Applied Discrete-Time Queues

Attahiru S. Alfa

Applied Discrete-Time Queues

Second Edition

Springer

Attahiru S. Alfa
Department of Electrical and
Computer Engineering
University of Manitoba
Winnipeg, Manitoba, Canada

Department of Electrical, Electronic and
Computer Engineering
University of Pretoria
Pretoria, South Africa

ISBN 978-1-4939-3418-8 ISBN 978-1-4939-3420-1 (eBook)
DOI 10.1007/978-1-4939-3420-1

Library of Congress Control Number: 2015955725

Springer New York Heidelberg Dordrecht London

Printed on acid-free paper

Springer Science+Business Media LLC New York is part of Springer Science+Business Media (www.springer.com)

This is dedicated to all who have supported me in one way or another during my academic journey, and specifically to all my parents, my wife, my children and my grandchildren.

Preface

This book started as a combination of two of my ideas

- A revision to the first book, which I now call the first edition; *Queueing Theory for Telecommunications: Discrete Time Modelling of a Single Node System*, including rearrangement of material and corrections of errors that were found, and
- Some new and significant additions that make the book more complete. The additions are;
  1. A new chapter on Special single node queueing models, which include; LCFS, SIRO, Pair formation, and Finite source queue
  2. A new chapter on Queues with Time Varying Parameters
  3. A new chapter on Tandem Queues and Queueing Networks
  4. A new chapter on Optimization and Control of Discrete Time Queues
  5. A new subsection on Review of Probability for Discrete Random Variables and Matrices
  6. A new subsection on Censored Markov Chains
  7. A new subsection on Special Types of QBDs
  8. A new subsection on PH/PH/1 System with Start-up Times
  9. A new subsection on Queues with Very Large Buffers
  10. A new subsection on Geo/G/∞ Systems
  11. An Appendix for $z-$transforms and Kronecker product
  12. Finally, the Introduction is fully expanded.

In the end, it has become a completely new book or a new edition; hence a new title to reflect the contents of the new edition. In the process of making some new additions to the material I have come to further realize that modelling a queueing system using discrete time approach is so practical, very realistic, very important in many systems such as telecommunications, and on top of all very exciting. There is always something interesting and subtle about each model in discrete time, especially the "book-keeping" aspect regarding order of event occurrences. For example when arrivals and service are the only two main events taking place in a discrete-time queue we know that it is important to first decide how we order the events since theoretically they can occur at the same time. So, when the two events occur at the same time do we recognize arrivals before service completions or vice versa? This presents two choices, known in the literature as *Late Arrivals* and *Early Arrivals*. One can practically develop a model for each one of them, even though it is not necessary. However, when we add some other events, for example the occurrence of vacations in vacation types of queues, then we have six possible combinations. Try and figure out the six! You can see that it becomes complex, a bit challenging but actually realistic. Would one then need to develop six different models? The answer is NO. A model should try and capture which of the sequences is actually used in the system considered. In this book we only select one sequence each time and present the associated model. For a different sequence one can use the same procedure developed in this book.

Most data are collected in discrete time, so to speak. In the end implementing queueing models for real life problems are carried out through numerical computations, which are based on digital operations. Even if continuous time models are created we can see that at the two supporting ends of the model, i.e. data collection and implementation on computers, are both based on discrete process. Continuous time models are nevertheless still very important especially when we want to understand the behaviours of a system, usually those that can be modelled nicely (possibly in closed form) to assist in revealing subtle behaviours that sometimes computational approaches can not bring out. Although this can also be achieved through discrete-time modelling of queueing systems, continuous time modelling seem to be more successful than discrete time in many of such cases. Hence this book is entitled "Applied Discrete Time Queues". It is written for those who are looking for models they can apply to solving practical queueing problems or need some ideas regarding how to develop such models.

The list of objectives expected to be achieved in writing this book include providing:

1. basic material for modelling commonly occurring discrete-time queueing models.
2. sufficient theoretical background to allow building the models and understanding the tools needed to build them.
3. numerical tools needed to compute the measures used to assess queueing systems
4. ideas that lead to further questions to researchers who wish to pursue this area of work.

The same philosophy that the first book was based on still applies in this new book; trying to show how to set up most queueing models using the concept of discrete time Markov chains (DTMC) and then guiding the reader to the tools in the literature that can be used for the analysis. As such a new chapter on numerical computations for DTMC has been created to avoid mixing the theory and computations of DTMC into the same chapter. Since the book is based on matrix analysis a new chapter on matrix representation of distributions has also been introduced.

Any corrections to errors found in the book will be posted to the following website:**http://home.cc.umanitoba.ca/~alfa/adtq/index.html**.

Winnipeg, Manitoba, Canada Attahiru S. Alfa
July 2015

The Preface to the first edition (blended with new edition chapter organization)

Again, I repeat, it is not as if there are not many books already on queueing theory! So, why this one – an additional book? Well, generally queueing models and their analyses can be presented in several different ways, ranging from the highly theoretical to the very applied approach. I have noticed that most queueing books are written based on the former approach, with mathematical rigour and as a result sometimes overlooking the needs of an application oriented reader who simply wants to apply the results to a real life problem. In other words, I am referring to that reader who has a practical queueing problem in mind and needs an approach to use for modelling such a system and obtain numerical measures to assist in understanding the system behaviour. Other books that take the purely applied approach usually oversimplify queueing models and analyses to the extent that a reader is not able to find results useful enough for the intended applications. For example, in an attempt to obtain very simple analysis, authors present models that are limited to simple unrealistic arrival and service processes and rules of operation, which may not capture the key behaviour of the system. One of the challenges is to present queueing models and clearly guide the reader on how to use the available mathematical tools to help in analyzing the problems associated with the models. To this end this book is mainly presented in a form that makes very easy to see how to apply the results. As such it is written to make queueing analysis easier to follow and also to make the new developments and results in the field much easier to follow for pragmatic users of the models.

Nearly every queueing model can be set up as a Markov chain, even though some of them may end up with huge state space. Markov chains have been very well studied in the field of applied stochastic processes. There are a class of Markov chains for which the matrix-analytic methods (MAM) are very suitable to their analysis. Most queueing models fall in that category. Hence there is definitely value in developing queueing models based on Markov chains and then using MAM for their analyses. The philosophy behind this book is that for most practical queueing problems that can be reduced to single node systems, we can set them up as Markov chains. If so, then we can apply the results from the rich literature on Markov chains and MAM. Two whole chapters are devoted to Markov chains in this book. Those chapters focus on Markov chain results and those specific to MAM that will be needed for the single node analysis. So most of the analysis techniques employed in this book are based on MAM. Occasionally the z-transform equivalent of some of the results are developed as alternatives. However, the use of transforms is not a major focus of this book. The book by Bruneel and Kim (1993) focused on the transform approach and readers who have interest in that topic are referred to that book. Additional information on discrete time approach for queues using z-transforms can be found by visiting Professor Bruneel Herwig's website.

It is well known that single node queues give very good insight into even the very complex non-single node queues if properly approximated. So single node queues are very important in the field of queueing theory and are encountered very frequently. This book focuses considerably on single node queues. Tandem queues are moderately covered and additional references given on the subject. As for queueing networks, a method that shows how the single node and tandem queues are used to approximate them is presented. For more on queueing networks in discrete time the reader is referred to the book by Daduna (2001).

One important area where queueing theory is applied very frequently is in the field of telecommunications. Telecommunication systems are currently analyzed in discrete time because they are based mainly on discrete technology (i.e. time is slotted and the system has shifted from analogue to discrete technology). Hence discrete time queueing models need special consideration in the fields of queueing and telecommunications. The material in this book has wider applications beyond telecommunications. They can be applied in transportation and traffic systems, manufacturing systems, healthcare systems, risk analysis and several other areas. These are discussed in Chapter 1.

The first chapter of this new edition introduces single node queueing systems and discusses discrete time briefly. We believe, and also want to show in this queueing book, that most single node queues in discrete time can be set up as discrete time Markov chains, and all we need to do is apply the results from Markov chain. Since most of the tools used will be based on matrix approach we present, in Chapter 2, matrix representation of probability distributions together with the arrival and service processes that usually characterize most of our queueing systems. The tools presented in the book are based on discrete time Markov chains, hence we present, in Chapter 3 the key results from discrete-time Markov chains, especially those related to MAM, that will be used to analyze our single node queueing systems. Chapter 4 focusses on infinite Markov chains.Generally, for most practical queueing problems we find ourselves turning to recursive schemes and hence needing to use some special computational procedures. Emphasis is placed on computational approaches for analyzing some classes of Markov chains in chapters 3 and 4. As we get into chapters 5 and 6 where basic single node queues are dealt with the reader will see that the real creative work in developing the queuing models is in setting them up as Markov chains. Chapter 7 deals with single node multi server systems, Chapter 8 with single node vacations queues and Chapter 9 with single node priority queues. Chapters 5 to 9 show that once the model is set up then all that is needed is to apply the results presented in Chapters 4 and 5 in analyzing the problem. The special philosophy of this book is that all we need is to understand some key results in discrete time Markov chains and we can analyze most single node queues. To summarize, the philosophy used in this book is such that chapters 3 and 4 actually contain the tools needed to analyze single node and some tandem queues considered in this book, and the materials presented in chapters 5 to 9 are mainly developing and also showing how to develop some key single server queueing models. For each of the models presented in chapters 5 and 9, the reader is pointed to the Markov chain results in chapters 3 and 4 that can be used for the queueing models' analyses. Chapter 2 presents general material that characterize queueing systems, including the very common distributions used to characterize the inter-arrival and service times. Chapter 10 deals with special class of single node queues. Other new additions to this new book include; Chapter 11 on queues with time-varying parameters, Chapter 12 on tandem queues with partial discussions on queueing network approximations, Chapter 13 on optimization of queues.

Winnipeg, Manitoba, Canada
September 2010

Attahiru S. Alfa
(revised July 2015)

References

Bruneel H, Kim BG (1993) Discrete-time Models for Communication systems Including ATM. Kluwer Academic Pub., Boston

Daduna H (2001) Queueing networks with discrete time scale: Explicit expressions for the steady state behaviour of discrete time stochastic networks. In: Springer Lecture Notes in Computer Science, Springer

Acknowledgements

I thank all those people who have provided me with the opportunity to complete this edition of the book and the first one. The list is very long to write out. However I wish to mention few people.

First I wish to thank all the people that have contributed to my learning of queueing theory, people like Do Le Minh, W. R. Blunden and Marcel Neuts, and several colleagues such as Srinivas Chakravarthy, V. Ramaswami, Qi-Ming He, and the list goes on. My sincere thanks goes to several of my graduate students over the years – they made me continue to maintain my keen interest in queueing theory. I sincerely thank Telex Magloire Ngatched Nkouatchah and Haitham Abu Ghazaleh for helping with going through the final stage draft of the first edition, to Hongqiao Tian for assistance with figures for this second edition and building the webpage for posting the errata and to Michelle Alfa for reading through the Preface and the Introduction for this second edition. Thanks to Mount-first for assisting with the rearrangement of the references by chapters in the second edition and Amy Dario for her help in printing several drafts of the second edition for me. Professor Bong Dae Choi of Korea University, Professors Vicente Casares Giner and Vicent Pla, both of the Universidad Politécnica de Valencia were kind enough to let me spend portions of my sabbatical leave of 2009-2010 in their respective universities with fundings and facilities to support my visit, during a period that I was trying to complete the first edition; and Professor Sunil Maharaj of the University of Pretoria and Professor Wei Li of the Texas Southern University who were kind enough to host me, in their respective universities, for portions of my sabbatical leave of 2014 while completing this second edition. To them I am most grateful because the timings were perfect in both cases.

Contents

Chapter 1
Introduction

1.1 Introduction

In a queueing system items arrive to be processed. The words "item" and "processed" are generic terms. An "item" could refer to customers. They could be customers arriving at a system such as a bank to receive service (be "processed"). Another example is in a manufacturing setting where an item could be a partially completed component that needs to be machined and is thus sent to a station where it is duly processed. Most of us go through city streets driving vehicles or as passengers in a vehicle. A vehicle going through a signalized intersection is an item that needs to be served and the service is provided by the intersection signal light in the form of giving the customer a green light to go through. The types of queueing systems that motivated me to write this book are the types encountered in a communication system. In such a system, an item could be a data packet arriving at a router (or at a switch) and needs its headers read so it can be forwarded to the appropriate next router or final destination. However, in the end, the systems considered in the book are generic and include any systems where items arrive to seek service (i.e. be processed). In this introductory chapter we will use all the similies of queueing system characteristics interchangeably so as to emphasize that we are dealing with a phenomenon that occurs in diverse walks of life. Generally in a queueing system there is a physical or a virtual location (sometimes moving) where items arrive to be processed. An item may be processed immediately after arrival if the processor is available, otherwise it has to wait if the processor is busy. Somehow, the waiting component that may occur if the processor is busy seems to dominate in the descriptor of the system, i.e. "queueing" or "waiting". In reality, this system can also be called a service system – sometimes a better descriptor. But we will stay with queueing system to maintain tradition.

Nearly all of us experience queueing systems directly or indirectly; directly through waiting in line ourselves or indirectly through some of our items waiting in line, such as a print job waiting in the printer buffer queue, or a packet belonging to a person waiting at a router node for processing. In any case, we all want our

A.S. Alfa, *Applied Discrete-Time Queues*, DOI 10.1007/978-1-4939-3420-1_1

delays to be minimum and also not be turned away from the system as a result of the buffer space not being available. There are additional types of factors that concern people, but these two are the major ones. Of course we all know that if those managing the system can provide "sufficient" resources to process the items then the delays or denial of service possibilities will be small. However, the system manager that provides the service is limited by the cost of the resources required. A system manager wishes to keep to a minimum the percentage of times the server (processor) is idle. Understanding how queueing systems behave under different scenarios is the key reason why we model queueing systems. This book intends to continue in that vein, but it focusses mainly on what we call a single node queueing system. Later in the book tandem queues and queueing networks are discussed in one chapter. In the following sections we give a few examples of everyday queueing systems which most people probably have experienced.

1.1.1 Examples of Transparent Queues

1.1.1.1 Cafeteria

Most people at some point in time during the day will go to grab a quick lunch at a cafeteria style restaurant. It is a service system. The usual practice is that a person (customer) looking for service picks up a tray and joins a line (queue) or a set of lines (queues) based on the local set up of the cafeteria, to pick what food items she/he wants and then continues to join another queue to pay for the items. The order of the queues all depend on the sequence and arrangement of the food service sites at the cafeteria. How quickly one goes through the system before starting to eat depends on many factors, such as the number of food stations, the number of checkout counters, how fast the counters operate, the number of other customers arriving, etc.

1.1.1.2 Bus Stop

Consider a bus stop. You arrive to catch a bus. If the bus is there waiting (and presumably there is space in the bus) you get on and the bus departs to its destination. If there is no bus yet, then you have to wait, together with other waiting passengers, for a period of time for the bus to arrive. That waiting time could be as long as a portion of the bus inter-arrival times, if you are able to get on the next bus that arrives. If the next bus that arrives does not have enough space for all those waiting then some passengers may have to wait for the next bus. You may be one of those who have to wait for the next bus. If you have to wait for the next bus then your waiting times would have been a portion of the inter-arrival time between two buses (first one and the one before that) plus one full inter-arrival time of the bus. From here on you can imagine the waiting time and queueing process involved. Queueing models can be used to study this system and also to develop optimal strategies for minimizing passenger waiting times and total operating cost for buses.

1.1.1.3 Internet Cafe

Most of us, when we travel, at some point in time have gone to an Internet cafe to try and send emails to our loved ones. There is usually a limited number of computers available for use at such cafes. If when we arrive at the cafe there is at least one computer free we use it, after we have paid. However, if they are all busy then we join a queue and wait our turn. Alternatively we may decide to go to another Internet cafe or just abort the plan and try again at some other time, depending on how urgent our need is. This system can be modelled as a simple queueing system.

1.1.1.4 Healthcare

There are all sorts of queueing experiences that we go through in the healthcare system, ranging from going to see the family physician to waiting for a major surgery. Here we give three examples.

1. We make an appointment to see our family physician. On the day of the appointment we arrive *early*, at least most people do. After we arrive we register with the front desk person and from then on we usually have to wait until the front desk person calls us. For most clinics after that call we are moved to the examination room where we wait again for awhile. We are now in our second stage of waiting. Then finally the physician comes to see us and we get the "service" (i.e. medical advice) we are seeking.
2. Suppose at the visit mentioned in the item above, the physician feels that one needs to go for a procedure, which could be a further test such as EKG or even a minor surgery at a particular establishment. The family physician's office then calls the establishment to make an appointment for you. At that point in time one goes into a it virtual waiting queue where they get notified after a spot is found for them, i.e. an appointment date. That is a waiting process in which we do not necessarily see the queue length, but it is there and we can ask and be told how long the queue is. On the appointed date one then goes to the establishment where they essentially go through the same queueing process as they did in the family physician's clinic. Once they arrive at the head of the queue they get the procedure they need.
3. The emergency clinic is the place where most queueing delays are encountered by many people. Different patients arrive at such facilities with different types of emergency needs. Some are life threatening, some are just painful and others are just "worried well". After a patient arrives they first go to an emergency admission clerk, if there is a clerk who is free, otherwise they have to wait until one is free. The emergency clerk takes information about the patients situation and based on the severity of the situation as assessed, the patient is placed in a position in the queueing system. That position could keep changing depending on whether the later arriving patients have problems that need more urgent attention. Anyone who has been to an emergency clinic knows the extent of the queueing problems associated with such clinics.

A reader is probably getting the message about the prevalence of queueing systems we go through in our daily lives and why it is important to make them efficient for customers queueing up while ensuring that it is cost effective for those that manage the operations.

1.1.1.5 Traffic Intersections

There are all sorts of queueing scenarios at traffic intersections. I will discuss the easiest one which is a STOP sign. Suppose I arrive at an intersection and I have a STOP sign regulating traffic on my lane. Let us assume that there no other lanes with STOP signs. That means I have to wait until the whole intersection clears and no vehicles ahead of me on my lane of the intersection before I can go through. Another queueing problem. I let a reader imagine other kinds of intersection controls and how the queueing system works.

1.1.1.6 Banks and Banking Machines

The queues at banks and banking machines are very obvious. They are very clearly transparent. Most of us use the banking system and are very familiar with this type of queues.

1.1.1.7 Getting Your Turn in a Conversation or at a Meeting

A type of queue that we all go through without realizing is in a conversation. Let us consider the simplest of all, when two people are having a conversation, usually each participant has to wait their turn, in a polite system. So if one has something to say, they have to wait until the other person finishes speaking before they can have their say. Now for a more complicated case, suppose there are $n \geq 2$ people having a casual conversation or are in a formal meeting. In a casual conversation, only one person can speak at a time and as soon as that person is finished the next person can then start to speak; however how the next person is selected (if there is more than one person who wants to speak) depends on the culture of the group, some usually unwritten code of conduct, or "who gets their voice out first"! In a formal meeting where there is a chairperson, that person selects who the next speaker would be using some criteria they feel is appropriate, or following rules that have been established for that type of meeting. Sometimes meetings may operate without a chairperson and then the selection of the next speaker is carried out just like the informal conversation situation. By now a queueing theorist can imagine all kinds of queueing models that can be used to describe all these.

1.1.2 *Examples of Non-Transparent Queues*

There are queueing systems which we go through and we are not aware that we do. I call these non-transparent queues.

1.1.2.1 Call Centers

At some point in time most people have probably been in communication with a call center, either by phone or through a computer chat. Usually a caller needs some kind of service and initiates the call. Very often, these days, within a short time there is a response from the other end to find out what one's call is about. Some times a response to one's queries are attended to immediately, but not too often. Usually the caller is asked to wait while an attempt is made to connect them to the right department. During this wait there could be music playing, advertisements going on to promote some products, etc. This is the first wait period, and the answers to one's queries may be obtained at the end of this stage, when the person on the other end of the communication returns to the conversation. If the person cannot answer the queries, the caller is passed on to another queue, and so on until the caller is reasonably satisfied, i.e. all their queries are adequately responded to. In these types of queues, usually a caller would not know their position in the queue. These days however, the system service has improved and callers are often either told their position in the queue or an estimated wait time. Call centers depend heavily on queueing models and statistics resulting from them in order to plan their system and manage them. In fact the modern call centers that tell a caller how long their wait is base the projections mainly on queueing models and the associated analysis.

1.1.2.2 Access to a Website

Trying to get information from a website involves a lot of queues in series as well as in parallel. The information being sought often resides at a server somewhere. When one sends a request for information on the web, the request goes through the Internet, most times, which consists of several routers where the requests, as well as the returned information, queue up along the way. Even the request leaving one's computer and information coming back are all carried by what is know as TCP/IP (transmission control protocol/Internet Protocol). Interestingly enough one knows that their request and information do queue (wait) somewhere along the route, but they as individuals do not actually see the queue. They feel though when the queue is long in terms of their own waiting times.

1.1.2.3 Access to a Communication System

Communication systems are typical examples of systems that employ different types of queueing models for their efficient running. All communication systems whether wireless or wired need to move information (in the form of packets these days) between several sources and destinations. Even while processing the information to be sent out, at the origin device, there is a considerable amount of queueing that goes on. Then when the packets leave the device to travel through some form of medium (wired or wireless) they compete with other packets getting transmitted from other sources to other destinations). They all have to go through some switches (routers) like an intersection and the same kind of queueing that is observed at road intersections also take place at routers (it is just a bit more complex with routers). Queueing theory actually got its officially documented beginning from telecommunication analysis in 1907.

1.1.2.4 Waiting for the Processing of Income Tax Returns

Nearly everyone files an income tax return in one form or another every year. For those who expect some refund from the government, it seems to take too long to receive it; even though that may only be a perception. The completed forms have to be processed and as such each form has to wait its turn since the number of staff working on income tax files is disproportionally small compared to the whole population making submission. So there is a set of queues which these documents go through for processing.

There are queueing systems which do not involve people queueing directly as customers but rather their items are queueing while the customer waits. They involve items waiting.

1.1.3 Examples of Queues Not Involving Humans Directly

1.1.3.1 Jobs Waiting at an Assembly Plant

Anyone who has ever worked in a factory or gone to visit a factory has seen some assembly lines where several parts arrive, from different sources inside or outside the plant, to be assembled into a single product. Unless there is a "perfect" coordination that can take place (which is doubtful), some parts have to wait for other parts to arrive in order to complete assembly. One also comes across parts waiting to be processed on machines, etc. These are items, rather than humans, that wait for processing of some sort. Even though they do not "feel" the waiting times, the customers waiting for those products in the end experience the waiting times. Queueing theory did receive significant attention in the manufacturing industry in the eighties.

1.1.3.2 Aircraft Waiting for take-off or Landing

At most airports, especially the big ones, sometimes aircraft do experience delays in take off and landing. There are usually limited number of runways and aircraft get scheduled for their departure and arrival times. In essence if all aircraft were able to be ready in time for take off and their requests to land were timed properly, then in theory, they should not experience delays. But departures sometimes get delayed due to other factors such as late arriving passengers (transferring from another flight), aircraft maintenance issues that have taken longer than expected, delays due to ground crew operations, etc. The result is that some aircraft do not meet their exact departure time and this affects other aircraft thereby leading to some long delays, because when they are finally ready to depart they have lost their spot in the scheduled queue leading to an extended wait (delay) on the ground. Equally leading to landing delays are things like aircraft taking longer time for the flight due to some natural situations and thereby arriving late. When they want to land late the runway may be occupied or even their gate may be already taken by another aircraft, making them wait in the air and circling the airport until a position is available. Even though humans themselves do not queue, their aircraft queues and this affects them. They may end up missing their connections, or missing their meetings for the day, and so many other possible ramifications.

There are so many types of queueing systems in life. One could confidently say nearly everyone goes through a number of queueing systems everyday. Later, in this book, there will be more detailed examples of queueing systems that have a major impact on us psychologically, on the economy, or on our personal life and how content we feel. Specifically we discus their mechanisms, how they operate and the different alternative ways they could be managed.

1.2 A single node queue

A single node queue is a queueing system in which a packet arrives to be processed at one location, such as at a router. After the packet has been processed it does not proceed to another location for further processing, or rather we are not interested in what happens to it at other locations after the service in the current location is completed. However the packet may re-enter the same location for processing immediately after its processing is completed. If it proceeds to another location for another processing and later returns to the first location for processing then we consider this return as a new arrival and the two requests are not directly connected. So a single node system has only one service location and service is provided by the same set of servers in that location even if a packet re-enters for another service immediately after finishing initial one. We give examples of single node queues in Figures 1.1, 1.2, 1.3 and 1.4. In Figure 1.1 we have the top portion showing a simple single server queue, where the big circle represents the server, the arrow leading to the server shows arrivals of items and the arrow leaving the server shows departure of items that have completed receiving service.

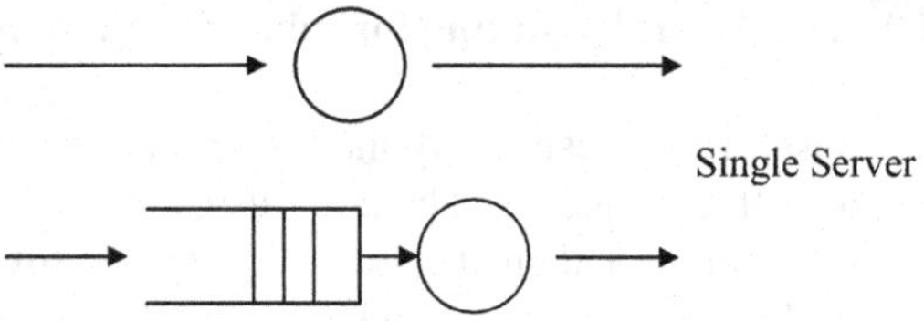

Fig. 1.1 A Single Node, Single Server Queue

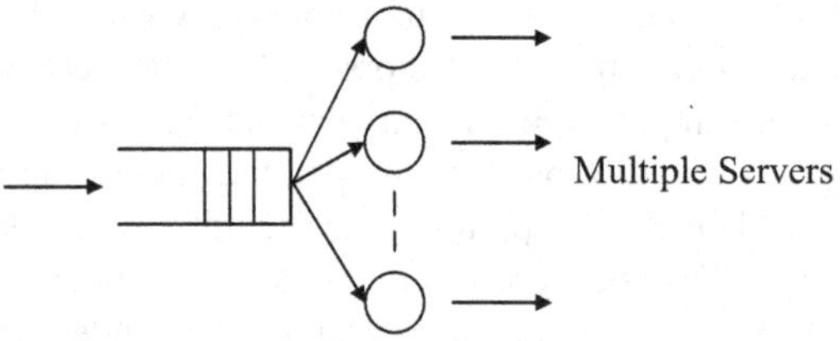

Fig. 1.2 A Single Node, Multiple Parallel Servers Queue

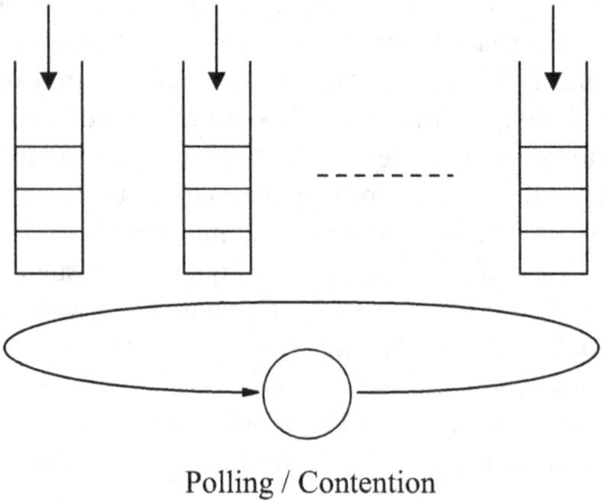

Fig. 1.3 A Single Node, Multiple Parallel Queues and One Server

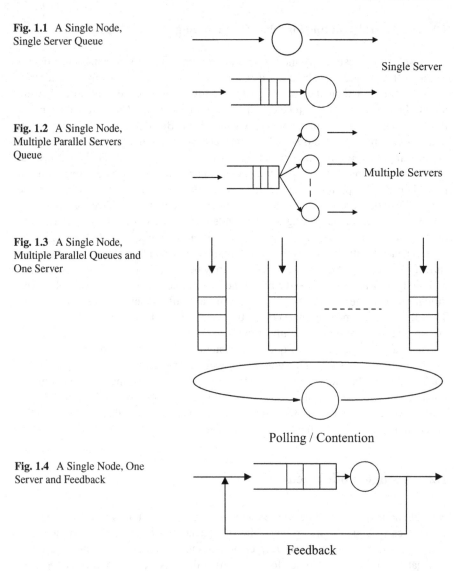

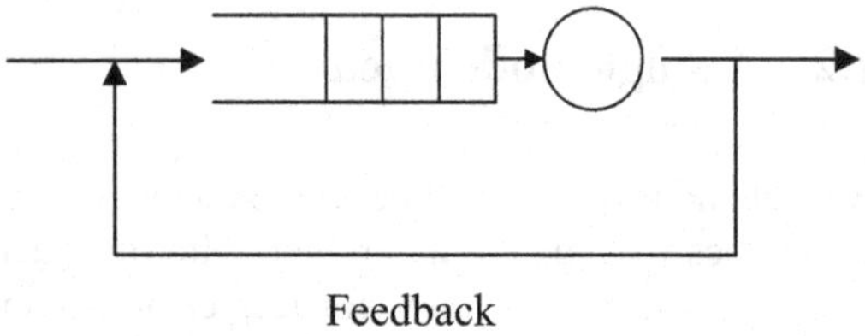

Fig. 1.4 A Single Node, One Server and Feedback

Figure 1.1 (the lower part) is an example of a simple single node queue. There is only one server (represented by the circle) at this node. Packets arrive into the buffer (represented by the open rectangle) and receive service when it is their turn. When they finish receiving service they leave and may even return for another service later. A queueing system that can be represented in this form is a bank with drive-in window where there is only one ATM machine to be used and vehicles queue up to use it.

In Figure 1.2 we have a single node queue with several servers in parallel. A packet may be sent to any of the servers for processing. This type of queueing system is very common at the in-buffer queue of a switch in communication networks, the check-out counter in a grocery store, at a bank where tellers attend to customers, a commercial parking lot where the parking bays are the servers and the service times are durations of parking, etc.

Figure 1.3 is a single node queue in which there are several buffers for different queues and a server attends to each queue based on a pre-scheduled rule, e.g. round-robin, table polling, etc. This is called a polling system. This type of queueing system is very common in telecommunication systems in the medium access control (MAC) where packets from different sources arrive at a router and need to be processed and given access to the medium. The server (router) attends to each source according to a pre-scheduled rule. Another example of this is a traffic intersection with lights – each arm of the intersection is a queue of its own and the server, which is the light, gives a green light according to a pre-scheduled rule to each arm. The example diagram of Figure 1.3 shows only one server, but we could have more than one server providing the same service.

Finally Figure 1.4 is a single node queue in which an item, after completing a service, may return to the same system immediately for another service. This is a feedback queue. For example consider a manufacturing system in which an item after going through a manufacturing process is inspected instantaneously and placed back in the same manufacturing queue if it is deemed to be defective and needs to be re-worked. These are just a few examples of single node queues.

What other types of queueing system exist and what makes them different from single node queues? Single node queues are the smallest units that can be classified as a queueing system. Other types of queueing systems are configurations of multiple single node queues.

1.3 A tandem queueing systems

Consider several single node queues in series, say N of them labelled $Q_1, Q_2, \cdots, Q_N$. Items that arrive at Q_1 get processed and once their service is completed they may return to Q_1 for another service, leave the system completely or proceed to Q_2 for an additional service. Arriving at Q_2 are items from Q_1 that need processing in Q_2 and some new items that join Q_2 from an external source. All these items arriving in Q_2 get processed at Q_2. Arriving items go through the N queues in series according to their service requirements. We usually reserve the term tandem queues for queues in which items flow in one direction only, i.e. $Q_1 \rightarrow Q_2 \rightarrow \cdots \rightarrow Q_N$, i.e. we say $p_{i,j}$ is the probability that an item that completes service at node Q_i proceeds to node Q_j, with $p_{i,j} \geq 0,\ j = i, i+1$ and $p_{i,j} = 0,\ j < i,\ j > i+1$, keeping in mind that $\sum_{i=1}^{N} p_{i,j} \leq 1$, since an item may leave the queue after service at node Q_i with probability $1 - \sum_{i=1}^{N} p_{i,j} \leq 1$, i.e. $1 - p_{i,i+1} \leq 1$. An example of a tandem queueing system is illustrated in Fig 1.5.

As one can see, a tandem queueing system can physically be decomposed into N single node queues. Even though the analysis as a set of single node queues is

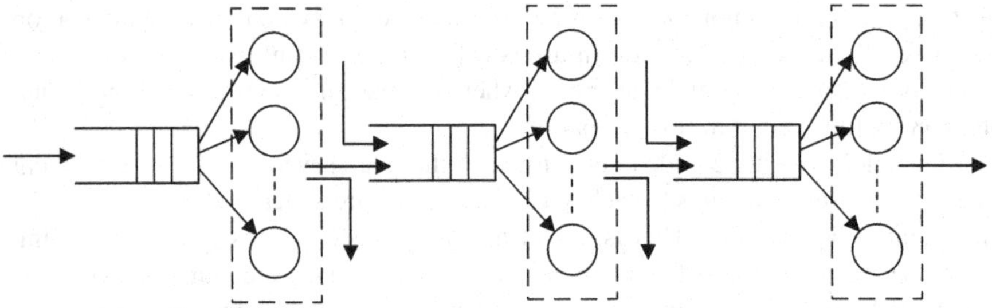

Fig. 1.5 Example of a Tandem Queue

involved and may also end up only as an approximation, this approach has been widely adopted in the literature Perros and Altiok (1989). For other results on discrete-time analysis of tandem queues, see Park et al (1994). A real life example of a tandem queueing system is a virtual circuit in a communication system. Another one is a linear production line in a manufacturing system. One example most of us come across nearly everyday is that of a path of travel. Once we have decided on our route of automobile travel in a city, then we see this as a linear path and each of the intersections along the path form a set of single node queues and all together form a tandem queueing system.

1.4 A network system of queues

Finally, we have a network system of queues which is a more general class of queueing systems. In such a system we may have N nodes. Items can arrive at any of the N nodes, e.g. node Q_i for processing. After processing at node Q_i an item may leave the system or proceed to another node Q_j for further processing. Here we have $p_{i,j} \geq 0,\ \forall(i,j)$, and $1 - \sum_{j=1}^{N} p_{i,j} \leq 1$ is the probability that a job leaves the system after processing is completed at node Q_i. There is no restriction as to which node an item may proceed to after completing service at a particular node, unlike in the case of the tandem queueing systems that has restrictions in its definition. Figure 1.6 illustrates an example of what a network of queues looks like.

Most real life queueing systems are usually networks. Examples are the communication packet switching system, road networks, etc. Once more one can see that a network system of queues is simply a collection of several single node queues. Sometimes a queueing network is decomposed into several single node queues as an approximation for analysis purposes. This is because analyzing queueing networks or even tandem networks could be very involved except in the simpler unrealistic cases. As a result decomposing tandem queues and queueing networks has received a considerable amount of attention in the literature. This is why we decided to focus considerably on single node queues and develop the proper analysis for them in this book. We also discuss tandem queues with considerable details and show how they can be approximated by single node queues. Work on discrete-time queueing networks in the literature is limited because it is comparatively more difficult to

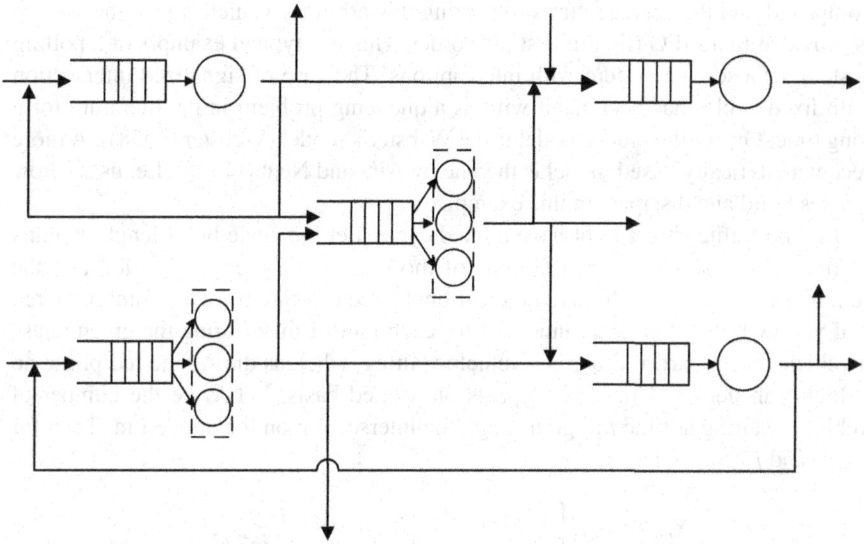

Fig. 1.6 Example of a network of Queues

study than its continuous time counterpart. The book by Daduna (2001) treats a class of discrete-time queueing networks with geometric inter-arrival times and geometric services. However the results in this model are limited when it comes to trying to develop computational aspects. Hence, we feel that this area of work is still full of unsolved problems. We discuss this under the Open Problems chapter.

1.5 Some Well-known Application Areas for Queueing Models

In this section we select a few applications of queueing models and discuss their mechanism in bit more detail.

1.5.1 Queues in Transportation and Traffic Systems

Consider a traffic intersection, whether signalized, non-signalized, or a traffic circle type. Let there be four approaches to the intersection. For the signalized intersection a vehicle may go through the intersection only when the light shows green for it, otherwise it waits its turn. Even then, if the light shows green for that vehicle's lane, the vehicles ahead of it in its lane get to go through the intersection first at the green signal. So in essence we can see that for an arbitrary vehicle arriving at the intersection, all the vehicles in its lane see green light as service and red as the server is away on vacation (serving other lanes of the intersection). When the vacation is completed and the server returns to serving this arbitrary vehicle's lane the service is provided in a FIFO (first in first out) order. This is a typical example of a polling system, or a service system with interruptions. The case of signalized intersection with fixed cycles has been dealt with as a queueing problem in the literature for a long time. One of the oldest model is the Webster's model (Webster (1958)). A more recent analytically based model is the one by Alfa and Neuts (1995). Let us see how queues build and dissipate in this example.

Let the traffic signal light have a fixed cycle. Let the cycle be of length C units of time. Suppose $g,\ 1 \leq g < C$ units of those are for a green indication and the remaining $C-g=r$ are for a red indication; by red indication we are combining red and yellow lights. Let us assume that for each unit of time during the green phase a vehicle can depart, if there is a vehicle waiting, whereas during the red phase no vehicle can depart. This cycle repeats on a fixed basis. Let X_t be the number of vehicles waiting at time t to go through the intersection on the target arm. Then for $n \geq 0$ and $j \geq 0$, we have

$$X_{nC+j+1} = \begin{cases} (X_{nC+j}-1)^{+}+A, & 1 \leq j \leq g, \\ X_{nC+j}+A, & g+1 \leq j \leq C. \end{cases},$$

where A is the number of arrivals during one interval of time, assuming that the arrival process is stationary, and $(a)^{+}$ stands for $max(0,a)$. The value X_t being the number of waiting vehicles at time t, is a measure of performance of the queueing system. Several queueing characteristics and performance measures can be derived from it after a very rigorous analysis.

1.5.2 *Queues in Manufacturing Systems*

In a manufacturing environment items go through some machines for processing and some times end up waiting or getting delayed due to the nature of product mix, service time and inter-arrival time fluctuations. This delay can probably be managed within reason with proper scheduling. However this situation can even be made worse if the machines have to wait for repairs due to unscheduled breakdowns. In this example we consider the model presented by Wu et al (2008).

Let the service time of a job be S, S_G its generalized service time which reflects the capacity of a workstation under the influence of interruptions, N_I the number of preemptive events such as breakdowns, d_k the duration of the kth downtime during the service time S, and T the duration of run-based non-preemptive product-induced events (this includes set-up times caused by product-induced events). Based on this we see that the departure time S_G for a product is given as

$$S_G = S + \sum_{i=1}^{N_I} d_i + T,$$

which represents how long a job is in the system (before departure) from when it first claims the service capacity of the workstation. Note that the inverse of S_G is a reflection of the capacity of the workstation, and of course the smaller it is the longer the delays.

1.5.3 Queues in the Health Care Systems

Application of queueing theory in the healthcare system is growing very fast. As a simple example, consider a case where people have to go for flu vaccinations. Suppose there are K healthcare workers available to give the vaccination and assume there is enough vaccine to cover all who come for it. Let us assume that each patient takes exactly one unit of time to be vaccinated. Let X_n be the number of people currently receiving or waiting for vaccination at time n and A be the number of arriving patients at any time interval, then we have

$$X_{n+1} = \begin{cases} A, & X_n \leq K \\ X_n - K + A, & X_n > K \end{cases}.$$

This is a very simple example. We can see that the queue build up can be captured by studying X_n, $n = 1, 2, 3, \cdots$.

Consider the case where we have K beds in an operating theater, of which $K_e < K$ are reserved for emergency cases. Again we assume that our unit of time is such that each patient requires one unit of time for surgery (more realistic time requirements will be covered later). Let $X(e)_n$ and $X(r)_n$ be the number of emergency and regular patients in the system waiting for, or currently undergoing, surgery at time n. We assume there is no preemption and emergency patients are not allowed to occupy any of the regular $K - K_e$ beds even if not occupied by regular patients; we assume that $K - K_e < \delta$, where δ is a very small integer. Let A_e and A_r be the number of arrivals of emergency and regular patients, respectively. Let

$$\mathscr{I}_1 := (X_n(e) \leq K_e, X_n(r) \leq K - K_e), \quad \mathscr{I}_2 := (X_n(e) \leq K_e, X_n(r) \geq K - K_e),$$

$$\mathscr{I}_3 := (X_n(e) \geq K_e, X_n(r) \leq K - K_e), \quad \mathscr{I}_4 := (X_n(e) \geq K_e, X_n(r) \geq K - K_e),$$

where $A := C$ implies that A is equivalent to C. Then we have

$$(X_{n+1}(e), X_{n+1}(r)) = \begin{cases} (A_e, A_r), & \mathscr{I}_1, \\ (A_e, ((X_n(r) - K + K_e)^+ + A_r)), & \mathscr{I}_2, \\ ((X_n(e) - K_e)^+ + A_e, A_r), & \mathscr{I}_3, \\ ((X_n(e) - K_e)^+ + A_e, (X_n(r) - K + K_e)^+ + A_r)), & \mathscr{I}_4. \end{cases}$$

Again from this we can study how the queues of regular and emergency patients grow with time.

1.5.4 Queues in Communication networks

As mentioned earlier in this Introduction, communication networks have more queueing systems occurring than most other types of queues. One simple example is in the data network area where packets are sent through a series of routers from the originating node to the destination node. Usually also we have different classes of packets. For the example to be discussed here we focus only on the case of one class of packets. Let there be two routers in series between the origin and the destination nodes. We make two other simplifying assumptions that the buffers between the router nodes have infinite sizes and also a packet only needs one unit of ime for processing at a router. Packets that leave the first node proceed to the second one for further processing and transmission. At the intermediate nodes (second node in this case) there are external packet arrivals, and the number of arrivals at a time interval is A_2, with arrivals at the first node given as A_1. So there will be two queues, one at each of the router nodes. Let $X_n(1)$ and $X_n(2)$ be the number of packets being processed or waiting at nodes 1 and 2, respectively at time n, we have

$$(X_{n+1}(1), X_{n+1}(2)) = ((X_n(1)-1)^+ + A_1, (X_n(2)-1)^+ + A_2 + U_{X_n(1)}),$$

where $U_v = \begin{cases} 1, & v \geq 1, \\ 0, & \text{otherwise.} \end{cases}$

There are a multitude of queueing systems in communication networks. In order to provide an effective and efficient communication system understanding the underlying system that governs the performance of the communication is only second to the development of the equipment needed for the system. In what follows we present just a few examples of communication networks, ranging from the very old to the very new systems.

- **Old POTS:** The plain old telephone system (POTS) is one of the oldest systems that actually established the need for queueing models. The works of the pioneering queueing analyst, A. K. Erlang (Erlang (2009), Erlang (1917)), were the results of trying to effectively understand the switching system of the POTS and assess how to determine the number of trunks required to effectively and efficiently operate such a system. That is an example of a queueing system at a high level. After switching is carried out at each node, a circuit is established the connect to communicating entities. Within the POTS there are several "mini" queueing systems. For example, trying to get a dial tone actually involved a queueing process.

  Queueing problems as seen in the POTS had to do with voice traffic – a more homogeneous type of traffic. Only one class of traffic was considered as important and was the main driver of the system. Things have changed significantly since then, These days telephones do a lot more.
- **Computer communication networks:** In the computer communication world data was moved in the form of packets. For data packets, instead of circuit

switching as in the POTS case, packet switching was employed. Circuits were not established, but packets were given headers and sent through the network via different paths and later assembled at the end (destination) node. Still packets go through a queueing system and depending on the types of packets being sent they may have some priorities set. The modern computer communication networks have also changed significantly over the years.

- **Integrated Services Digital Networks (ISDN):** In the late 1970s and early 1980s a considerable effort was focussed on trying to integrate both the voice and data networks into one so as to maximize resources. The ISDN project was started and out of it came also the Asynchronous Transfer Mode (ATM), an idea that packetizes both voice and data to small size ($48+5$kb) cells and sends them through the system either through a virtual circuit or as packet switched data. With ATM came serious considerations for different classes of traffic and priority queues became very important in telecommunication queues. In fact, the advent of ISDN, ATM and the fact that communication systems started to be operated as digital systems (rather than analogue) led to discrete time queueing systems becoming more prevalent.
- **Wireless networks:**When wireless communication networks became accessible to the paying public the modelling of the queueing systems associated with them opened up a new direction in queueing systems. With the wireless system came mobility of customers. So customers or mobile stations that are receiving service while they are moving could end up switching to a different server (when they change location) without completing their service at the previous server, i.e. mobile stations go through a handoff process from one location to another.
- **Cognitive radio networks:** Cognitive radio network is currently one of the recent topics of interest in communication systems that opens up the study of a new type of queueing systems. In this system we have several channels - used by primary users (*aka* licensees of the channel) that can only be used by secondary users when the channels are free. Here the secondary users keep searching for free channels that are available for use at some points in time. So this is a queueing system in which the server is available only intermittently and the customer (secondary user) has to find which channel (server) and when it is available, keeping in mind that other secondary users may also be attempting to use the same channel.

These are just a few of the queueing systems that have historically been associated with telecommunication systems.

Trying to present material in this book specifically for telecommunication systems would be limiting the capabilities of the analysis techniques in it. Despite the fast changing world of telecommunications the fundamental queueing tools available seem to keep pace or sometimes are way ahead of the communication technologies. As such, I decided to keep the book mainly as a tool for telecommunication systems analysis rather than a book on telecommunications systems queues. I present the material as a fundamental discrete time single node queue analysis. The results can be used for different telecommunication systems as they arise or appropriately modified to suit the analysis of the new systems. As we see from this section, telecommunication systems technologies ranging from the POTS to cognitive radio network all use queueing models and these models are conceptually based on the same stochastic models. I hope the results and analysis techniques in this book can be applied for a long time as the telecommunication system evolves and that they can be applied to several other systems. As pointed out in this book real world queueing situations are everywhere. I have used mostly telecommunication examples and terminologies in the book just to give a reference point and allow consistent use of terminologies as much as is possible.

1.6 Why are queueing systems of interest?

Queueing systems are of interest to everyone, but mainly to two categories of people; 1) the users of the systems (customers) and 2) the service providers. The customers need (want) to use the system and minimize their delays, loss (or blocking) probabilities, etc. On the other hand the service provider wishes to minimize its cost for providing the service to customers while ensuring that the customers are "reasonably" satisfied. In the end, customers will access a service only if they feel they get at least their "money's worth" for the service they receive. Granted there are situations where customers are captive of a system and a measure of level of service does not come into play. In such cases customers just want to know how long it would take to get served and accept it. In general, the service providers do not want to assign too many servers when this would result in a high percentage of them becoming idle most of the time. Also if buffering is possible and relevant, service providers do not want to provide an extremely large buffer size which is often unoccupied, especially if it costs more to increase the buffer size. Service providers generally plan without knowing exactly the workload to be submitted by customers because that component is stochastic, even though they learn the patterns of the traffic later. Sometimes the rules used to operate the system is a major factor in achieving the proper goals. For example, just by changing the operation rule from first-come-first-served to a priority system will change how the system is perceived by the customers. In such cases a service provider may classify customers by categories (how much they pay for a service) and give priorities to some classes. Several of these factors (delay, loss, preference service, more servers, larger buffers, etc) are inter-related and difficult to control by the service provider without incurring major extra costs. So the need to understand how to operate a queueing system and how all these factors are inter-related is necessary and needs a very good understanding of the systems That is why we need models for understanding the system behaviour!

In order to articulate this further we select an example of a queueing system for discussion. Consider the Internet system. People send packets through this system to be processed and forwarded to a destination. One could consider the packets as the items that need to be processed for a customer. The Internet service provider (ISP) ensures that the customer is provided with the resources to get the packets processed and sent to the correct destination with minimum delay and a low loss probability. The customer pays for a service and expects a particular Quality of Service (QoS) commensurate with how much is paid. In an utopian world the customer would want the packet processed immediately and the ISP would want to obtain the maximum revenue possible without incurring any costs! In the real world the resources needed to process the packets costs money and the ISP needs to make a profit so it wishes to provide the most cost effective amount of resources while the customer who pays for the service defines the QoS target that needs to be met by the ISP.

In designing a queueing system we need to find the optimal configurations and rules that will optimize the profit for the ISP and meet the QoS of customers. In order to do this we need to understand how the queueing system will perform under different configurations and rules.

1.7 Performance Measures

It is clear that both service system managers and customers have ways of assessing the performance of a queueing system. What would make a customer extremely happy going through a queueing system is if they do not have to wait at all, even though they know that is most unlikely! Nevertheless, they would like to know how long one has to wait to receive service from the system. So waiting time or delay is one typical performance measure as seen by customers. Even though the service provider would like to provide a service with reasonably low waiting time, it knows that to achieve this a good amount of service resources would have to be provided which also could be expensive as more resources are used. However, the service provider is also aware that a customer's subscription to its system depends on how well the waiting time scales compared to the price charged to the customer for use the system. So waiting time or delay is equally important to a customer as it is to a service system provider.

The number of customers waiting for service is another common measure of performance as seen by both the customer and the service system provider. The number waiting, if seen by customers, is a major discouraging factor in joining the queue for service. Also the service provider would need to make sure there is a sufficient buffer space to accommodate the waiting customers. If the buffer space is limited, then there could be losses of customers when the buffer is full; this is a dillema for both service provider and customers. Finally, the number waiting in the queue is strongly related to the waiting time through the service times.

Another very important measure is how idle or busy the system is. If it is too busy, customers will not be happy as that is a sign of possible long waits. However for the service provider a long idle time of servers is a waste of resources and uneconomical. These two measures are conflicting, as expected.

1.7.1 Performance Measures

1. **Queue Length:** The queue length refers to the number of items, in this case packets, that are waiting in some location, or buffer, to be processed. This is often an indication of how well a queueing system is performing. The longer the queue length the worse its performance from a user's point of view, even though this may be argued not to be quite correct all the time.
2. **Loss Probability:** If the buffer space where items have to wait is limited, which often is the case for real systems, then some items may not be able to wait if they arrive to find the waiting space to be full. We usually assume that such items are lost and may retry by arriving again at a later point in time. In data packet systems the loss of a packet may be very inconvenient and customers are concerned about this loss probability. The higher its value the worse is the system performance from the customer's perspective.
3. **Waiting Times:** waiting time or delay as referred to by some customers is the duration of the time between an item's arrival and when it starts to receive service. This is the most used measure of system performance by customers. Of course the longer this performance measure is the worse is the perception of the system from a customer's point of view.
4. **System Time:** This is simply the waiting time plus the the time to receive service. It is perceived in the same way as the waiting time except when dealing with a preemptive system where an item may have its service interrupted sometimes.
5. **Work Load:** Whenever items are waiting to be processed a server or the system manager is interested in knowing how much time is required to complete processing the items in the system. Of course, if there are no items in the system the workload will be zero. Work load is a measure of how much time is needed to process the waiting items and it is the sum of the remaining service time of the item in service and the sum of the service times of all the waiting items in a work conserving system. In a work conserving system no completed processing is repeated and no work is wasted. A queueing system becomes empty and the server becomes idle once the workload reduces to zero.
6. **Age Process:** In the age process we are interested in how long the item receiving service has been in the system, i.e. the age of the item in process. By age we imply that the item in service was "born" at its arrival time. The age is a measure that is very useful in studying queues where our interest is usually only in the waiting times. This is also very useful in analyzing queues with heterogeneous items that are processed in a first come first served (FCFS) order, also known as first in first out (FIFO) order.
7. **Busy Period:** This is a measure of the time from when a server begins to process after an empty queue to the first time when the queue becomes empty again. This measure is more of interest to the ISP, who wishes to keep their resources fully utilized. So the higher this value the happier an ISP. However if the resource that is used to provide service is human such as in a bank, grocery store, etc, then there is a limit to how long a service provider wishes to keep a server busy before the server becomes ineffective.

8. **Idle Period:** This is a measure of the time that elapses from when the last item is processed leaving the queue empty to when the service begins again usually after the first item arrival. This is like the "dual" of the busy period. The longer the idle period is the more wasteful are the resources. A service provider wishes to keep the idle period as small as possible.
9. **Departure Times:** This is a measure of the time an item completes service in a single node system. This measure is usually important in cases where an item has a departure time constraint and also where a departing item has to go into another queue for processing. When a single node queue serves an input queue to another queue downstream, the departure process forms the arrival process into the next queue. How this departure process is characterized determines the performance of the downstream queue.

In the end our interest is to be able to develop tools for assessing the performance measures of different queueing systems. That is what this book is about.

1.8 Characterization of Queues

It was pointed out in the last section of this book what performance measures are of interest in queueing. Let us briefly mention the characteristics of queueing systems that determine or govern those measures. It is clear that both arrival and service processes have a lot to do with it, and can be used to characterize a queueing system. Common sense tells us that if the arrival rate increases without anything else changing then both the waiting times and number waiting in the system would increase and conversely increasing service rate decreases those measures. Adding servers in parallel would also increase the effective service rate.

The size of the buffer determines the loss rate; as the buffer size increases the loss rate would decrease. But one has to remember that even though the loss rate decreases, the waiting time for service would increase on the average.

Finally, another very important characteristic of a queueing system is the service discipline, whether first come first served, last come first served, etc.

In summary, some of the very popular parameters used for characterizing a queueing system include; arrival and service processes, buffer size, queue discipline such as FIFO (first in first out), SIRO (service in random order), LIFO (last in first out), PRIO (priority), etc.

These are just few of the fundamental measures of interest to a queue analyst. There are several others that are specific to some queues that will be discussed as they appear in the book. However, not all of the measures mentioned above will be presented for all queues in the book.

1.9 Discrete time vs Continuous time analyses

Most queueing models in the literature before the early 1990s were developed in continuous time. Only the models such as the M/G/1 and GI/M/1 that were based on the embedded Markov chains studied the systems in discrete times. Such models were well studied in the 1950s (Kendall (1951) Kendall (1953)). The discrete time models developed before then were few and far between. The other discrete time models that were later studied are those by Galliher and Wheeler (1958), Dafermos and Neuts (1971), Neuts and Klimko (1973), and Minh (1978), just to name a few. However, researchers then did not see any major reasons to study queues in discrete time, except when it was felt that by doing so made difficult models easier to analyze. Examples of such cases include the embedded systems and the queues with time-varying parameters.

Modelling of a communication system is an area that uses queueing models significantly and continuous time models were seen as adequate for the purpose. However now that communication systems are more digital than analogue, and we work in time slots, discrete time modelling has become more appropriate. Hence new results for discrete time analysis and books specializing on this approach are required.

Time is a continuous quantity. However, for practical measurement related purposes, this quantity is sometimes considered as discrete, especially in modern communication systems where we work in time slots. We often hear people say a system is observed every minute, every second, every half a second, etc. If for example we observe a system every minute, how do we relate the event occurrence times with those events that occur in between our two consecutive observation time points? This is what makes discrete time systems slightly different from continuous time systems. Besides most models, for practical large systems, end up getting solved using numerical methods. Most numerical methods usually require some form of discretization. Let us clarify our understanding of continuous and discrete view of time.

1.9.1 Time

Consider time represented by a real line. Since time is continuous we are not able to observe any system at time t. We can only observe it between time t and $t+\delta t$, where δt is infinitesimally small. In other words, it takes time to observe an event, irrespective of how small this observation time is. This is because by the time we finish observing the system, some additional infinitesimal time δt would have gone by. Hence, in continuous time, we can only observe a system for a time interval, i.e. between times t and time $t+\delta t$. Note that δt could be as small as one wishes.

In discrete time, on the other hand, our thinking is slightly different. We divide time up into finite intervals. These intervals do not have to be of equal lengths and

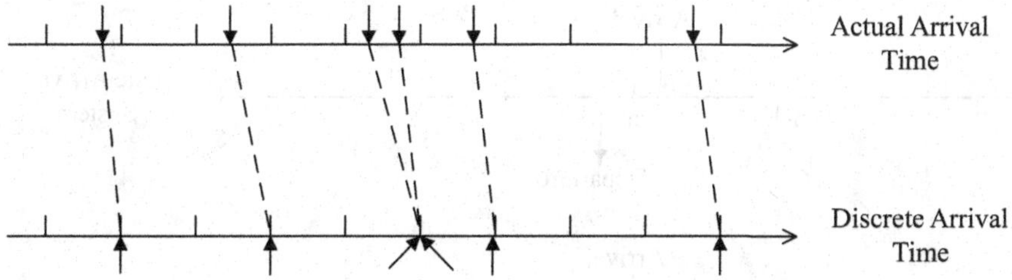

Fig. 1.7 Example of Discrete Time Concept

the time points may or may not represent specific events. We can now observe the system at these points in time. For example, we observe the system at time t_1, t_2, t_3, ... , t_n, t_{n+1}, ..., implying that everything that occurs after time t_n and throughout the interval up to time t_{n+1}, is observable at time t_{n+1}; for example all arrivals to a system that occur between the interval $(t_n, t_{n+1}]$ are observed, and assume to take place, at time epoch t_{n+1}. Another way to look at it is that nothing occurs between the interval $(t_n, t_{n+1}]$ that is not observable at time epoch t_{n+1}. By this reasoning we assume the picture of the system is frozen at these time points t_1, t_2, $\cdots$, t_n, $\cdots$, and also assume that it takes no time to "freeze" the picture. Fig. 1.7 illustrates this idea.

In addition, we only know what we observe at the time points and they constitute the only knowledge we have about the system. Whatever activities occur between our last observation of the system at time t_n, and our next observation at time t_{n+1}, are assumed to occur at time t_{n+1}. We normally require that an item is in the system for at least one unit of time, e.g. a data packet requires at least one time slot for processing, and hence it will be in the system for at least that duration before departing. Hence, our discrete time analysis by definition will be what Hunter (1983) calls the *late arrival with - delayed access*, i.e. arrivals occur just before the end of an epoch, services terminate at the end of an epoch and the arriving customer is blocked from entering an empty service facility until the servicing interval terminates. This is discrete time analysis of queueing systems as proposed by Dafermos and Neuts (1971), Neuts and Klimko (1973), Klimko and Neuts (1973), Galliher and Wheeler (1958), and Minh (1978). We are simply taking a snapshot of the system at these time epochs. We will not consider the other two discrete time arrangements described by Hunter (1983) and Takagi (1993), since the modelling aspects will be repetitious with only minor modifications. Hunter Hunter (1983) explained three possible arrangements of arrival and service processes in discrete time. They are: 1) Early arrival 2) Late arrival with immediate access and 3) Late arrival with delayed access. In this book we work with Hunter's Case 3) i.e. Late arrival with delayed access. The first one and the third one are more realistic and are illustrated in Fig. 1.8. The other two types of arrangements of discrete time views described by Hunter will not be considered in this book. However, it is very simple to extend the discussions in this book to those other two arrangements.

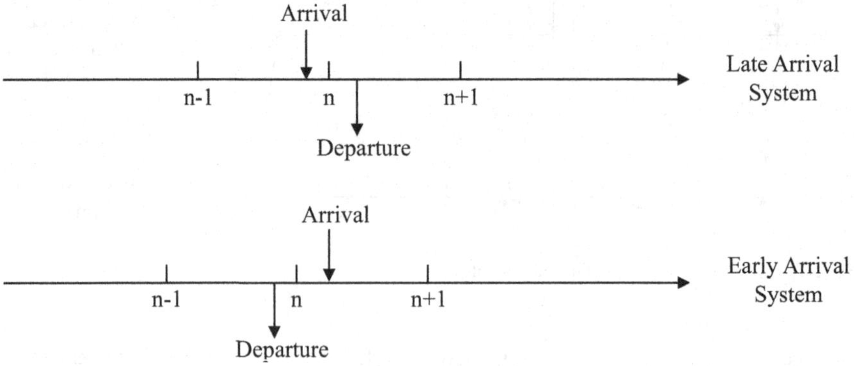

Fig. 1.8 Arrival and Service Completion Orders in Discrete Time Concept

References

Alfa AS, Neuts MF (1995) Modelling vehicular traffic using the discrete time Markovian arrival process. Transportation Science 29:109–117

Daduna H (2001) Queueing networks with discrete time scale: Explicit expressions for the steady state behaviour of discrete time stochastic networks. In: Springer Lecture Notes in Computer Science, Springer

Dafermos SC, Neuts MF (1971) A single server queue in discrete time. Cahiers Centre d'Etudes Rech Oper 13:23–40

Erlang AK (1917) Solution of some problems in the theory of probabilities of significance in automatic telephone exchanges. Electrotejnikeren (Danish), 13, 5-13 (English Translation, P O Elec Eng J 10, 189-197 (1917-1918) 13:5–13

Erlang AK (2009) The theory of probabilities and telephone conversations. vol 20, pp 33–39

Galliher HP, Wheeler RC (1958) Nonstationary queuing probabilities for landing congestion of aircraft. Operations Research 6:264–275

Hunter JJ (1983) Mathematical Techniques of Applied Probability - Volume 2, Discrete Time Model: Techniques and Applications. Academic Press, New York

Kendall DG (1951) Some problems in the theory of queues. J Roy Statist Soc Ser B 13:151–185

Kendall DG (1953) Stochastic processes occurring in the theory of queues and their analysis by the method of imbedded Markov chains. Ann Math Statist 24:338–354

Klimko MF, Neuts MF (1973) The single server queue in discrete time - numerical analysis, II. Naval Res Logist Quart 20:304–319

Minh DL (1978) The discrete-time single-server queue with time-inhomogeneous compound Poisson input. Journal of Applied Probability 15:590–601

Neuts MF, Klimko MF (1973) The single server queue in discrete time - numerical analysis, III. Naval Res Logist Quart 20:297–304

Park D, H P, Yamashita H (1994) Approximate analysis of discrete-time tandem queuing networks with bursty and correlated input traffic and customer loss. Oper Res Letters 15:95–104

Perros HG, Altiok T (1989) Queueing Networks with Blocking. North, Amsterdam

Takagi H (1993) Queuing analysis, a foundation of performance analysis - volume 3. In: Discrete-Time Systems, North Holland, p Amsterdam

Webster FV (1958) Traffic signal settings. Road Research Technical paper 39, Department of Scientific & Industrial Research, Her majesty's Stationary Office, London, England

Wu K, McGinnis LF, Zwart B (2008) Queueing models for single machine manufacturing systems with interruptions. In: SJMason, Hill RR, Monch L, Rose O, Jefferson T, Fowler JW (eds) Proc. 2008 Winter Simulation Conference, pp 2083–2092

Chapter 2
Arrival and Service Processes

2.1 Introduction

For every queuing system we have to specify both the arrival and the service processes very clearly. The stochastic process that characterize arrival and service are key to explaining how queues build, fluctuate and dissipate. They are some of the key characteristics that determine the performance of a queueing system. If the times between arrivals of items are generally short and the service times are long it is clear that the queue will build up. Most important is how those times vary, i.e. the distribution of the inter-arrival times and service times. We will present the arrival and service processes that commonly occur in most queuing systems. However before we do that let us briefly review some important probability axioms and other aspects associated with probability of discrete event since a matrix approach of representing probability distributions will be used in this chapter a brief review of

2.2 Review of Probability for Discrete Random Variables and Matrices

We define a discrete state space $\mathscr{X} = \{a_0, a_1, a_2, a_3, a_4, \cdots\}$ as a countable sequence, where $a_i \neq a_j$, $i \neq j$. Throughout this book we will consider special discrete state space where $\mathscr{X} = \{0, 1, 2, 3, 4, \cdots\}$. Consider any random collection of discrete variables we call the set $\mathscr{X}_s = \{a_1, a_2, \cdots, a_N\}$. Let $\mathscr{X}_s \subseteq \mathscr{X}$. As an example $\mathscr{X}_s$ could be the possible outcome of tossing a six-sided die, i.e. $\mathscr{X}_s = \{1, 2, 3, 4, 5, 6\}$, or the number of heads showing up when three unbiased coins are tossed in which case $\mathscr{X}_s = \{0, 1, 2, 3, \}$, etc. Let $\mathscr{A}$ be a random variable that is in the set $\mathscr{X}$ and $P(\mathscr{A})$ be the probability of the event $\mathscr{A}$. We define p_i as the probability that $\mathscr{A}$ assumes the value $i \in \mathscr{X}$, then we write

A.S. Alfa, *Applied Discrete-Time Queues*, DOI 10.1007/978-1-4939-3420-1_2

1. $P(\mathscr{A} = i) = p_i,\ i \in \mathscr{Z}$,
2. $0 \leq p_i \leq 1$,
3. $\sum_{i\in\mathscr{Z}} p_i = 1$.

The expectation of a random variable can be written as

1. $E[\mathscr{A}]$ as the expected value of $\mathscr{A}$.
2. For the example above we can write $E[\mathscr{A}] = \sum_{i\in\mathscr{Z}} i p_i = \sum_{i>0} \sum_{k=i}^{\infty} p_k$, and
3. in general we have $E[g(\mathscr{A})] = \sum_{i\in\mathscr{Z}} g(i) p_i$.

2.2.1 *The z transform*

For $|z| \leq 1$, we define the z-transform or the probability generating function *(pgf)* of $\mathscr{A}$ as

$$A(z) = \sum_{i\in\mathscr{Z}} z^i p_i.$$

We have

$$E[\mathscr{A}] = \frac{dA(z)}{dz}|_{z\to 1} = \sum_{i\geq 1} i p_i,$$

and in general we have the nth factorial moment of $\mathscr{A}$ written as

$$E_f^n[\mathscr{A}] = \frac{d^n A(z)}{dz^n}|_{z\to 1} = \sum_{i\geq n}^{\infty} \frac{i!}{(i-n)!} p_i.$$

2.2.2 *Bivariate Cases*

Consider two random variables $\mathscr{A}$ and $\mathscr{B}$, with $\mathscr{A} \in \mathscr{Z}$ and $\mathscr{B} \in \mathscr{Z}$, and let $\mathscr{A} \cup \mathscr{B}$and $\mathscr{A} \cap \mathscr{B}$ represent the union and intersection, respectively, of $\mathscr{A}$ and $\mathscr{B}$, we have

1. $P(\mathscr{A} \cup \mathscr{B}) = P(\mathscr{A}) + P(\mathscr{B}) - P(\mathscr{A} \cap \mathscr{B})$, where $P(\mathscr{A} \cap \mathscr{B})$ is the probability of both event $\mathscr{A}$ and $\mathscr{B}$ occurring.
2. $P(\mathscr{A} \cap \mathscr{B}) = P(\mathscr{A})P(\mathscr{B}|\mathscr{A})$, where $P(\mathscr{B}|\mathscr{A})$ is the conditional probability that $\mathscr{B}$ occurs given that $\mathscr{A}$ has occurred.
3. $P(\mathscr{A} \cap \mathscr{B}) = P(\mathscr{A})P(\mathscr{B})$, if $\mathscr{A}$ and $\mathscr{B}$ are independent
4. $P(\mathscr{B}|\mathscr{A}) = \frac{P(\mathscr{A}\cap\mathscr{B})}{P(\mathscr{A})},\ P(\mathscr{A}) > 0$.
5. Let us write $p_{i,j} = Pr\{\mathscr{A} = i, \mathscr{B} = j\}$ then we have $P(z_1, z_2) = \sum_{i\in\mathscr{Z}} \sum_{j\in\mathscr{Z}} z_1^i z_2^j$ $p_{i,j},\ |z_k| \leq 1,\ k = 1,2$.

All these results can be written for multivariate cases of 3 or more random variables. For example, consider n random variables $\mathscr{A}_1, \mathscr{A}_2, \cdots, \mathscr{A}_n$ and let $p_{i_1,i_2,\cdots,i_n} = Pr\{\mathscr{A}_1 = i_1, \mathscr{A}_2 = i_2, \cdots, \mathscr{A}_n = i_n\}$, then we can write

$$P(z_1, z_2, \cdots, z_n) = \sum_{i_1 \in \mathscr{X}} \sum_{i_2 \in \mathscr{X}} \cdots \sum_{i_n \in \mathscr{X}} z_1^{i_1} z_2^{i_2} \cdots z^{i_n} p_{i_1,i_2,\cdots,i_n}, \quad |z_k| \leq 1, \forall k \geq 1.$$

2.2.3 *Some very Common Discrete Distributions*

2.2.3.1 Bernoulli Process

The Bernoulli process is one of the most important processes in probability and especially when carrying out discrete time analysis of queues. Consider a random variable or an event $\mathscr{A}$ that has only two possible outcomes, success (S) or failure (F). Let $P(\mathscr{A} = S) = p$ and $P(\mathscr{A} = F) = q = 1 - p$, then this process is called a Bernoulli process, if each trial is independent of the previous one and the outcomes have the same probability.

The $z-$transform or probability generating function *(pgf)* associated with Bernoulli distribution $B(z)$ is stated as

$$B(z) = q + pz, \quad |z| \leq 1.$$

As an example, consider taking well shuffled deck of playing cards (52 of them). If we draw a card from the deck at random and we consider drawing a King as success then

$$P(\mathscr{A} = S) = p = \frac{4}{52} = \frac{1}{13},$$

and of course

$$P(\mathscr{A} = F) = \frac{48}{52} = \frac{12}{13} = 1 - \frac{1}{13}.$$

If we replace the drawn card into the deck, reshuffle the deck and draw again at random, the probability p remains the same. If we take this example further and say customers arrive at every time interval according to a Bernoulli process with probability p, then we can determine, probabilistically how the number of customers grow in the system at any time.

2.2.3.2 Geometric Distribution

Geometric distribution is derived from the Bernoulli process. It captures how many repetitive tries it takes for a success to occur. For example, let us consider the example case of drawing a card from the deck as given above. Suppose we draw from the deck and replace the drawn card each time back to the deck and we want to know the probability of how long it takes to draw the first King (first success). If we were successful the first time then the probability that it took us one draw to be successful is p. The probability that it took us two draws to be successful will imply that the first draw was unsuccessful while the second was successful. The probability that it took two draws for the first success is qp. Let the $\mathscr{X}$ be the number of attempts for first success, then it follows that

$$P(\mathscr{X} = i) = q^{i-1}p, \; i \geq 1.$$

The pgf associated with the geometric distribution, $G(z)$, is written as

$$G(z) = \sum_{i=1}^{\infty} q^{i-1}pz^i = \frac{pz}{1-qz}, \; |z| \leq 1.$$

In some other disciplines or books people study, instead, the number of failures before a success. In that case we speak of a number $\tilde{\mathscr{X}} = \mathscr{X} - 1$. Hence we have

$$P(\tilde{\mathscr{X}} = i) = q^i p, \; i \geq 0.$$

In this case the *pgf*, $\widetilde{G(z)}$ is written as

$$\widetilde{G(z)} = \sum_{i=0}^{\infty} q^i pz^i = \frac{p}{1-qz}, \; |z| \leq 1.$$

We will be using the first definition in this book, because our interest is in queueing systems for which both inter-arrival and service times are always at least one time unit.

2.2.3.3 Binomial Distribution

If on the other hand the question of interest to us is given that we are allowed to draw $\mathscr{Y}$ times from the deck (with replacement), how many successes do we expect to have? Let $\mathscr{X}$ be the number of draws and $\mathscr{Y}$ the number of successes out of the $\mathscr{X}$ draws ($\mathscr{Y} \leq \mathscr{X}$) then

$$P(\mathscr{Y} = j)|\mathscr{X} = n) = \binom{n}{j} p^j q^{n-j}, \; 0 \leq j \leq n.$$

The pgf associated with the Binomial distribution, $\hat{B}(z)$, is written as

$$\hat{B}(z) = \sum_{j=0}^{n} \binom{n}{j} p^j q^{n-j} z^j = (pz+q)^n = (B(z))^n, \; |z| \leq 1.$$

The Poisson distribution can be derived as a limiting distribution of the Binomial under certain conditions.

2.2.3.4 Negative Binomial Distribution

Another distribution that is very related to the Bernoulli process is the negative binomial distribution. Here the interest is in knowing how many draws are needed to have exactly a particular number of successes. Let $\mathscr{Z}$ be the number of draws required for the $\mathscr{Y}^{th}$ success to occur. We have

$$P(\mathscr{Y}=j|\mathscr{Z}=n,)=\binom{n-1}{j-1}p^jq^{n-j},\ 1\leq j\leq n.$$

It is clear that when $j=1$ we have that the Negative Binomial is a special case of the Geometric distribution. The *pgf* associated with the Negative Binomial distribution, $NB(z)$, is written as

$$NB(z)=\sum_{j=1}^{n-1}\binom{n-1}{j-1}p^jq^{n-j}z^j=pz(pz+q)^{n-1}=pz(B(z))^{n-1},\ |z|\leq 1.$$

This is the discrete analog of the Erlang distribution.

2.2.3.5 Mixture of Geometric Distributions

Finally we present another distribution that is also related to the Bernoulli process and Geometric distribution – a mixture of geometric distributions. Consider n geometric distributions with parameters (p_k,q_k), $k=1,2,\cdots,n$. Suppose we are given a probability distribution $(\theta_1,\ \theta_2,\ \cdots,\ \theta_n)$ with $\sum_{i=1}^{n}\theta_i=1$. There is a distribution called a mixture of geometric distributions. If the number of draws for first success as $\mathscr{X}$ is of the mixture of geometric distribution, then

$$P(\mathscr{X}=i)=\sum_{k=1}^{n}\theta_k q_k^{i-1}p_k,\ k\geq 1.$$

This is the discrete equivalent of the hyper-exponential distribution.

For detailed exposition to these distributions, the reader is referred to basic probability books. Most of these distributions will be covered in detail later with regards to queueing theory. Specifically how they arise in discrete time queueing models will be further discussed later.

2.2.4 Brief Summary of Required Material from Matrices

In this section we briefly present a few properties of a square matrix that will be needed in this chapter. Generally a matrix D of dimension $m \times n$ can be written as
$$D = \begin{bmatrix} d_{11} & d_{12} & \cdots & d_{1n} \\ d_{21} & d_{22} & \cdots & d_{2n} \\ \vdots & \vdots & \cdots & \vdots \\ d_{m1} & d_{m2} & \cdots & d_{mn} \end{bmatrix}.$$ Consider

1. a non-negative finite square matrix A of dimension n, i.e. each element $a_{i,j}$ of this matrix has the property that $0 \leq a_{i,j} < \infty$,
2. an n row vector $\mathbf{c} = [c_1,\ c_2,\ \cdots,\ c_n]$, and
3. an n column vector $\mathbf{b} = \begin{bmatrix} b_1 \\ b_2 \\ \vdots \\ b_n \end{bmatrix}$.
4. A column vector $\mathbf{1} = [1\ 1\ \cdots\ 1]^T$, where A^T is the transpose of a matrix A
5. $\mathbf{e}_j(n)$ which is an n column vector of zeros in all locations except at location j where there is a 1.
6. An identity matrix I. We will usually assume the dimension of I is obvious and when it is not we write it as $I(n)$, i.e. of dimension n.

We have the following properties of matrices that will be used in this book

1.
$$A^2 = \begin{bmatrix} \sum_{j=1}^n a_{1,j}a_{j,1} & \sum_{j=1}^n a_{1,j}a_{j,2} & \cdots & \sum_{j=1}^n a_{1,j}a_{j,n} \\ \sum_{j=1}^n a_{2,j}a_{j,1} & \sum_{j=1}^n a_{2,j}a_{j,2} & \cdots & \sum_{j=1}^n a_{2,j}a_{j,n} \\ \vdots & \vdots & \cdots & \vdots \\ \sum_{j=1}^n a_{n,j}a_{j,1} & \sum_{j=1}^n a_{n,j}a_{j,2} & \cdots & \sum_{j=1}^n a_{n,j}a_{j,n} \end{bmatrix},$$
2. From the above we can infer $A^k,\ k = 1, 2, 3, \cdots,$
3. By definition $A^0 = I$, the identity matrix, where $I_{i,j}$ are the elements of I and $I_{j,j} = 1$ with $I_{j,k} = 0$, for $j \neq k$.
4.
$$\mathbf{c}A = \left[\sum_{i=1}^n c_i a_{i,1},\ \sum_{i=1}^n c_i a_{i,2},\ \cdots \sum_{i=1}^n c_i a_{i,n}\right],$$
5.
$$A\mathbf{b} = \begin{bmatrix} \sum_{j=1}^n a_{1,j}b_j \\ \sum_{j=1}^n a_{2,j}b_j \\ \vdots \\ \sum_{j=1}^n a_{n,j}b_j \end{bmatrix},$$

6. The matrix A has n eigenvalues $\lambda_i,\ i=1,2,\cdots,n$ and associated left eigenvectors $\mathbf{a}_i$ such that

$$\lambda_i \mathbf{a}_i = \mathbf{a}_i A.$$

7. The spectral radius of A denoted by $sp(A)$ is given as

$$sp(A) = max\{|\lambda_i|,\ i=1,2,\cdots,n\}.$$

Next we give some simple but very interesting examples of matrix representations of discrete distributions.

2.2.5 Examples of Simple Representations of Discrete Distributions Using Matrices

We will consider three simple examples of representations.

1. Consider a variable $\theta,\ 0 \le \theta \le 1$, and an integer $n < \infty$. Let $b(n,\theta,i) = \binom{n}{i} \theta^i (1-\theta)^{n-i},\ 0 \le i \le n$. Define

- an n row vector $\boldsymbol{\alpha} = [\alpha_1,\ \alpha_2,\ \cdots,\ \alpha_n]$, where $\alpha_i = b(n,\theta,i)$,
- a column vector $\mathbf{t} = \begin{bmatrix} 1 \\ 0 \\ \vdots \\ 0 \end{bmatrix}$ and
- a matrix $T = \begin{bmatrix} 0 & 0 & \cdots & 0 \\ 1 & 0 & \cdots & 0 \\ 0 & 1 & \cdots & 0 \\ \vdots & \vdots & \ddots & \vdots \end{bmatrix}$.

If we define a random variable $\mathscr{A} = \{0,1,2,\cdots,n\}$, for which we have

$$Pr\{\mathscr{A} = 0\} = 1 - \boldsymbol{\alpha}\mathbf{1},$$

$$Pr\{\mathscr{A} = i\} = \boldsymbol{\alpha} T^{i-1}\mathbf{t},\ i=1,2,\cdots,n.$$

It is immediately obvious that what we have done is written the Binomial distribution in a matrix form (be it in efficiently). As an example,

$$Pr\{\mathscr{A} = j\} = \binom{n}{j} (1-\theta)^j \theta^{n-j}.$$

2. Consider an n row vector $\boldsymbol{\alpha} = [1,\ 0,\ 0,\ \cdots,\ 0]$, an n column vector $\mathbf{t} = \begin{bmatrix} 0 \\ 0 \\ \vdots \\ 1 \end{bmatrix}$

   and an n square matrix $T = \begin{bmatrix} 0\,1\,0 \cdots 0 \\ 0\,0\,1 \cdots 0 \\ \vdots\ \vdots\ \vdots\ \ddots\ \vdots \\ 0\,0\,0 \cdots 1 \end{bmatrix}$. If we define a random variable $\mathscr{A} = \{0,1,2,\cdots,n\}$, for which we have

$$Pr\{\mathscr{A} = i\} = \boldsymbol{\alpha} T^{i-1}\mathbf{t},\ i = 1,2,\cdots,n,$$

   then one sees that what we have is

$$Pr\{A = n\} = 1, \text{ and } Pr\{\mathscr{A} = j\} = 0,\ j \neq n.$$

   This is a constant random variable.
3. Consider

   - a variable θ,
   - an n row vector $\boldsymbol{\alpha} = [1,\ 0,\ 0,\ \cdots,\ 0]$,
   - an n column vector $\mathbf{t} = \begin{bmatrix} 0 \\ 0 \\ 0 \\ \vdots \\ \theta \end{bmatrix}$ and
   - an n square matrix $T = \begin{bmatrix} 1-\theta & \theta & 0 & \cdots & 0 & \\ 0 & 1-\theta & \theta & \cdots & 0 & \\ 0 & 0 & 1-\theta & \theta & \cdots & 0 \\ \vdots & \vdots & \vdots & \vdots & \ddots & \vdots \\ 0 & 0 & 0 & 0 & \cdots & 1-\theta \end{bmatrix}$.

   If we define a random variable $A = \{0,1,2,\cdots,n\}$, for which we have

$$Pr\{\mathscr{A} = i\} = \boldsymbol{\alpha} T^{i-1}\mathbf{t},\ i = 1,2,\cdots,n.$$

   One sees that this random variable has the negative binomial distribution. For example,

$$Pr\{\mathscr{A} = j\} = \binom{n-1}{j-1}(1-\theta)^{j-1}\theta^{n-j}.$$

These are just simple examples of representation of some well known discrete distributions using the matrix approach. The key here is that we have demonstrated, through examples, that at least some well-known discrete distributions can be represented using matrix form. This approach becomes very useful later when we discuss some very important and basic queueing models.

2.2.6 Matrix Representation

Consider a square matrix T of dimension $1 < m < \infty$ with the following properties.

- its elements $T_{i,j}$ satisfy the condition: $0 \leq T_{i,j} < \infty$
- there is constant κ such that the spectral radius $\rho = sp(\kappa T) < 1$. Usually $0 < \kappa \leq 1$.

Theorem 2.1 Then there exists a row vector $\mathbf{a}$, with elements $a_i > 0$ and $\mathbf{a}\mathbf{1} \leq 1$, and a column vector $\mathbf{w}$, with elements $w_i > 0$, such that

$$0 \leq \mathbf{a}T^{k-1}\mathbf{w} \leq 1, \ \forall k = 1, 2, \cdots$$

and

$$0 \leq \mathbf{a}(I-T)^{-1}\mathbf{w} \leq 1.$$

Proof: If $0 \leq T_{ij} < \infty$ and $sp(T) < 1$, then it is known from matrix analysis that $0 \leq (T^n)_{ij} < \infty, \ \forall k \geq 0$, hence it is possible to find two vectors $\mathbf{a}$ and $\mathbf{w}$ such that

$$0 \leq \mathbf{a}T^{k-1}\mathbf{w} \leq 1, \text{ and } \sum_{k=1}^{\infty} \mathbf{a}T^{k-1}\mathbf{w} \leq 1..$$

Definition: Let $\mathscr{A}$ be a discrete random variable, and p_k be defined as $p_k = Pr\{\mathscr{A} = k\}$, then

$$p_k = \mathbf{a}T^{k-1}\mathbf{w}, \ k = 1, 2, 3, \cdots.$$

$$p_0 = 1 - \mathbf{a}(I-T)^{-1}\mathbf{w}.$$

These matrix representations will be used later in Section 2.5.

2.3 Arrival and Service Processes

These two processes can probably be discussed separately. But because the distributions used for service processes can also be used for the inter-arrival times, we will combine them in the same section.

Consider a situation where packets arrive to a router at discrete time points. Let J_n be the arrival time of the nth packet. Further let A_n be the inter-arrival time between the nth and $(n-1)^{th}$ packet, i.e. $A_n = J_n - J_{n-1},\ n = 1,2,\cdots$. We let J_0 be the start of our system and assume an imaginary zeroth packet so that $A_1 = J_1 - J_0$. If $A_1,\ A_2,\ A_3,\ \ldots$ are independent random variables then the time points $J_1, J_2, J_3,\ \ldots$ are said to form renewal points. If also $A_n = A,\ \forall n$, then we say the inter-arrival times have identical distributions. We present the distributions commonly assumed by A. We will not be concerned at this point about A_1 since it may have a special behaviour. We assume it is a well behaved interval and not a residual. Dealing with the residuals at this point would distract from the main discussion in this chapter.

Let t_n^0 be the start time of the service time of the nth packet and let t_n^f be the completion time of its service time duration, i.e. $S_n = t_n^f - t_n^0$. If $S_n = S,\ \forall n$, then we say that the service times have identical distributions. We can also consider a system that is saturated with a queue of packets needing service. Let C_n be the service completion time of the nth packet. Then $S_n = C_n - C_{n-1}$.

First let us briefly introduce the renewal process, which is related to some of these arrival and service processes which are independent.

2.4 Renewal Process

Let X_n be the inter-event time between the $(n-1)$th and nth events such as arrivals. In our context X_n could be A_n or S_n. If $Y_0 = 0$ represents the start of our observing the system, then $Y_n = \sum_{i=0}^n X_i$ is the time of occurrence of the nth event. For example, $J_n = \sum_{i=0}^n A_i$ is the arrival time of the nth customer. We shall assume for simplicity that $X_0 = 0$ and $X_1, X_2, \ldots$ are independent. Later we may also assume that they are identically distributed variables (note that it is more general to assume that X_0 has a value and it has a different distribution than the rest, but that is not very critical for our purpose here). If they are independent and identically distributed we say they are *iid*.

If we consider the duration of X_n as how long the nth inter-event lasts, then Y_n is the point in time when the nth process starts and then lasts for X_n time units. The time Y_n is the regeneration point or the renewal epoch. We assume that $P\{X_n > 0\} > 0,\ \forall n > 1$, and we let $\mathbf{p}_n$ be the collection $\mathbf{p}_n = (p_1^{(n)}, p_2^{(n)} \ldots)$, where $p_i^{(n)} = Pr\{X_n = i\},\ \forall n \geq 1$. We define the $z-$transform of this inter-event times as

$$p_n^*(z) = \sum_{j=1}^{\infty} p_j^{(n)} z^j,\ |z| \leq 1,\ \forall n. \tag{2.1}$$

Let the mean number of events per unit time be given by μ then the mean interevent time

$$E[X_n] = \frac{dp_n^*(z)}{dz}|_{z\to 1} = \mu^{-1} > 0,\ \forall n. \tag{2.2}$$

Since $X_1, X_2, \ldots$ are independent we can obtain the distribution of Y_n as the convolution sum of $X_1, X_2, \ldots$ as follows

$$\Pr\{Y_n\} = \Pr\{X_1 * X_2 * \ldots * X_n\}. \tag{2.3}$$

Letting $q_j^{(n)} = Pr\{Y_n = j\}$ and $q_n^*(z) = \sum_{j=1}^{\infty} q_j^{(n)} z^j, \ |z| \leq 1$ we have

$$q_n^*(z) = p_1^*(z) p_2^*(z) \cdots p_n^*(z). \tag{2.4}$$

2.4.1 *Example:*

$$Pr\{Y_2 = j\} = \sum_{v=1}^{j-1} Pr\{X_1 = v\} Pr\{X_2 = j - v\}.$$

If $X_1 = X_2 = X$, then $p_1^*(z) = p_2^*(z) = p^*(z)$ and hence we have

$$q_2^*(z) = (p^*(z))^2.$$

Further let

$$p_1 = 0.1,\ p_2 = 0.3,\ p_3 = 0.6,\ p_j = 0, j \geq 4,$$

then

$$p^*(z) = 0.1z + 0.3z^2 + 0.6z^3$$

and

$$q_2^*(z) = 0.01z^2 + 0.06z^3 + 0.21z^4 + 0.36z^5 + 0.36z^6.$$

From this we obtain

$$q_1^{(2)} = 0,\ q_2^{(2)} = 0.01,\ q_3^{(2)} = 0.06,\ q_4^{(2)} = 0.21,\ q_5^{(2)} = 0.36,\ q_6^{(2)} = 0.36,\ q_j^{(2)} = 0, \forall j \geq 7.$$

In this case, for example, X could be the length of time it takes to process a job. If we have two jobs in a system the total time it takes to process the two is given by Y_2. It is easy to show that if we have K jobs then the z-transform of the total time to process all of them will be $(p(z))^K$.

2.4.2 *Number of renewals*

Let Z_n be the number of renewals in $[0,n]$ then

$$Z_n = \max\{m \geq 0 | Y_m \leq n\}. \tag{2.5}$$

Hence,

$$P\{Z_n \geq j\} = P\{Y_j \leq n\} = \sum_{v=1}^{n} P\{Y_j = v\}. \tag{2.6}$$

If we let $m_n = E[Z_n]$ be the expected number of renewals in $[0,n)$, it is straightforward to show that

$$m_n = \sum_{j=1}^{n} \sum_{v=j}^{n} P\{Y_j = v\} \tag{2.7}$$

m_n is called the *renewal function*. It is a well known elementary renewal theorem that the mean renewal rate μ is given by

$$\mu = \lim|_{n \to \infty} \frac{m_n}{n}. \tag{2.8}$$

For a detailed treatment of renewal process, the reader is referred to Wolff (1989). Here we only give a skeletal proof.

In what follows, we present distributions that are commonly used to describe arrival and service processes. Some of them are of the renewal types.

2.5 Special Arrival and Service Processes in Discrete Time

2.5.1 *Bernoulli Process*

The Bernoulli process in general terms was presented in a previous section of the chapter. Let us consider it in the context of a queueing system. If time is slotted, i.e. if we consider discrete time of equal intervals, and assume the arrival of packets at each interval could only be singly with probability $p > 0$ and we have $q = 1 - p$ and are equally likely at each time. Then at any interval between time t_n and t_{n+1} the probability of one packet arrival is p and no packet arrival is q.

The idea can be extended to service completion probability. Suppose a job (packet) is receiving service (being processed) and at each time when it is in service that service could end with probability p or continue with probability q. Then the service process is based on a Bernoulli process.

2.5.2 Geometric Distribution

Geometric distribution is the most commonly used discrete time inter-arrival or service time distribution. Its attraction in queuing theory is its lack of *memory property*.

Consider a random variable X which has only two possible outcomes - *success* or *failure*, represented by the state space $\{0,1\}$, i.e. 0 is failure and 1 is success. Let q be the probability that an outcome is a failure, i.e. $Pr\{X=0\}=q$ and $Pr\{X=1\}=p=1-q$ be the probability that it is a success.

The mean number of successes in one trial is given as

$$E[X]=0\times q+1\times p=p.$$

Let $\theta^*(z)$ be the $z-$transform of this random variable and given as

$$\theta^*(z)=q+pz.$$

Also we have

$$E[X]=\frac{d(q+pz)}{dz}|_{z\to 1}=p. \tag{2.9}$$

Suppose we carry out an experiment which has only two possible outcomes 0 or 1 and each experiment is independent of the previous and they all have the same outcomes X, we say this is a Bernoulli process,and $\theta(z)$ is the $z-$transform of this Bernoulli process. Further, let τ be a random variable that represents the time (number of trials) by which the first success occurs, where all the trials are independent of each other. It is simple to show that if the first success occurs at the τth trial then the first $\tau-1$ trials must have been failures. The random variable τ has a geometric distribution and

$$\Pr\{\tau=n\}=q^{n-1}p,\ n\geq 1. \tag{2.10}$$

The mean interval for a success is

$$E[X]=\sum_{n=1}^{\infty} nq^{n-1}p=p^{-1}. \tag{2.11}$$

We chose to impose the condition that $n\geq 1$ because in the context of discrete time queueing systems our inter-arrival times and service times have to be at least one unit of time long, respectively. Let the $z-$ transform of this geometric distribution be $T^*(z)$, we have

$$T^*(z)=pz(1-qz)^{-1},\ |z|\leq 1. \tag{2.12}$$

We can also obtain the mean time to success from the $z-$transform as

$$E[X] = \frac{T^*(z)}{dz}|_{z\to 1} = p^{-1}. \tag{2.13}$$

In the context of arrival process, T is the inter-arrival time. Success implies an arrival. Hence, at any time, the probability of an arrival is p and no arrival is q. This is known as the Bernoulli process.

Let Z_m be the number of arrivals in the time interval $[0,m]$, with $Pr\{Z_m = j\} = a_{m,j},\ j \geq 0,\ m \geq 1$ and $a_m^*(z) = \sum_{j=0}^m a_{m,j} z^j$. Since the outcome of time trial follows a Bernoulli process then we have

$$a_m^*(z) = (q+pz)^m,\ m \geq 1,\ |z| \leq 1. \tag{2.14}$$

The distribution of Z_m is given by the Binomial distribution as follows:

$$\Pr\{Z_m = i\} = a_{m,i} = \binom{m}{i} q^{m-i} p^i,\ 0 \leq i \leq m \tag{2.15}$$

The variable $a_{m,i}$ is simply the coefficient of z^i in the term $a_m^*(z)$. The mean number of successes in m trials is given as

$$E[Z_m] = \frac{a_m^*(z)}{dz}|_{z\to 1} = mp. \tag{2.16}$$

In the context of service times, T is the service time of a customer. Success implies the completion of a service. Hence, at any time the probability of completing an ongoing service is p and no service completion is q.

This is what is known as the lack of memory property.

2.5.2.1 Lack of Memory Property:

This lack of memory property is a feature that makes geometric distribution very appealing for use in discrete stochastic modelling. It is shown as follows:

$$\Pr\{T = n+1|T > n\} = \frac{\Pr\{T = n+1\}}{\Pr\{T > n\}} = \frac{q^n p}{\sum_{m=n+1}^{\infty} q^{m-1} p} = \frac{q^n p}{q^n} = p \tag{2.17}$$

which implies that the duration of the remaining portion of a service time is independent of how long the service had been going on, i.e. *lack of memory*.

The mean inter-arrival time or mean service time is given by p^{-1}.

Throughout the rest of this book when we say a distribution is geometric with parameter p we imply that it is governed by a Bernoulli process that has the success probability of p.

While there are several discrete distributions of interest to us in queueing theory, we find that most of them can be studied under the general structure of what is known as the Phase type distribution.

2.5.3 Phase Type Distribution

Phase type distributions are getting to be very commonly used these days after Neuts (1981) made them very popular and easily accessible. They are often referred to as the PH distribution. The PH distribution has become very popular in stochastic modelling because it allows numerical tractability of some difficult problems and in addition several distributions encountered in queueing seem to resemble the PH distribution. In fact, Johnson and Taaffe (1989) have shown that most of the commonly occurring distributions can be approximated by the phase type distributions using moment matching approach based on three moments. The approach is based on using mixtures of two Erlang distributions - not necessarily of common order. They seem to obtain very good fit for most of the cases which they studied. Other works of fitting phase-type distributions include those of Asmussen and Nerman (1991), Bobbio and Telek (1992), and Bobbio and Cumani (1992). The data fitting works by most of these authors are for the continuous PH distributions. There are several other works by Telek and his team for the fitting discrete PH (Bobbio et al (2004)).

Phase type distributions are distributions of the time until absorption in an absorbing Markov chain. If after an absorption the chain is restarted, then it represents the distribution of a renewal process.

Consider an (n_t+1) absorbing discrete time Markov chain (DTMC) with state space $\{0,1,2,\cdots,n_t\}$ and let state 0 be the absorbing state. Let T be an m-dimension sub-stochastic matrix with entries

$$T = \begin{bmatrix} T_{1,1} & \cdots & T_{1,n_t} \\ \vdots & \vdots & \vdots \\ T_{n_t,1} & \cdots & T_{n_t,n_t} \end{bmatrix}$$

and also let

$$\boldsymbol{\alpha} = [\alpha_1, \alpha_2, \ldots, \alpha_{n_t}], \text{ and } \boldsymbol{\alpha}\mathbf{1} \le 1.$$

An example of a PH with $n = 2$ states is shown in Fig. 2.1.

In this context, α_i is the probability that the system starts from a transient state i, $1 \le i \le n_t$, and T_{ij} is the probability of transition from a transient state i to a transient state j. We say the phase type distribution is characterized by $(\boldsymbol{\alpha}, T)$ of dimension n_t. We also define α_0 such that $\boldsymbol{\alpha}\mathbf{1} + \alpha_0 = 1$.

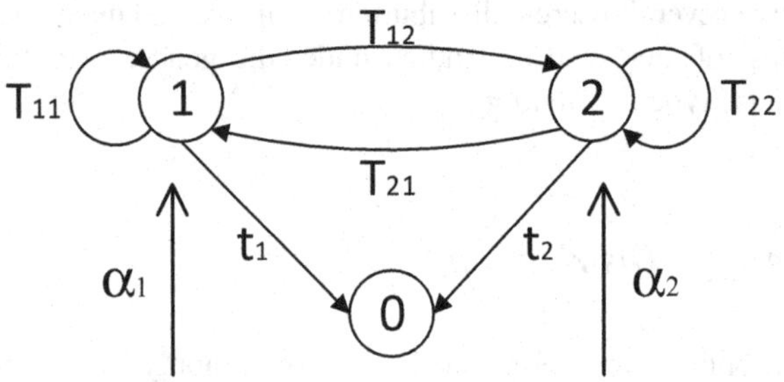

Fig. 2.1 A Two State PH Distribution Diagram

The transition matrix P of this absorbing Markov chain is given as

$$P = \begin{bmatrix} 1 & 0 \\ \mathbf{t} & T \end{bmatrix} \tag{2.18}$$

where $\boldsymbol{t} = \mathbf{1} - T\mathbf{1}$.

The phase type distribution with parameters $\boldsymbol{\alpha}$ and T is usually written as PH distribution with representation $(\boldsymbol{\alpha}, T)$. The matrix T and vector $\boldsymbol{\alpha}$ satisfy the following conditions:

Conditions:

1. Every element of T is between 0 and 1, i.e. $0 \leq T_{i,j} \leq 1$.
2. At least for one row i of T we have $\sum_{j=1}^{n_t} T_{i,j}\mathbf{1} < 1$.
3. The matrix $T + \mathbf{t}\boldsymbol{\alpha}$ is irreducible.
4. $\boldsymbol{\alpha}\mathbf{1} \leq 1$ and $\alpha_0 = 1 - \boldsymbol{\alpha}\mathbf{1}$.

Throughout this book, whenever we write a PH distribution $(\boldsymbol{\alpha}, T)$ there is always a bolded lower case column vector associated with it; in this case $\mathbf{t}$. As another example, if we have a PH distribution $(\boldsymbol{\beta}, S)$ then there is a column vector $\mathbf{s}$ which is given as $\mathbf{s} = \mathbf{1} - S\mathbf{1}$.

If we now define p_i as the probability that the time to absorption into state $n_t + 1$ is i, then we have

$$p_0 = \alpha_0, \tag{2.19}$$

$$p_i = \alpha T^{i-1}\mathbf{t}, \;\; i \geq 1. \tag{2.20}$$

Let $p^*(z)$ be the $z-$transform of this PH distribution, then

$$p^*(z) = \alpha_0 + z\boldsymbol{\alpha}(I - zT)^{-1}\mathbf{t}, \; |z| \leq 1. \tag{2.21}$$

Then nth factorial moment of the time to absorption is given as

$$\mu_n' = n!\boldsymbol{\alpha} T^{n-1}(I-T)^{-n}\mathbf{1}. \tag{2.22}$$

Specifically the mean time to absorption is

$$\mu_1' = E[X] = \boldsymbol{\alpha}(I-T)^{-1}\mathbf{1}. \tag{2.23}$$

We will show later in the study of the phase renewal process that also

$$\mu_1' = E[X] = (\boldsymbol{\pi}\mathbf{t})^{-1}, \tag{2.24}$$

where

$$\boldsymbol{\pi} = \boldsymbol{\pi}(T+\mathbf{t}\boldsymbol{\alpha}), \quad \boldsymbol{\pi}\mathbf{1} = 1.$$

Example:
Consider a phase type distribution with representation $(\boldsymbol{\alpha}, T)$ given as

$$T = \begin{bmatrix} 0.1 & 0.2 & 0.05 \\ 0.3 & 0.15 & 0.1 \\ 0.2 & 0.5 & 0.1 \end{bmatrix}, \quad \boldsymbol{\alpha} = [0.3\ 0.5\ 0.2],\ \alpha_0 = 0.$$

For this $\alpha_0 = 0$ and $\mathbf{t} = [0.65\ 0.45\ 0.2]^T$. We have

$$p_1 = 0.46,\ p_2 = 0.2658,\ p_3 = 0.1346,\ p_4 = 0.0686,\ p_5 = 0.0349,$$

$$p_6 = 0.0178,\ p_7 = 0.009,\ p_8 = 0.0046,\ p_9 = 0.0023,\ p_{10} = 0.0012,$$

$$p_{11} = 0.00060758,\ p_{12} = 0.0003093,\ \cdots$$

Alternatively we may report our results as the complement of the cumulative distribution, i.e.

$$P_k = Pr\{X \geq k\} = 1 - Pr\{X \leq k\} = \boldsymbol{\alpha} T^{k-1}(I-T)^{-1}\mathbf{t}.$$

This is given as

$$P_1 = 1.0,\ P_2 = 0.54,\ P_3 = 0.2743,\ P_4 = 0.1397,\ P_5 = 0.0711,\ P_6 = 0.0362,\ \cdots.$$

2.5.3.1 Two very important closure properties of phase type distributions:

Consider two discrete random variables X and Y that have phase type distributions with representations $(\boldsymbol{\alpha},T)$ and $(\boldsymbol{\beta},S)$.

1. **Sum:** Their sum $Z = X + Y$ has a phase type distribution with representation $(\boldsymbol{\delta},D)$ with

$$D = \begin{bmatrix} T & \mathbf{t}\boldsymbol{\beta} \\ \mathbf{0} & S \end{bmatrix}, \quad \boldsymbol{\delta} = [\boldsymbol{\alpha} \;\; \alpha_0\boldsymbol{\beta}].$$

2. **Mixture:** Their mixture with $[\theta_1, \theta_2]$, $0 \le \theta_i \le 1$, $i = 1,2$, mixing density $(\theta_1 + \theta_2 = 1)$ has a phase type distribution with representation $(\boldsymbol{\delta},D)$ with

$$D = \begin{bmatrix} T & \mathbf{0} \\ \mathbf{0} & S \end{bmatrix}, \quad \boldsymbol{\delta} = [\theta_1\boldsymbol{\alpha} \;\; \theta_2\boldsymbol{\beta}].$$

3. **Minimum:** Their minimum $W = min(X,Y)$ has a phase type distribution with representation $(\boldsymbol{\delta},D)$ with

$$D = T \otimes S, \quad \boldsymbol{\delta} = [\boldsymbol{\alpha} \otimes \boldsymbol{\beta}].$$

4. **Maximum:** Their maximum $U = max(X,Y)$ has a phase type distribution with representation $(\boldsymbol{\delta},D)$ with

$$D = \begin{bmatrix} T \otimes S & \mathbf{t} \otimes S & T \otimes \mathbf{s} & \mathbf{t} \otimes \mathbf{s} \\ \mathbf{0} & T & \mathbf{0} & \mathbf{0} \\ \mathbf{0} & \mathbf{0} & S & \mathbf{0} \\ \mathbf{0} & \mathbf{0} & \mathbf{0} & \mathbf{0} \end{bmatrix}, \quad \boldsymbol{\delta} = [\boldsymbol{\alpha} \otimes \boldsymbol{\beta}, \; \alpha_0\boldsymbol{\beta}, \; \boldsymbol{\alpha}\beta_0, \; \mathbf{0}].$$

These results can be extended directly to the case of more than two PH distributions.

2.5.3.2 Minimal coefficient of variation of a discrete PH distribution

The coefficient of variation (cv) of a discrete PH distribution has a different behaviour compared to its continuous counterpart. For example, for some integer $K < \infty$ a random variable X with

$$Pr\{X = k\} = \begin{cases} 1, & k = K, \\ 0, & k \neq K, \end{cases} \tag{2.25}$$

can be represented by the discrete PH distribution. This PH distribution has a cv of zero. This type of case with cv of zero is not encountered in the continuous PH. This information can sometimes be used to an advantage when trying to fit a dataset to a discrete PH distribution. In general for the discrete PH the coefficient of variation is a function of its mean.

A general inequality for the minimal cv of a discrete PH distribution was obtained by Telek (2000) as follows. Consider the discrete PH distribution $(\boldsymbol{\alpha}, T)$ of order n_t. Its mean is given by μ'. Let us write μ' as $\mu' = \lfloor \mu' \rfloor + \langle \mu' \rangle$ where $\lfloor \mu' \rfloor$ is the integer part of μ' and $\langle \mu' \rangle$ is the fractional part with $0 \leq \langle \mu' \rangle < 1$. Telek (2000) proved that the cv of this discrete PH distribution written as $cv(\boldsymbol{\alpha}, T)$ satisfies the inequality

$$cv(\boldsymbol{\alpha}, T) \geq \begin{cases} \frac{\langle \mu' \rangle (1 - \langle \mu' \rangle)}{(\mu')^2}, & \mu' < n_t, \\ \frac{1}{n_t} - \frac{1}{\mu'}, & \mu' \geq n_t. \end{cases} \tag{2.26}$$

The proof for this can be found in Telek Telek (2000).

Throughout this book we will assume that $\alpha_0 = 0$, since we are dealing with queueing systems for which we do not allow inter-arrival times or service times to be zero.

2.5.3.3 Examples of special phase type distributions

Some special cases of discrete phase type distributions include:

1. **Geometric distribution** with $\alpha = 1$, $T = q$, $\mathbf{t} = p = 1 - q$. Then $p_0 = 0$, and $p_i = q^{i-1}p$, $i \geq 1$ and $n_t = 1$. For this distribution, the mean $\mu' = 1/p$ and the $cv(1, q) = 1 - 1/p$
2. **Negative binomial distribution** with

$$\boldsymbol{\alpha} = [1, 0, 0, \ldots, 0], \text{ and } T = \begin{bmatrix} q & p & & \\ & q & p & \\ & & \ddots & \ddots \\ & & & q \end{bmatrix},$$

and n_t is the number of successes we are looking for occurring at the ith trial. It is easy to show that

$$p_i = \boldsymbol{\alpha} T^{i-1} \mathbf{t} = \binom{i-1}{n_t - 1} p^{n_t} q^{i-n_t}, \; i \geq n_t.$$

3. **Mixed Geometric distribution** with

$$\boldsymbol{\alpha} = [\theta_1, \; \theta_2, \; \ldots, \; \theta_n], \; 0 \leq \theta_i \leq 1, \; i = 1, 2, \cdots, n \text{ and } T = \begin{bmatrix} q_1 & & & \\ & q_2 & & \\ & & \ddots & \\ & & & q_n \end{bmatrix},$$

with $\sum_{i=1}^{n} \theta_i = 1$ and $0 < q_i < 1, \forall i$.

4. **Multiple-time-scaled PH Distribution**: Consider a random variable X with PH distribution $(\boldsymbol{\alpha}, T)$ of dimension n. Now consider another random variable Y such that

$$Pr\{X = j\} = Pr\{Y = kj\}, \; j = 1, 2, \cdots; \; k = 1, 2, \cdots,$$

with

$$E[Y] = kE[X].$$

We find that Y has a PH distribution $(\boldsymbol{\beta}, S)$ of dimension kn and its parameters are given as

$$\boldsymbol{\beta} = \boldsymbol{\alpha} \otimes \mathbf{e}_{k-1}^T(k), \;\; S = \begin{bmatrix} S_{11} & \cdots & S_{1n} \\ \vdots & \cdots & \vdots \\ S_{n1} & \cdots & S_{nn} \end{bmatrix},$$

where $S_{ii} = \begin{bmatrix} \mathbf{0} & I_{k-1} \\ T_{ii} & \mathbf{0}^T \end{bmatrix}$ and $S_{ij} = T_{ij}\mathbf{e}_k(k) \otimes \mathbf{e}_k(k)^T, \;\; i \neq j$.

5. **General discrete distribution with finite** support can be represented by a discrete phase type distribution with $\alpha = [1, 0, 0, \ldots, 0]$ and $0 < t_{ij} \leq 1, \; j = i+1$, and $t_{ij} = 0, \; j \neq i+1$ where n_t is the length of the support. For example, for a constant inter-arrival time with value of $n_t = 4$ we have $\boldsymbol{\alpha} = [1, \; 0, \; 0, \; 0]$ and

$$T = \begin{bmatrix} 0 & 1 & & \\ 0 & 0 & 1 & \\ 0 & 0 & 0 & 1 \\ 0 & 0 & 0 & 0 \end{bmatrix}.$$

Note that the general distribution with finite support can also be represented as a phase type distribution with $\boldsymbol{\alpha} = [\alpha_1, \; \alpha_2, \; \cdots, \; \alpha_{n_t}]$ and $t_{ij} = 1, \; i = j-1$, and $t_{ij} = 0, \; i \neq j-1$. For the example of constant inter-arrival time with value of $n_t = 4$ we have $\boldsymbol{\alpha} = [0, \; 0, \; 0, \; 1]$ and

$$T = \begin{bmatrix} 0 & & & \\ 1 & 0 & & \\ 0 & 1 & 0 & \\ 0 & 0 & 1 & 0 \end{bmatrix}.$$

The case of general discrete distribution will be discussed in detail later.

2.5.3.4 Analogy between PH and Geometric distributions

Essentially discrete phase type distribution is simply a matrix version of the geometric distribution. The geometric distribution has parameters p for success (or arrival) and q for failure (no arrival). The discrete phase type with representation $(\boldsymbol{\alpha}, T)$ has $\mathbf{t}\boldsymbol{\alpha}$ for success (arrival) and T for failure (no arrival). So if we consider this in the context of a Bernoulli process we have the $z-$transform of this process as $\theta^*(z) = T + z\mathbf{t}\boldsymbol{\alpha}$. Next we discuss the Phase renewal process.

2.5.3.5 Phase Renewal Process:

Consider an inter-event time X which is described by a phase type distribution with the representation $(\boldsymbol{\alpha}, T)$. The matrix T records transitions with no event occurring and the matrix $\mathbf{t}\boldsymbol{\alpha}$ records the occurrence of an event and the re-start of the renewal process. If we consider the interval $(0,n)$ and define the number of renewals in this interval as $N(n)$ and the phase of the PH distribution at phase $J(n)$ at time n, and define

$$P_{i,j}(k,n) = Pr\{N(n) = k, J(n) = j | N(0) = 0, J(0) = i\},$$

and the associated matrix $P(k,n)$ such that $(P(k,n))_{i,j} = P_{i,j}(k,n)$, we have

$$P(0,n+1) = TP(0,n),\ n \geq 0 \tag{2.27}$$

$$P(k,n+1) = TP(k,n) + (\mathbf{t}\boldsymbol{\alpha})P(k-1,n),\ k = 1,2,\cdots; n \geq 0. \tag{2.28}$$

Define

$$P^*(z,n) = \sum_{k=0}^{n} z^k P(k,n),\ n \geq 0.$$

We have

$$P^*(z,n) = (T + z(\mathbf{t}\boldsymbol{\alpha}))^n. \tag{2.29}$$

This is analogous to the $(p+zq)^m$ for the $z-$ transform of the Binomial distribution presented earlier. It is immediately clear that the matrix $T^* = T + \mathbf{t}\boldsymbol{\alpha}$ is a stochastic matrix that represent the transition matrix of the phase process associated with this process.

Secondly, we have T as the matrix analogue of p in the Bernoulli process while $\mathbf{t}\boldsymbol{\alpha}$ is the analogue of q in the Bernoulli process. Hence for one time epoch $T + z(\mathbf{t}\boldsymbol{\alpha})$ is the $z-$transform of an arrival. This phase renewal process can be found in state i in the long run with probability π_i where $\boldsymbol{\pi} = [\pi_1,\ \pi_2,\ \cdots,\ \pi_{n_t}]$ and it is given by

$$\boldsymbol{\pi} = \boldsymbol{\pi}(T + \mathbf{t}\boldsymbol{\alpha}), \tag{2.30}$$

and

$$\boldsymbol{\pi}\mathbf{1} = 1. \tag{2.31}$$

Hence the average number of arrivals in one time unit is

$$E[Z_1] = \boldsymbol{\pi}(0 \times T + 1 \times (\mathbf{t}\boldsymbol{\alpha}))\mathbf{1} = \boldsymbol{\pi}(\mathbf{t}\boldsymbol{\alpha}))\mathbf{1} = \boldsymbol{\pi}\mathbf{t}. \tag{2.32}$$

This is the arrival rate and its inverse is the mean inter-arrival time of the corresponding phase type distribution, as pointed out earlier on.

Define r_n as the probability of a renewal at time n, then we have

$$r_n = \boldsymbol{\alpha}(T + \mathbf{t}\boldsymbol{\alpha})^{k-1}\mathbf{t}, \;\; k \geq 1. \tag{2.33}$$

Keep in mind that $\boldsymbol{\pi}$ is the solution to the equations

$$\boldsymbol{\pi} = \boldsymbol{\pi}(T + \mathbf{t}\boldsymbol{\alpha}), \;\; \boldsymbol{\pi}\mathbf{1} = 1.$$

This $\boldsymbol{\pi}$ represents the probability vector of the PH renewal process being found in a particular phase in the long term. If we observe this phase process at arbitrary times given that it has been running for a while, then the remaining time before an event (or for that matter the elapsed time since an event) also has a phase type distribution with representation $(\boldsymbol{\pi}, T)$. It was shown in Latouche and Ramaswami (1999) that

$$\boldsymbol{\pi} = (\boldsymbol{\alpha}(I - T)^{-1}\mathbf{1})^{-1}\boldsymbol{\alpha}(I - T)^{-1}. \tag{2.34}$$

Both the inter-arrival and service times can be represented by the phase type distributions. Generally, phase type distributions are associated with cases in which the number of phases are finite. However, recently, the case with infinite number of phases has started to receive attention.

2.5.4 *The infinite phase distribution (IPH)*

The IPH is a discrete phase type distribution with infinite number of phases. It is still represented as $(\boldsymbol{\alpha}, T)$, except that now we have the number of phases $n_t = \infty$. The IPH was introduced by Shi and Liu (1998). One very important requirement is that the matrix T be irreducible and $T\mathbf{1} \leq 1$, with at least one row being strictly less than 1.

If we now define p_i as the probability that the time to absorption into a state we label as $*$ is i, then we have

$$p_0 = \alpha_0$$
$$p_i = \alpha T^{i-1}\mathbf{t}, \;\; i \geq 1$$

Every other measure carried out for the PH can be easily derived for the IPH. However, we have to be cautious in many instances. For example, the inverse of $I-T$ may not be unique and as such we have to define it as appropriate for the situation under consideration. Secondly computing p_i above requires special techniques at times depending on the structure of the matrix T. The rectangular iteration was proposed by Shi et al (1996) for such computations.

2.5.5 General Inter-event Times

General types of distributions, other than of the phase types, can be used to describe both inter-arrival and service times. In continuous times it is well known that general distributions encountered in queueing systems can be approximated by continuous time PH distributions. This is also true for discrete distributions. However, discrete distributions have an added advantage in that if the distribution has a finite support then it can be represented exactly by discrete PH. We proceed to show how this is true by using a general inter-event time X with finite support and a general distribution given as

$$Pr\{X=j\}=a_j,\ j=1,2,\cdots,n_t<\infty.$$

There are at least two exact PH representations for this distribution, one based on remaining time and the other on elapsed time.

2.5.5.1 Remaining Time Representation

Consider a PH distribution (α,T) of dimension n_t. Let

$$\alpha=[a_1,\ a_2,\ \cdots,\ a_{n_t}], \tag{2.35}$$

and

$$T_{i,j}=\begin{cases}1, & j=i-1\\ 0, & \text{otherwise},\end{cases} \tag{2.36}$$

Then the distribution of this PH is given as

$$\alpha T^{k-1}\mathbf{t}=a_k;\ k=1,2,\cdots,n_t. \tag{2.37}$$

For detailed discussion on this see Alfa (2004).

In general, even if the support is not finite we can represent the general inter-event times with the IPH, by letting $n_t\to\infty$. For example, consider the inter-event times $\mathscr{A}$ which assume values in the set $\{1,2,3,\cdots\}$ with $a_i=Pr\{\mathscr{A}=i\}$, $i=1,2,3,\cdots$. It is easy to represent this distribution as an IPH with $\alpha=[a_1,\ a_2,\ \cdots]$ and $T=\begin{bmatrix}0 & 0\\ I_\infty & 0\end{bmatrix}$.

2.5.5.2 Elapsed Time Representation

Consider a PH distribution $(\boldsymbol{\alpha}, T)$ of dimension n_t. Let

$$\boldsymbol{\alpha} = [1,\ 0,\ \cdots,\ 0], \tag{2.38}$$

and

$$T_{i,j} = \begin{cases} \tilde{a}_i, & j = i+1 \\ 0, & \text{otherwise}, \end{cases} \tag{2.39}$$

where

$$\tilde{a}_i = \frac{u_i}{u_{i-1}},\ u_i = 1 - \sum_{v=1}^{i} a_v,\ u_0 = 1,\ \tilde{a}_{n_t} = 0.$$

Then the distribution of this PH is given as

$$\boldsymbol{\alpha} T^{k-1} \mathbf{t} = a_k;\ k = 1, 2, \cdots, n_t. \tag{2.40}$$

For detailed discussion on this see Alfa (2004). Similarly we can use IPH to represent this distribution using elapsed time by allowing $n_t \to \infty$.

2.5.6 Markovian Arrival Process

All the inter-event times discussed so far are of the renewal types and are assumed to be independent and identically distributed (*iid*). However, in telecommunication queueing systems and most other traffic queueing systems for that matter, inter-arrival times are usually correlated. So the assumption of independent inter-arrival times is not valid in some instances.

Earlier Neuts (1979, 1992) and Lucantoni (1991) presented the Markovian arrival process (MAP) which can handle correlated arrivals and is also tractable mathematically. In what follows, we first describe the single arrival MAP and then briefly present the batch MAP.

Define two sub-stochastic matrices D_0 and D_1, both of the dimensions n. The elements $(D_0)_{ij}$ refer to transition from state i to state j without an (event) arrival because the transitions are all within the n_t transient states. The elements $(D_1)_{ij}$ refer to transition from state i into the absorbing state 0 with an instantaneous restart from the transient state j with an (event) arrival during the absorption. We note that the phase from which an absorption occurred and the one from which the next process starts are connected and hence this captures the correlation between inter-arrival times. The matrix $D = D_0 + D_1$ is a stochastic matrix, and we assume it is irreducible. Note that $D\mathbf{1} = \mathbf{1}$. If we define $\{(N_n, J_n), n \geq 0\}$ as the total number of arrivals and the phase of the MAP at time n, then the transition matrix representing this system is

$$P = \begin{bmatrix} D_0 & D_1 & & & \\ & D_0 & D_1 & & \\ & & D_0 & D_1 & \\ & & & \ddots & \ddots \end{bmatrix}. \tag{2.41}$$

Consider the discrete-time Markov renewal process embedded at the arrival epochs and with transition probabilities defined by the sequence of matrices

$$Q(k) = [D_0]^{k-1} D_1, \; k \geq 1. \tag{2.42}$$

The MAP is a discrete-time point process generated by the transition epochs of that Markov renewal process.

Once more let N_m be the number of arrivals at time epochs $1, 2, \ldots, m$, and J_m the state of the Markov process at time m. Let $P_{r,s}(n,m) = \Pr\{N_m = n,\; J_m = s \mid N_0 = 0,\; J_0 = r\}$ be the (r,s) entry of a matrix $P(n,m)$. The matrices $P(n,m)$ satisfy the following discrete Chapman-Kolmogorov difference equations:

$$P(n, m+1) = P(n,m)D_0 + P(n-1,m)D_1, \; n \geq 1, \; m \geq 0 \tag{2.43}$$

$$P(0, m+1) = P(0,m)D_0 \tag{2.44}$$

$$P(0,0) = I \tag{2.45}$$

where I is the identity matrix and $P(u,v) = 0$ for $u \geq v+1$.

The matrix generating function

$$P^*(z,m) = \sum_{n=0}^{m} P(n,m) z^n, \; |z| \leq 1 \tag{2.46}$$

is given by

$$P^*(z,m) = (D_0 + zD_1)^m, \; m \geq 0 \tag{2.47}$$

If the stationary vector π of the Markov chain described by D satisfies the equation

$$\pi D = \pi, \tag{2.48}$$

and

$$\pi \mathbf{1} = 1 \tag{2.49}$$

then $\lambda' = \boldsymbol{\pi} D_1 \mathbf{1}$ is the probability that, in the stationary version of the arrival process, there is an arrival at an arbitrary point. The parameter λ' is the expected number of arrivals at an arbitrary time epoch or the discrete arrival rate of the MAP.

It is clear that its kth moments about zero are all the same for $k \geq 1$, because $1^j \boldsymbol{\pi} D_1 \mathbf{1} = \boldsymbol{\pi} D_1 \mathbf{1} = \lambda'$, $\forall j \geq 1$, since its jth moment about zero is given as $\sum_{k=0}^{\infty} k^j \boldsymbol{\pi} D_k \mathbf{1}$. Hence its variance σ_X^2 is given as

$$\sigma_X^2 = \lambda'' - (\lambda')^2 = \lambda' - (\lambda')^2 = \lambda'(1 - \lambda'). \tag{2.50}$$

Its jth autocorrelation factor for the number of arrivals, $ACF(j)$, is given as

$$ACF(j) = \frac{1}{\sigma_X^2}[\boldsymbol{\pi} D_1 D^{j-1} D_1 \mathbf{1} - (\lambda')^2], \; j \geq 1. \tag{2.51}$$

The autocorrelation between inter-arrival times is captured as follows. Let X_i be the ith inter-arrival time then we can say that

$$Pr\{X_i = k\} = \boldsymbol{\pi} D_0^{k-1} D_1 \mathbf{1}, \; k \geq 1, \tag{2.52}$$

where $\boldsymbol{\pi} = \boldsymbol{\pi}(I - D_0)^{-1} D_1$, $\boldsymbol{\pi}\mathbf{1} = 1$. Then $E[X_i]$, the expected value of X_i, $\forall i$, is given as

$$E[X_i] = \sum_{j=1}^{\infty} j \boldsymbol{\pi} D_0^{j-1} D_1 \mathbf{1} = \boldsymbol{\pi}(I - D_0)^{-1} \mathbf{1}. \tag{2.53}$$

The autocorrelation sequence r_k, $k = 1, 2, \cdots$ for the inter-arrival times is thus given as

$$r_k = \frac{E[X_\ell X_k] - E[X_\ell]E[X_k]}{E[X_\ell^2] - (E[X_\ell])^2}, \; \ell = 0, 1, 2, \cdots. \tag{2.54}$$

Special Cases:

The simplest MAP is the Bernoulli process with $D_1 = q$ and $D_0 = p = 1 - q$. The discrete phase type distribution is also a MAP with $D_1 = \mathbf{t}\boldsymbol{\alpha}$ and $D_0 = T$.

Another interesting example of MAP is the Markov modulated Bernoulli process (MMBP) which is controlled by a Markov chain which has a transition matrix P and rates given by the $\theta = diag(\theta_1, \theta_1, \ldots, \theta_{n_t})$, $0 \leq \theta_i \leq$, $i = 1, 2, \cdots, n$. In this case, $D_0 = (I - \theta)P$ and $D_1 = \theta P$.

Several examples of discrete MAP can be found in Alfa and Neuts (1995), Alfa and Chakravarthy (1994), Alfa et al (1995), Liu95, Park et al (1994) and Blondia (1992).

2.5.6.1 Platoon Arrival Process (PAP)

The PAP is a special case of MAP which occurs in several traffic situations mainly in telecommunications and vehicular types of traffic. It captures distinctly, in addition to correlation in arrival process, the bursts in traffic arrival process termed platoons here.

The PAP is an arrival process with two regimes of traffic. There is a platoon of traffic (group of packets) which has intra-platoon intervals of arrival times that are identical. The number of arrivals in a platoon is random. At the end of a platoon there is an inter-platoon interval between the end of one platoon and the start of the next platoon arrivals. The inter-platoon intervals are different from the intra-platoon intervals. A good example is if one observes the departure of packets from a router queue, one observes a platoon departure that consists of traffic departing as part of a busy period. The inter-platoon times are essentially the service times. The last packet that departs at the end of a busy period marks the end of a platoon. The next departure will be the head of a new platoon. The interval between the last packet of a platoon and the first packet of the next platoon is the inter-platoon time which in this case is the sum of the service time and the residual inter-arrival time of a packet into the router. If we consider this departure from the router as an input to another queueing system, then the arrival to this other queueing system is a PAP. This type of arrival pattern is also noticed at a signalized road traffic intersection, where departing vehicles up form some kind of platoon when the light turns green. This platoon eventually dissipates, depending on how long the green light lasts.

A PAP is described as follows. Let packets arrive in platoons of random sizes with probability mass function (*pmf*) given as $\{p_k,\ k \geq 1\}$, i.e. p_k is the probability that the number in a platoon is k. Time intervals between arrivals of packets in the same platoon denoted as intra-platoon inter-arrival times have a *pmf* denoted as $\{p_1(j),\ j \geq 1\}$. The time intervals between the arrival of the last packet in a platoon and the first packet of the next platoon is referred to as the inter-platoon inter-arrival times and has the *pmf* denoted by $\{p_2(j), j \geq 1\}$. Let the size of platoon be distributed according to the PH distribution $(\boldsymbol{\delta}, F)$. For this PH let us allow $\delta_0 = 1 - \boldsymbol{\delta}\mathbf{1} \geq 0$, and define $\mathbf{f} = \mathbf{1} - F\mathbf{1}$. Here we have

$$p_k = \begin{cases} \delta_0,\ k = 1 \\ \boldsymbol{\delta} F^{k-1}\mathbf{f},\ k = 1, 2, \cdots \end{cases}. \tag{2.55}$$

The platooned arrival process (PAP) is a discrete-time Markov renewal process whose transition probabilities are described by the sequence of matrices

$$f(j) = \begin{bmatrix} \delta_0 p_2(j) & \boldsymbol{\delta} p_2(j) \\ \mathbf{f} p_1(j) & F p_1(j) \end{bmatrix},\ j \geq 1. \tag{2.56}$$

If we let the intra-platoon and inter-platoon inter-arrival times assume PH distribution then we actually end up with a PAP that is a special case of a MAP.

Let $(\boldsymbol{\alpha}_1, T_1)$ be a PH distribution that describes intra-platoon times and $(\boldsymbol{\alpha}_2, T_2)$ a PH distribution describing the inter-platoon times. Then the PAP is described by two matrices D_0 and D_1, with

$$D_0 = \begin{bmatrix} T_2 & \mathbf{0} \\ \mathbf{0} & I \otimes T_1 \end{bmatrix}, \tag{2.57}$$

and

$$D_1 = \begin{bmatrix} \delta_0 \mathbf{t}_2 \boldsymbol{\alpha}_2 & \boldsymbol{\delta} \otimes \mathbf{t}_2 \boldsymbol{\alpha}_1 \\ \mathbf{f} \otimes \mathbf{t}_1 \boldsymbol{\alpha}_2 & F \otimes \mathbf{t}_1 \boldsymbol{\alpha}_1 \end{bmatrix}. \tag{2.58}$$

Note that $\mathbf{f} = \mathbf{1} - F\mathbf{1}, \ \mathbf{t}_k = \mathbf{1} - T_k\mathbf{1}, \ k = 1, 2.$

Next we explain the elements of the matrices D_0 and D_1.

For the matrix D_0 we have

- the matrix T_2 captures the phase transitions during an inter-platoon arrival time, and
- the matrix $I \otimes T_1$ captures the transitions during an inter-arrival time within a platoon.

For the matrix D_1 we have

- the matrix $\delta_0 \mathbf{t}_2 \boldsymbol{\alpha}_2$ captures the arrival of a platoon of a single packet type, with an end to the arrival of a platoon and the initiation of an inter-platoon inter-arrival time process, whilst
- the matrix $\boldsymbol{\delta} \otimes \mathbf{t}_2 \boldsymbol{\alpha}_1$ captures the arrival of a platoon consisting of at least two packets, with an end to the inter-platoon inter-arrival and the initiation of an intra-platoon inter-arrival time process.
- The matrix $F \otimes \mathbf{t}_1 \boldsymbol{\alpha}_1$ captures the arrival of an intra-platoon packet, with an end to the intra-platoon inter-arrival and the initiation of a new intra-platoon inter-arrival time process, whilst
- the matrix $\mathbf{f} \otimes \mathbf{t}_1 \boldsymbol{\alpha}_2$ captures the arrival of the last packet in a platoon, with an end to the intra-platoon inter-arrival and the initiation of an inter-platoon inter-arrival time process.

A simple example of this is where intra-platoon and inter-platoon inter-arrival times follow geometric distributions, with $p_1(j) = (1-a_1)^{j-1}a_1, \ p_2(j) = (1-a_2)^{j-1}a_2, \ j \geq 1$, and the distribution of platoon size is geometric with $p_0 = 1-b, \ p_k = b(1-c)^{k-1}c, \ k \geq 1$, where $0 < (a_1, a_2, b, c) < 1$. In this case we have

$$D_0 = \begin{bmatrix} 1-a_2 & 0 \\ 0 & 1-a_1 \end{bmatrix}, \quad D_1 = \begin{bmatrix} ba_2 & (1-b)a_2 \\ ba_1 & (1-b)a_1 \end{bmatrix}.$$

For a detail discussion of the discrete PAP see Alfa and Neuts (1995).

2.5.6.2 Batch Markovian Arrival Process (BMAP)

Define substochastic matrices $D_k,\ k \geq 0$, such that $D = \sum_{k=0}^{\infty} D_k$ is stochastic. The elements $(D_k)_{ij}$ refer to a transition from state i to state j with $k \geq 0$ arrivals.

If we define π such that

$$\pi D = \pi, \quad \pi \mathbf{1} = 1$$

then the arrival rate $\lambda' = E[X] = \pi \sum_{k=1}^{\infty} k D_k \mathbf{1}$. Let $\lambda'' = E[X^2] = \pi \sum_{k=1}^{\infty} k^2 D_k \mathbf{1}$ be its second moment about zero, then its variance σ_X^2 is given as

$$\sigma_X^2 = \lambda'' - (\lambda')^2. \tag{2.59}$$

Its jth autocorrelation factor $ACF(j)$ is given as

$$ACF(j) = \frac{1}{\sigma_X^2}[\pi(\sum_{k=1}^{\infty} kD_k)D^{j-1}(\sum_{k=1}^{\infty} kD_k)\mathbf{1} - (\lambda')^2], \ j \geq 1. \tag{2.60}$$

2.5.7 *Marked Markovian Arrival Process*

Another class of arrival process of interest in telecommunication is the marked Markovian arrival process (MMAP[K]), with K classes. It is represented by $K+1$ matrices $D_0,\ D_1,\ \cdots,\ D_K$ all of dimension $n_t \times n_t$. The elements $(D_k)_{i,j},\ k = 1, 2, \cdots, K$, represent transitions from state i to state j with type k packet arrival and $(D_0)_{i,j}$ represents no arrival. Let $D = \sum_{v=0}^{K} D_v$, then D is a stochastic matrix. For a detailed discussion of this class of arrival process see He and Neuts (1998).

An example application of this type of arrival process is in the telecommunications where we may have different paying classes of customers (users) who demand different types of service. The arrivals of these customers may be correlated and a MMAP can capture this process.

2.5.8 *Semi Markov Processes*

Service times with correlations can be described by a class of semi-Markov processes, which are the analogue of the MAP. For such examples, see Alfa and Chakravarthy (1994) and Lucantoni and Neuts (1994).

2.5.9 Data Fitting for PH and MAP

Fitting of PH distribution is both a science and an art. This is because PH representation of a given probability distribution function is not unique. As a trivial example, the geometric distribution $r_k = q^{k-1}p,\ k = 1, 2, \cdots$, can be represented as a PH distribution with parameters $(1, q)$ or even as $(\boldsymbol{\alpha}, T)$, where $\boldsymbol{\alpha} = [0,\ 1,\ 0]$ and $T = \begin{bmatrix} b_1 & b_2 & b_3 \\ 0 & q & 0 \\ c_1 & c_2 & c_3 \end{bmatrix}$, where $0 \leq (b_i, c_i) \leq 1$ and $b_1 + b_2 + b_3 \leq 1,\ \ c_1 + c_2 + c_3 \leq 1$. Several examples of this form can be presented. It is also known that PH representation is non-minimal, as demonstrated by the last example. So, fitting a PH distribution usually involves selecting the number of phases in advance and then finding the best fit using standard statistical methods such as the maximum likelihood method of moments. Alternatively one may select the structure of the PH and then find the best fitting order and parameters. So the art is in the pre-selection process which is often guided by what the PH distribution is going to be used for in the end. For example if the PH is for representing service times in a queueing problem we want to have a small dimension so as to reduce the computational load associated with the queueing model. In some other instances the structure of the PH may be more important if a specific structure will make computation easier.

In general assume we are given a set of N observations of inter-event times $y_1,\ y_2,\ \cdots,\ y_N$, with $\mathbf{y} = [y_1,\ y_2,\ \cdots,\ y_N]$. If we want to fit a PH distribution $(\boldsymbol{\alpha}, T)$ of order n and/or known structure to this set of observations we can proceed by of the following two methods:

- **Method of moments:** We need to estimate a maximum of $n^2 + n - 1$ parameters. This is because we need a maximum of n^2 parameters for the matrix T and $n - 1$ parameters for the vector $\boldsymbol{\alpha}$. If we have a specific structure in mind then the number of unknowns that need to be determined can be reduced. For example if we want to fit a negative binomial distribution of order n then all we need is one parameter. But if it is a general negative binomial then we need n parameters. Let the number of parameters needed to be determined by $m \leq n^2 + n - 1$. Then we need to compute m moments of the dataset $\mathbf{y}$, which we write as $\tilde{\mu}_k$, $k = 1, 2, \cdots, m$. With this knowledge and also knowing that the factorial moments of the PH distribution are given as $\mu_k^{'} = k!\boldsymbol{\alpha}(I - T)^{-k}\mathbf{1}$ we can then obtain the moments of the PH μ_k from the factorial, equate them to the moments from the observed data. This will lead to a set of m non-linear equations, which need to be solved to obtain the best fitting parameters.
- **Maximum likelihood (ML) method:** This second approach is more popularly used. It is based on the likelihoodness of observing the dataset. Let $f_i(y_i, \boldsymbol{\alpha}, T) = \boldsymbol{\alpha}T^{y_i - 1}\mathbf{t},\ i = 1, 2, \cdots, N$. Then the likelihood function is

$$\mathscr{L}(\mathbf{y}, \boldsymbol{\alpha}, T) = \prod_{i=1}^{N} f_i(y_i, \boldsymbol{\alpha}, T).$$

  This function is usually converted to its log form and then different methods can be used to find the best fitting parameters.

Data fitting of the PH distribution is outside the scope of this book. However, several methods can be found in Telek (2000) and Bobbio et al (2004). Bobbio et al (2004) specifically presented methods for fitting acyclic PH. Acyclic PH is a special PH (α, T) for which $T_{i,j} = 0, \forall i \geq j$ and $\alpha_1 = 1$.

There has not been much work with regards to the fitting of discrete MAP. However, Breuer and Alfa (2005) did present the ME algorithm based on ML for estimating parameters for the PAP.

2.6 Service Times: What does this really mean?

We conclude this chapter with a point that has not been addressed much in queueing. What does service distribution really imply? Is the distribution because a server cannot guarantee to process a job that requires k units of work time in k units of time due to its own lack of consistency? Or is it that the server can complete k units of work in k units of time but the arriving customers come with different service time requirements? This has a major effect when considering telecommunication problems.

2.7 Problems

1. Question 1: Packets arrive at a router for processing. Each of the packets requires 2 units of time for set up before processing; this is a set up time which we call S_s. After the set up time of 2 units a packet requires additional units of service and this additional units of service has a geometric distribution with parameter a; i.e. if the additional service is S_a then $Pr\{S_a = j\} = (1-a)^{j-1}a, j \geq 1$. Therefore the total service time of a packet is $S = S_s + S_a$. Show how this S can be represented by a discrete phase type distribution, and give the representation. What is the mean of S, i.e. $E[S]$? Find the z-transform of this processing time. What is the probability that the processing time of two packets is 8 units of time?
2. Question 2: Consider a service system with each item's service time following a geometric distribution with parameter p. Suppose the server can take a break after serving n items. Let X be the length of time the server works before taking a break and $x_k = Pr\{X = k\}$. Write down the expression for x_k, $k \geq n$. If the service time of an individual item follows a phase type distribution (α, T), write down the expression for x_k.
3. Question 3: Consider three boxes with infinite capacities. Packets are generated according to the phase type distribution (α, T). Supposes these items are placed in the boxes according to the following rule; items $1, 4, 7, 10, \cdots$, go into first box; items $2, 5, 8, 11, \cdots$, go into the second box and items $3, 6, 9, 12, \cdots$, go into the third box. Develop an expression that captures the arrival process of items into any of the boxes. Extend the idea to the case of $n \geq 1$ boxes.

4. Question 4: Consider two classes of items A and B. Items A arrive according to the phase type distribution with parameter (α, T) and item B according to phase type distribution (β, S). Write down an expression for the probability that item A arrives before item B. What is the probability that the first arrival of an item is at time n?

References

Alfa AS (2004) Markov chain representations of discrete distributions applied to queueing models. Computers and Operations Research 31:2365–2385

Alfa AS, Chakravarthy S (1994) A discrete queue with Markovian arrival processes and phase type primary and secondary services. Stochastic Models 10:437–451

Alfa AS, Neuts MF (1995) Modelling vehicular traffic using the discrete time Markovian arrival process. Transportation Science 29:109–117

Alfa AS, Dolhun KL, Chakravarthy S (1995) A discrete single server queue with Markovian arrivals and phase type group service. Journal of Applied Mathematics and Stochastic Analysis 8:151–176

Asmussen S, Nerman O (1991) Fitting phase-type distributions via the EM algorithm. In: Symposium i Anvendt Statiskik, K. Vest Nielsen (ed.), Copenhagen, pp 335–346

Blondia C (1992) A discrete-time batch Markovian arrival process as B-ISDN traffic model. Belgian Journal of Operations Research, Statistics and Computer Science 32:3–23

Bobbio A, Cumani A (1992) ML estimation of the parameters of a PH distribution in triangular canonical form. Computer Performance Evaluation, G Balbo an G Serazzi (ed), Elsevier Science Pub BV pp 33–46

Bobbio A, Telek M (1992) Parameter estimation of phase type distributions. In: 20th European Meeting of Statisticians, Bath, UK, pp 335–346

Bobbio A, Hovarth A, Scarpa M, Telek M (2004) Acyclic discrete phase type distributions: properties and a parameter estimation algorithm. Performance Evaluation 54:1 –32

Breuer L, Alfa AS (2005) An EM algorithm for platoon arrival processes in discrete time. Operations Res Lett 33:535–543

He QM, Neuts MF (1998) Markov arrival process with marked transitions. Stochastic Processes and Applications 74:37–52

Johnson M, Taaffe M (1989) Matching moments to phase distributions: mixtures of erlang distributions of common order. Stochastic Models 5

Latouche G, Ramaswami V (1999) Introduction to matrix analytic methods in stochastic modeling. In: Pennsylvania, Philadelphia, Pennsylvania

Lucantoni DM (1991) New results on single server queue with a batch Markovian arrival process. Stochastic Models 7:1–46

Lucantoni DM, Neuts MF (1994) Some steady-state distributions of the MAP/SM/1 queue. Stochastic Models 10

Neuts MF (1979) A versatile Markovian point process. Journ Appl Prob 16:764–779

Neuts MF (1981) Matrix-geometric solutions in stochastic models - an algorithmic approach. In: The Johns Hopkins University Press, Baltimore

Neuts MF (1992) Models based on the Markovian arrival process. IEICE Trans Commun E75-B:1255–1265

Park D, H P, Yamashita H (1994) Approximate analysis of discrete-time tandem queuing networks with bursty and correlated input traffic and customer loss. Oper Res Letters 15:95–104

Shi D, Liu D (1998) Markovian models for non-negative random variables. In: Alfa AS, Chakravarthy S (eds) Advances in Matrix-analytic Methods in Stochastic Models, Notable Publications, pp 403–427

Shi D, Guo J, Liu L (1996) SPH-distributions and the rectangle-iterative algorithm. In: Chakravarthy S, Alfa AS (eds) Matrix-analytic Methods in Stochastic Models, Marcel Dekker, pp 207–224

Telek M (2000) The minimal coefficient of variation of discrete phase type distributions. 3rd Conf on MAM pp 391–400

Wolff RW (1989) Stochastic Modelling and the Theory of Queues. Prentice hall, Engelwood Cliffs, New Jersey

Chapter 3
Discrete-Time Markov Chains

3.1 Introduction

Markov processes are a special class of stochastic processes. In order to fully understand Markov processes we first need to introduce stochastic processes. However, to help us understand stochastic process we need to remember the basic probability theory associated with it, which was briefly reviewed in the last chapter.

3.2 Stochastic Processes

Consider a sample space $\mathscr{S}$ and a random variable $X(\omega)$ where $\omega \in \mathscr{S}$. A stochastic process may be defined as a collection of points $X(t,\omega)$, $t \geq 0$, indexed over the parameter t, usually representing time but may represent others such as space, etc. Put in a simpler form we can say a stochastic process is a collection of random variables $X(t)$, $t \geq 0$, which is observed at time t. Examples of a stochastic process could be the amount of rainfall $X(t)$ on day t, the remaining gasoline in the tank of an automobile at any time, etc. Most quantities studied in a queueing system involve a stochastic process. We will introduce the concept in the context of discrete time queues in the next section.

In this section, we consider integer valued stochastic processes in discrete times. Throughout this book we only consider non-negative integer valued processes. Consider a state space $\mathscr{I} = \{0,1,2,3,\cdots\}$ and a state space $\mathscr{I}_s$ which is a subset of $\mathscr{I}$, i.e. $\mathscr{I}_s \subseteq \mathscr{I}$. Now consider a random variable X_n which is observable at time $n = 0,1,2,\cdots$ in discrete time, where $X_n \in \mathscr{I}_s$.

The collection of these $X_0, X_1, X_2, \cdots$ form a stochastic process in discrete time; or written as $\{X_n, n = 0,1,2,\cdots\}$. A discrete parameter stochastic process is simply a family of random variables X_n, $n = 0,1,2,\cdots$ which evolves with parameter n, and in this case n could be time or any other quantity, such as an event count.

A.S. Alfa, *Applied Discrete-Time Queues*, DOI 10.1007/978-1-4939-3420-1_3

This collection describes how the system evolves with time. Our interest is to study the relations of these collections. For example X_n could be the number of data packets that arrive at a router at time n, $n = 0,1,2,\cdots$, or the number of packets left behind in the router by the nth departing packet. Examples of stochastic processes usually of interest to us in queueing are the number of packets in a queue at a particular time or between short intervals of time, the waiting time of packets which arrive in the system at particular times or between intervals of time, the waiting time of the nth packet, etc.

Of interest to us about a stochastic process is usually its probabilistic aspects; the probability that X_n assumes a particular value. Also of interest to us is the behaviour of a collection of a group of the random variables. For example the $Pr\{X_m = a_m, X_{m+1} = a_{m+1}, \cdots, X_{m+v} = a_{m+v}\}$. Finally we may be interested in the moments of the process at different times.

As an example, consider a communication processing system to which messages arrive at time $A_1, A_2, A_3, \ldots$, where the nth message arrives at time A_n. Let the nth message require S_n units of time for processing. We assume that the processor can only serve one message at a time in a first come first served order. Let there be an infinite waiting space (buffer) for messages waiting for service. Let X_t be the number of messages in the system at time t. X_t is a stochastic process. Further let D_n be the departure time of the nth message from the system after service completion.

In order to understand all the different ways in which this stochastic process can be observed in discrete time, we consider a particular realization of this system. For example, let us consider the process for up to 10 message arrivals for a particular realization as shown in Table (3.1) We can now study this process at any point in time.

In discrete time, we observe the system at times $t_1, t_2, t_3, \ldots$, where for simplicity, without loss of generality, we set $t_1 = 0$ to be the time of the first arrival, we have several choices of the values assumed by t_i, $i > 1$ in that we can study the system at:

Table 3.1 Example of a sample path

n	A_n	S_n	D_n
1	0	3	3
2	2	4	7
3	5	1	8
4	6	3	11
5	7	6	17
6	12	2	19
7	14	4	23
8	15	5	28
9	19	1	29
10	25	3	32
$\vdots$	$\vdots$	$\vdots$	$\vdots$

- Equally spaced time epochs such that $t_2 - t_1 = t_3 - t_2 = t_4 - t_3 = ... = \tau$. For example, if $\tau = 1$, then $(t_1, t_2, t_3, ...) = (0, 1, 2, ...)$. In which case, $X_{t_1} = X_0 = 1$, $X_{t_2} = X_1 = 1$, $X_{t_3} = X_2 = 2$, $X_{t_4} = X_3 = 1$, $X_{t_5} = X_4 = 1$, $X_{t_6} = X_5 = 2$, $X_{t_7} = X_6 = 3, \cdots$
- Arbitrarily spaced time epochs such that $t_2 - t_1 = \tau_1$, $t_3 - t_2 = \tau_2$, $t_4 - t_3 = \tau_3$, . . . with τ_1, τ_2, τ_3, . . . not necessarily being equal. For example, if $\tau_1 = 1$, $\tau_2 = 2$, $\tau_3 = 1$, $\tau_4 = 4$, . . . hence $(t_1, t_2, t_3, t_4, t_5, \cdots) = (0, 1, 3, 4, 8, \cdots)$ we have $X_{t_1} = X_0 = 1, X_{t_2} = X_1 = 1, X_{t_3} = X_3 = 1, X_{t_4} = X_4 = 1, X_{t_5} = X_8 = 2, \cdots$
- Points of event occurrences. For example if we observe the system at departure times of messages, i.e. $(t_1, t_2, t_3, ...) = (D_1, D_2, D_3, ...) = (3, 7, 8, \cdots)$. In this case we have $X_{t_1} = X_3 = 1, X_{t_2} = X_7 = 3, X_{t_3} = X_8 = 2, \cdots$

Any events that occur between the time epochs of our observations are not of interest to us in discrete time analysis. We are only taking "snap shots" of the system at the specific time epochs.

By this assumption, then any discrete time observation is contained in continuous time observations.

Note that we may also have multivariate stochastic processes. For example, let X_n be the number of packets in the system at time n and let $Y_n \leq X_n$ be how many of the X_n packets require special service. The stochastic process for this system is $\{(X_n, Y_n), n = 01, 2, 3 \cdots\}$, $X_n \in \mathscr{I}$, $Y_n \leq X_n$. We will discuss multivariate processes later when we consider Markov chains.

3.3 Markov Processes

In 1907, A. A. Markov published a paper Markov (1907) in which he defined and investigated the properties of what are now called as Markov processes. A Markov process is a special stochastic process in which the state of the system in the future is independent of the past history of the system but dependent on only the present. The consideration of Markov processes is very important and in fact, is central to the study of queuing systems because one usually attempts to reduce the study of nearly all queueing systems to the study of Markov process.

Consider a stochastic process X_n in discrete time $(n = 0, 1, 2, ...)$. The set of possible values that X_n takes on is referred as its state space. X_n is a Markov process if

$$Pr\{X_{n+1} \in A_{n+1} | X_n = x_n, X_{n-1} \in A_{n-1}, ..., X_0 \in A_0\}$$
$$= Pr\{X_{n+1} \in A_{n+1} | X_n = x_n\}$$

holds for each state integral n, each state x_n and the set of states A_{n+1}, A_{n-1}, A_0.

If the numbers in the state space of a Markov process are finite or countable, the Markov process is called a Markov chain. In this case, the state space is usually assumed to be the set of integers {0, 1, . . .}. In the case that the numbers in the state space of a Markov process are over a finite or infinite continuous interval (or set of such intervals), the Markov process is referred to as a Markov process. For most times in queueing applications, Markov chains are more common even though continuous space Markov processes, in the form of stochastic fluid models, are getting popular for applications to queueing systems. We focus mainly on Markov chains in this chapter.

3.4 Discrete Time Markov Chains

Consider a discrete time stochastic process X_0, X_1, . . ., with discrete (i.e. finite or countable) state space $\mathscr{I}_c = \{i_0, i_1, i_2, ...\} \subseteq \mathscr{I}$. If

$$\begin{aligned} &Pr\{X_{n+1} = i_{n+1} | X_n = i_n, X_{n-1} = i_{n-1}, ..., X_0 = i_0\} \\ &= Pr\{X_{n+1} = i_{n+1} | X_n = i_n\} \end{aligned}$$

holds for any integral n, and the states i_{n+1}, i_n, i_{n-1}, . . ., i_0, then X_n is said to be a discrete time Markov chain (DTMC). If further we have

$$Pr\{X_{n+m+1} = j | X_{n+m} = i\} = Pr\{X_{n+1} = j | X_n = i\}, \forall (i,j) \in \mathscr{I}, \forall (m,n) \geq 0,$$

we say the Markov chain is time-homogeneous or stationary. Throughout most of this book we shall be dealing with this class of DTMC.

Generally we say

$$p_{i,j}(n) = P\{X_{n+1} = j | X_n = i\}$$

is the transition probability from state i at time n to state j at time $n+1$. However, for most of our discussions in this book (until later chapters) we focus on the case where the transition is independent of the time, i.e. $p_{i,j}(n) = p_{i,j}$, $\forall n$. The matrix $P = (p_{i,j}), (i,j) \in \mathscr{I}$ is called the transition probability matrix of the Markov chain, in which

$$p_{i,j} = Pr\{X_{n+1} = j | X_n = i\}.$$

The basic properties of the elements of the transition probability matrix of a Markov chain are as follows:

Properties:

$$0 \leq p_{i,j} \leq 1, \quad \forall i \in \mathscr{I}, \forall j \in \mathscr{I}$$
$$\sum_{j \in \mathscr{I}} p_{ij} = 1$$

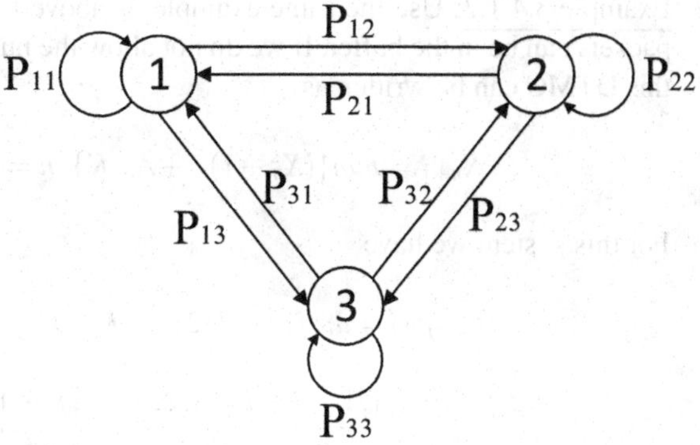

Fig. 3.1 Three States Markov Chain

For example consider a DTMC with the state space $\{1,2,3\}$. We show, in Fig. 3.1, its transition behaviour between time n and $n+1$.

3.4.1 Examples

1. Example 3.4.1.1: Consider a discrete stochastic process X_n which is the number of data packets in the buffer of a router at time n. At time n one packet is removed from the buffer if it is not empty and A_n packets are added at the same time. Let $\{A_n,\ n=0,1,2,\cdots\}$ be a discrete stochastic process which assumes values in the set $\mathscr{I}$. This DTMC can be written as

$$X_{n+1} = (X_n - 1)^+ + A_n, \quad n = 0,1,2,3,\cdots,$$

where $(z)^+ = max\{z,0\}$. For example, let $A_n = A,\ \forall n$ and $a_k = Pr\{A = k\}$, $k = 0,1,2,3,\cdots$, then

$$p_{0,j} = a_j,\ j = 0,1,2,\cdots,$$

$$p_{i,j} = a_{j-i+1}, 1 \le i \le j = 1,2,3,\cdots,$$

and

$$p_{i,j} = 0,\ j < i.$$

The transition matrix P associated with this DTMC can be written as

$$P = \begin{bmatrix} a_0 & a_1 & a_2 & a_3 & a_4 & \cdots \\ a_0 & a_1 & a_2 & a_3 & a_4 & \cdots \\ & a_0 & a_1 & a_2 & a_3 & \cdots \\ & & a_0 & a_1 & a_2 & \cdots \\ & & & \ddots & \ddots & \ddots \end{bmatrix}.$$

2. Example 3.4.1.2: Use the same example as above but put a limit on how many packets can be in the buffer. If we do not allow the number to exceed $K < \infty$, then this DTMC can be written as

$$X_{n+1} = min\{(X_n - 1)^+ + A_n, K\} \ \ n = 0, 1, 2, 3, \cdots.$$

For this system we have

$$p_{0,j} = a_j, \ j = 0, 1, 2, \cdots, K-1, \ \ p_{0,K} = \tilde{a}_K,$$

$$p_{i,j} = a_{j-i+1}, 1 \le i \le j = 1, 2, 3, \cdots, K-1, \ p_{i,K} = \tilde{a}_{K-i+1},$$

and

$$p_{i,j} = 0, \ j < i,$$

where $\tilde{a}_w = \sum_{v=w}^{\infty} a_v$.

3. Example 3.4.1.3: Consider a discrete stochastic process Y_n which is the number of packets in a buffer at time n. Suppose at time n, one packet is added to contents of the buffer and the minimum of $Y_n + 1$ and B_n items are removed from the buffer, where $\{B_n, \ n = 0, 1, 2, \cdots\}$ is a discrete stochastic process which assumes values in the set $\mathscr{I}$. This DTMC can be written as

$$Y_{n+1} = (Y_n + 1 - B_n)^+, \ \ n = 0, 1, 2, 3, \cdots,$$

where $(z)^+ = max\{z, 0\}$. For example, let $B_n = B, \ \forall n, \ b_k = Pr\{B = k\}, \ k = 0, 1, 2, 3, \cdots$, and $\tilde{b}_k = \sum_{v=k+1}^{\infty} b_v$, then

$$p_{i,0} = \tilde{b}_i, \text{ and } p_{i,i+1} = b_0, \ \ i = 0, 1, 2, \cdots,$$

$$p_{i,j} = b_{i+1-j}, 1 \le j \le i = 1, 2, 3, \cdots,$$

and

$$p_{i,j} = 0, \ i < j.$$

The transition matrix P associated with this DTMC can be written as

$$P = \begin{bmatrix} \tilde{b}_0 & b_0 & \cdots & & \\ \tilde{b}_1 & b_1 & b_0 & \cdots & \\ \tilde{b}_2 & b_2 & b_1 & b_0 & \cdots \\ \vdots & \vdots & \vdots & \vdots & \ddots \end{bmatrix}.$$

4. Example 3.4.1.4: Consider two types of items; type A and type B. There are N of each, i.e. a total of $2N$ items. Suppose these $2N$ items are randomly divided into two sets of equal numbers with N placed in one urn and the remaining N placed in the other urn. If there are X_n of type A items in the first urn at time n then we have $N-X_n$ of type B in the first urn. If we now take one item each at random from each urn and exchange their urns, then it is clear that X_n, $n = 0,1,2,3,\cdots$ is a DTMC represented as

$$X_{n+1} = \begin{cases} X_n - 1, \text{ with probability } \frac{X_n^2}{N^2} \\ X_n, \quad \text{ with probability } \frac{2X_n(N-X_n)}{N^2} \\ X_n + 1, \text{ with probability } \frac{(N-X_n)^2}{N^2} \end{cases}$$

This is the well known Bernoulli-Laplace diffusion model. We can write

$$p_{i,i-1} = (\frac{i}{N})^2,$$

$$p_{i,i} = 2\frac{i}{N}(1 - \frac{i}{N}),$$

$$p_{i,i+1} = (1 - \frac{i}{N})^2.$$

For the example of when $N = 4$ we can write the transition matrix as

$$P = \begin{bmatrix} 0 & 1 & & & \\ \frac{1}{16} & \frac{3}{8} & \frac{9}{16} & & \\ & \frac{1}{4} & \frac{1}{2} & \frac{1}{4} & \\ & & \frac{9}{16} & \frac{3}{8} & \frac{1}{16} \\ & & & 1 & 0 \end{bmatrix}.$$

5. Example 3.4.1.5: Consider a single server queueing system in which arrivals and service of packets are independent. Let A_n be the inter-arrival time between the nth and the $n-1^{st}$ packet, S_n the processing time of the nth packet and W_n the waiting time of the nth packet, then we can see that W_n, $n = 1,2,3,\cdots$ is a Markov chain given as

$$W_{n+1} = (W_n + S_n - A_{n+1})^+.$$

Here the parameter n is not time but a sequential label of a packet.

6. Example 3.4.1.6: The Gilbert-Elliot (G-E) channel model is a simple, but effective two-state DTMC. Consider a communication channel which is subject to noise. The G-E model assumes that at all times the channel is in one of two possible states: state 0 – channel is good, and state 1 – channel is bad. The state

of the channel at the next time slot depends only on the state at the current time slot and not on the previous times. The probability of the channel state going from good to bad is q and probability of it going from bad to good is p. With this system as a DTMC with state space $\{0,1\}$, its transition matrix is given as

$$P = \begin{bmatrix} 1-q & q \\ p & 1-p \end{bmatrix}.$$

7. Example 3.4.1.7: The Fritchman channel model is an extension of the G-E model in that it assumes that the channel could be in one of 2 or more possible states. Let the number of states be M with $2 \leq M < \infty$, which are numbered $\{0,1,2,\cdots,M-1\}$. We let state 0 correspond to the best channel state and state $M-1$ to the worst channel state. The Fritchman model restricts the channel changes from one time to the next time not to jump up or down by more than one state. The probability of going from state i to $i+1$ is given as u and going from i to $i-1$ is given as d, while that of remaining in i is given as a. It also assumes that the channel state at the next time depends only on the current state. The transition matrix for this model is written as

$$P = \begin{bmatrix} 1-u & u & & & & \\ d & a & u & & & \\ & d & a & u & & \\ & & \ddots & \ddots & \ddots & \\ & & & d & a & u \\ & & & & d & 1-d \end{bmatrix}.$$

3.4.2 *State of DTMC at arbitrary times*

One of our interests in DTMC is the state of the chain at an arbitrary time, i.e. $Pr\{X_n = i\}$. This we can obtain by using the total probability rule as

$$Pr\{X_{n+1} = i\} = \sum_{k \in \mathscr{I}} Pr\{X_{n+1} = i | X_n = k\} Pr\{X_n = k\}. \tag{3.1}$$

Let $\boldsymbol{x}^{(n)} = [x_0^{(n)}, x_1^{(n)}, ...,]$ be the state probability vector describing the state of the system at time n, where $x_i^{(n)}$ = Probability that the system is in the state i at time n. Then we can write

$$x_i^{(n+1)} = \sum_{k \in \mathscr{I}} x_k^{(n)} p_{k,i} \tag{3.2}$$

Fig. 3.2 Two States Markov Chain

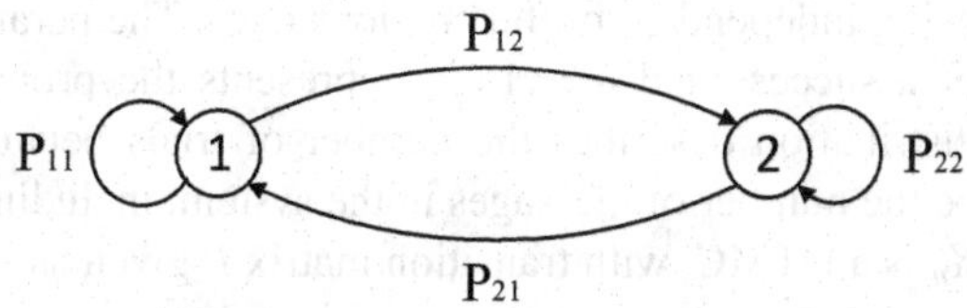

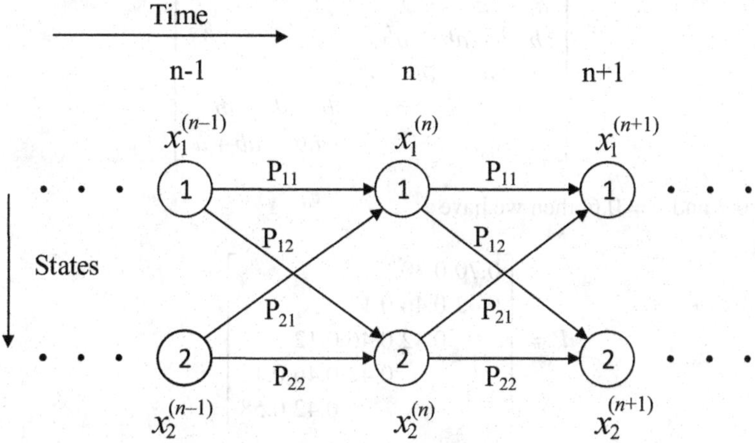

Fig. 3.3 State Probabilities at Arbitrary Times

and in matrix form we have

$$\boldsymbol{x}^{(n+1)} = \boldsymbol{x}^{(n)} P = \boldsymbol{x}^{(0)} P^n. \tag{3.3}$$

Consider a DTMC with the state space $\{1,2\}$, then the transition probabilities and the state of the system at any arbitrary times are shown in the Fig. 3.2.

Another way to view this process is to draw N nodes directed network as shown in Fig. 3.3, for the case of $N = 2$, and then connect the node pairs (i,j) that have $p_{ij} > 0$. If we now consider the case in which the DTMC goes from node i to node j with the probability p_{ij} at each time step, then $x_i^{(n)}$ is the probability that we find the DTMC at node i at time n.

We now introduce some examples as follows:

3.4.2.1 Example

Example 3.4.2.1.1: Consider a processor which serves messages one at a time in order of arrival. Let there be a finite waiting buffer of size 4 for messages. Assume messages arrive according to the Bernoulli process with parameter a ($\bar{a} = 1 - a$) and service is geometric with parameter b ($\bar{b} = 1 - b$). A Bernoulli process is a stochastic process with only two possible outcomes, viz: success or failure, with each trial being independent of the previous ones. The parameter a represents the probability of a success and $\bar{a} = 1 - a$ represents the probability of a failure. A geometric distribution describes the number of trials between successive successes. Let X_n be the number of messages in the system, including the one being served, at time n. X_n is a DTMC, with transition matrix P given as

$$P = \begin{bmatrix} \bar{a} & a & & & \\ \bar{a}b & \bar{a}\bar{b}+ab & a\bar{b} & & \\ & \bar{a}b & \bar{a}\bar{b}+ab & a\bar{b} & \\ & & \bar{a}b & \bar{a}\bar{b}+ab & a\bar{b} \\ & & & \bar{a}b & \bar{a}\bar{b}+a \end{bmatrix}$$

If $a = 0.3$ and $b = 0.6$, then we have

$$P = \begin{bmatrix} 0.70 & 0.30 & & & \\ 0.42 & 0.46 & 0.12 & & \\ & 0.42 & 0.46 & 0.12 & \\ & & 0.42 & 0.46 & 0.12 \\ & & & 0.42 & 0.58 \end{bmatrix}$$

If at time zero the system was empty, i.e. $\boldsymbol{x}^{(0)} = [1,0,0,0,0]$, then

$$\boldsymbol{x}^{(1)} = \boldsymbol{x}^{(0)}P = [0.7, 0.3, 0.0, 0.0, 0.0].$$

Similarly,

$$\boldsymbol{x}^{(2)} = \boldsymbol{x}^{(1)}P = \boldsymbol{x}^{(0)}P^2 = [0.616, 0.348, 0.036, 0.0, 0.0]$$

and

$$\boldsymbol{x}^{(5)} = \boldsymbol{x}^{(4)}P = \boldsymbol{x}^{(0)}P^5 = [0.5413, 0.3639, 0.0804, 0.0130, 0.0014].$$

We also obtain

$$\boldsymbol{x}^{(100)} = \boldsymbol{x}^{(99)}P = \boldsymbol{x}^{(0)}P^{100} = [0.5017, 0.3583, 0.1024, 0.0293, 0.0084].$$

We notice that

$$\boldsymbol{x}^{(101)} = \boldsymbol{x}^{(100)}P = \boldsymbol{x}^{(0)}P^{101} = [0.5017, 0.3583, 0.1024, 0.0293, 0.0084]$$
$$= \boldsymbol{x}^{(100)}, \text{ in this case.}$$

We will discuss this observation later.

Keeping in mind that a Markov chain can also be of infinite state space, consider this same example but where the buffer space is unlimited. In that case, the state size is countably infinite and the transition matrix P is given as

$$P = \begin{bmatrix} \bar{a} & a & & & \\ \bar{a}b & \bar{a}\bar{b}+ab & a\bar{b} & & \\ & \bar{a}b & \bar{a}\bar{b}+ab & a\bar{b} & \\ & & \bar{a}b & \bar{a}\bar{b}+ab & a\bar{b} \\ & & & \ddots & \ddots & \ddots \end{bmatrix}$$

We shall discuss the case of infinite space Markov chains later.

3.4.2.2 Chapman Kolmogorov Equations

Let $P^{(n)}$ be the matrix whose (i,j)th entry refers to the probability that the system is in state j at the nth transition, given that it was in state i at the start (i.e. at time zero), then $P^{(n)} = P^n$, i.e., the matrix $P^{(n)}$ is the nth order transition matrix of the system. In fact, let $p_{ij}^{(n)}$ be the (i,j)th element of $P^{(n)}$ then we have

$$\begin{aligned} p_{ij}^{(2)} &= \sum_{v\in\mathscr{I}} p_{iv}p_{vj} \\ p_{ij}^{(3)} &= \sum_{v\in\mathscr{I}} p_{iv}^{(2)}p_{vj} = \sum_{v\in\mathscr{I}} p_{iv}p_{vj}^{(2)} \\ \vdots &= \vdots \\ p_{ij}^{(n)} &= \sum_{v\in\mathscr{I}} p_{iv}^{(n-1)}p_{vj} = \sum_{v\in\mathscr{I}} p_{iv}p_{vj}^{(n-1)} \end{aligned}$$

which when written out in a more general form becomes

$$p_{ij}^{(n)} = \sum_{v\in\mathscr{I}} p_{iv}^{(m)}p_{vj}^{(n-m)}, \tag{3.4}$$

and in matrix form becomes

$$P^{(n)} = P^{(n-m)}P^{(m)}. \tag{3.5}$$

Since $P^{(2)} = P^2$ from the first of above equation, by a simple induction, we will know $P^{(n)} = P^n$ holds for any non-negative integer n. This is the well known Chapman-Kolmogorov equation.

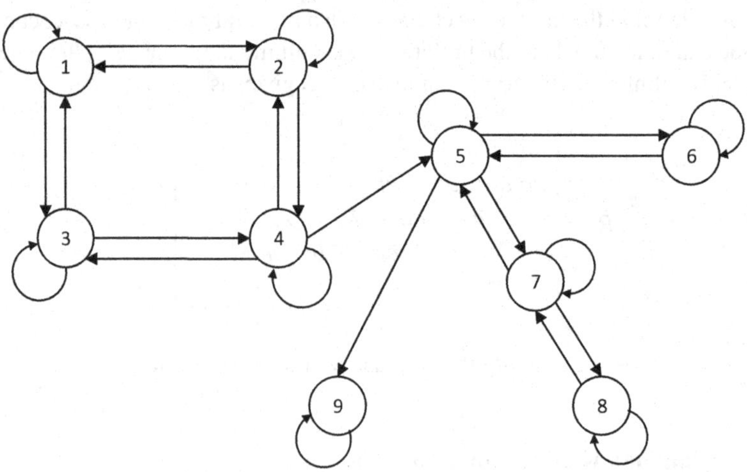

Fig. 3.4 Markov Chain Network Example

3.4.3 Classification of States

In this section, we explain how the states of a Markov chain are classified. Consider a 9 state DTMC shown in Fig. 3.4. We let an arrow from i to j imply that $p_{ij} > 0$, and assume that $p_{ii} > 0$ for all the 9 states. We use this figure to help us in explaining the classes of states of a DTMC.

Communicating States - State j is said to be accessible from state i if there exists a finite n such that $p^n_{i,j} > 0$. If two states i and j are accessible to each other, they are said to communicate. Consider the nine state DTMC with state space $\{1,2,3,4,5,6,7,8,9\}$ and the following transition matrix

$$P = \begin{bmatrix} p_{1,1} & p_{1,2} & p_{1,3} & & & & & & \\ p_{2,1} & p_{2,2} & & p_{2,4} & & & & & \\ p_{3,1} & & p_{3,3} & p_{3,4} & & & & & \\ & p_{4,2} & p_{4,3} & p_{4,4} & p_{4,5} & & & & \\ & & & & p_{5,5} & p_{5,6} & p_{5,7} & & p_{5,9} \\ & & & & p_{6,5} & p_{6,6} & & & \\ & & & & p_{7,5} & & p_{7,7} & p_{7,8} & \\ & & & & & & p_{8,7} & p_{8,8} & \\ & & & & & & & & p_{9,9} \end{bmatrix}.$$

We assume that all the cells with entries have values that are greater than zero.

For example, $p_{1,2} > 0$, whereas, $p_{1,3} = 0$. For example in this DTMC states 1 and 2 do communicate. However state 3 is not accessible from state 2 in one transition step, but it is accessible in two transition steps via state 4, i.e. we have $p^{(2)}_{2,3} \geq p_{2,4}p_{4,3} > 0$, also $p^{(2)}_{3,2} \geq p_{3,4}p_{4,2} > 0$, and hence state 3 is accessible from state 2. Therefore states 2 and 3 do communicate.

This transition matrix can also be shown as a network with each node representing a state and a directed arc joining two nodes signify a non-zero value, i.e. direct access (transition) between the two nodes (states) in one step. Fig. 3.4 represents this transition matrix. In general we say state j is accessible from state i if a path exists in the network from node i to node j, and the two states are said to communicate if there are paths from node $i(j)$ to node $j(i)$.

Absorbing States - a state i is said to be an absorbing state if once the system enters state i it never leaves it. Specifically, state i is an absorbing state if $p_{ii} = 1$. In Fig. 3.4 state 9 is an absorbing state. Also if we set $p_{6,5} = 0$, then state 6 becomes an absorbing state and as we can see $p_{6,6} = 1$ by virtue of it being the only non-zero element in that row.

Transient States - a state i is said to be a transient state if there exists a state j which can be reached in a finite number of steps from state i but state i cannot be reached in a finite number of steps from state j. Specifically, state i is transient if there exists a state j such that $p^{(n)}_{ij} > 0$ for some $n < \infty$ and $p^{(m)}_{ji} = 0, \ \forall m > 0$. In the above example, state 5 can be reached from state 3 in a finite number of transitions but state 3 can not be reached from state 5 in a finite number of transitions, hence state 3 is a transient state.

Recurrent States - a state i is said to be recurrent if the probability that the system ever returns to state i is 1. Every state which is not a transient state is a recurrent state. Consider a three state DTMC with states $\{0,1,2\}$ and probability transition matrix P given as

$$P = \begin{bmatrix} 0.2 & 0.7 & 0.1 \\ 0.1 & 0.3 & 0.6 \\ & 0.6 & 0.4 \end{bmatrix}.$$

All the states here are recurrent. If however, the matrix P is

$$P = \begin{bmatrix} 0.2 & 0.8 & \\ 0.1 & 0.3 & 0.6 \\ & & 1 \end{bmatrix},$$

then states 0 and 1 are transient with state 2 being an absorbing state and technically also being a recurrent state. Later we show that there are two types of recurrent states, viz: positive recurrent for which the mean recurrence time is finite and the null recurrent for which the mean recurrence time is infinite.

Periodic States - a state i is said to be periodic if the number of steps required to return to it is a multiple of k, $(k > 1)$. As an example the Markov chain with the following transition matrix

$$P = \begin{bmatrix} 0 & 1 & 0 \\ 0 & 0 & 1 \\ 1 & 0 & 0 \end{bmatrix}$$

has all the three states as periodic. It is show diagrammatically in Fig. 3.5.

Another example of periodic states are shown in the Markov chain represented as a network in Fig. 3.6. All the three states are periodic with period of $k = 3$. For this figure the associated transition matrix is

$$P = \begin{bmatrix} 0 & 0 & 1 \\ 1 & 0 & 0 \\ 0 & 1 & 0 \end{bmatrix}$$

If a state is not periodic, it is said to be aperiodic. In both Figures 3.5 and 3.6, to go from any state back to the same state requires hopping over three links, i.e. visiting two other nodes all the time.

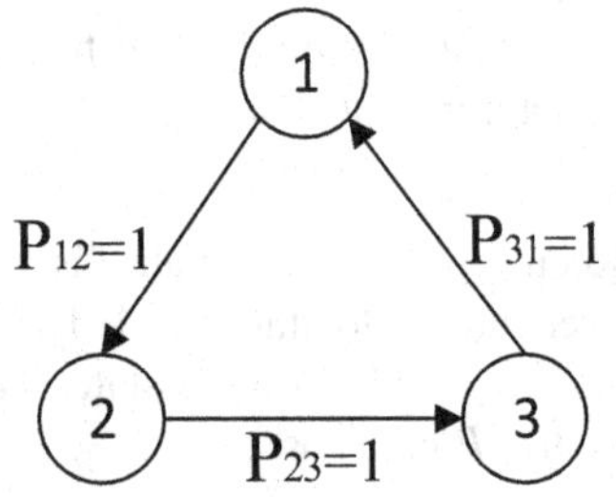

Fig. 3.5 Periodic States Example 1

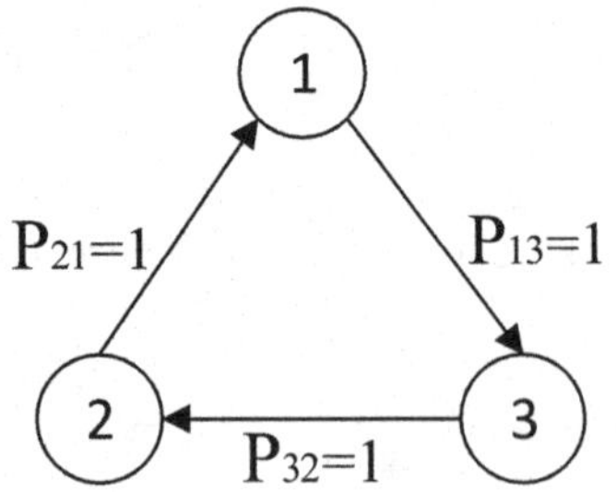

Fig. 3.6 Periodic States Example 2

3.4.4 *Classification of Markov Chains*

In this section, we show how Markov chains are classified.

Irreducible Chains - a Markov chain is said to be irreducible if all the states in the chain communicate with each other. As an example consider two Markov chains with five states S_1, S_2, S_3, S_4, S_5 and the following transition matrices

$$P_1 = \begin{bmatrix} 0.70 & 0.30 & & & \\ 0.40 & 0.30 & 0.15 & 0.15 & \\ & 0.60 & 0.10 & 0.30 & \\ & & 0.30 & 0.50 & 0.20 \\ & & & 0.20 & 0.80 \end{bmatrix} \quad P_2 = \begin{bmatrix} 0.70 & 0.30 & & & \\ 0.40 & 0.60 & & & \\ & & 0.10 & 0.90 & \\ & & 0.30 & 0.50 & 0.20 \\ & & & 0.20 & 0.80 \end{bmatrix}.$$

The DTMC with transition matrix P_1 is irreducible, but the one with P_2 is reducible into two sub-sets of states: the set of states in $A = \{S_1, S_2\}$ and the set of states in $B = \{S_3, S_4, S_5\}$. The set of states in A do not communicate with the set of states in B. Hence, this is a reducible Markov chain. The set of states in A are said to form a closed set and so also are the states in B. Note that states $\{1,2,3,4\}$ cannot be reached from states $\{5,6,7,8,9\}$.

Absorbing Chains - a Markov chain is said to be an absorbing chain if any of its states is an absorbing state, i.e. if at least for one of its states i we have $Pr\{X_{n+1} = i | X_n = i\} = p_{i,i} = 1$. As an example consider an absorbing DTMC with states $\{0,1,2\}$ in which state 0 is an absorbing state. Fig. 3.7 is a an example representation of such a chain.

Recurrent Markov Chain - a Markov chain is said to be recurrent if all the states of the chain are recurrent. It is often important to also specify if a chain is positive or null recurrent type based on if any of the states are positive recurrent or null recurrent. The conditions that govern the differences between these two recurrence will be presented later in the next subsection. Proving that a Markov chain is positive recurrent is usually involved and problem specific. We will not attempt to give a general method for proving recurrence in this book. However, we will present the method used for proving recurrence for some classes of Markov chains which are commonly encountered in queueing theory.

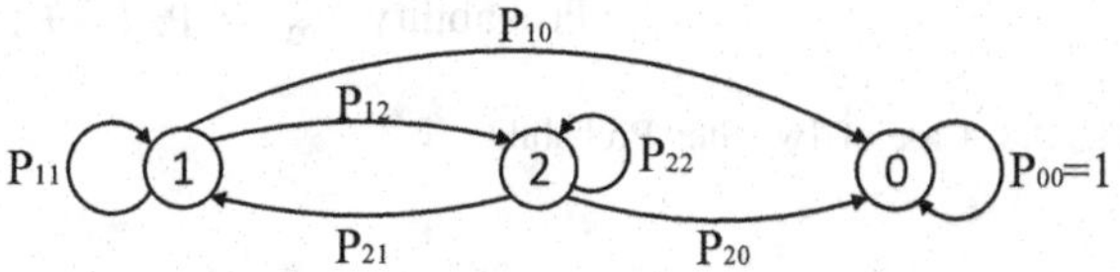

Fig. 3.7 An Absorbing Markov Chain

Ergodic Markov Chain - an irreducible Markov chain is said to be ergodic if all its states are aperiodic and positive recurrent. In addition, if the chain is finite, then aperiodicity is all it needs to be an ergodic chain.

3.5 First Passage Time

In this section we consider first passage time, a very important measure which is used in many situations to characterize a Markov chain. Consider a Markov chain with the state space $\mathscr{N} = 0, 1, 2, \ldots, N$, $\mathscr{N} \subseteq \mathscr{I}$ and associated transition matrix P. Given that the chain is in a particular state, say i at any arbitrary time, we want to find from then on when it reaches a state j for the first time. Define $f_{ij}^{(n)}$ as the probability that state j is visited for the first time at the nth transition, given that the system was in state i at the start, i.e. at time 0. When $j = i$ we have $f_{ii}^{(n)}$ as the probability that state i is visited for the first time at the nth transition, given that the system was in state i at the start, i.e. at time 0; this is known as first return probability. Note that $p_{ij}^{(n)}$ is different from $f_{ij}^{(n)}$ in that the former gives the probability that the system is in state j at the nth transition given that it was in state i at the start; the system could have entered state j by the 1st, 2nd, . . ., $(n\text{-}1)^{th}$ or n^{th} transition and remained in, or went into another state and returned to j by the nth transition. Note that $p_{ij}^{(n)} \geq f_{ij}^{(n)}$, $\forall n$ in general and $p_{ij} = f_{ij}^{(1)}$. Consider a DTMC with states $\{1, 2, 3\}$ we show in Figs. 3.8 and 3.9, the difference between $p_{2,3}^{(2)}$ and $f_{2,3}^{(2)}$, respectively.

We see that

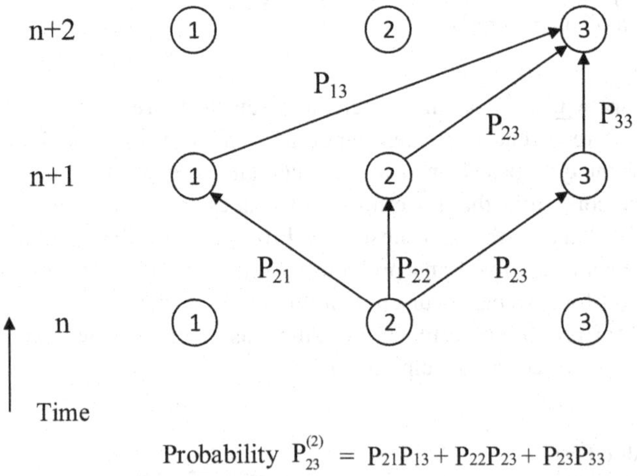

Fig. 3.8 Case of Two Step Probability

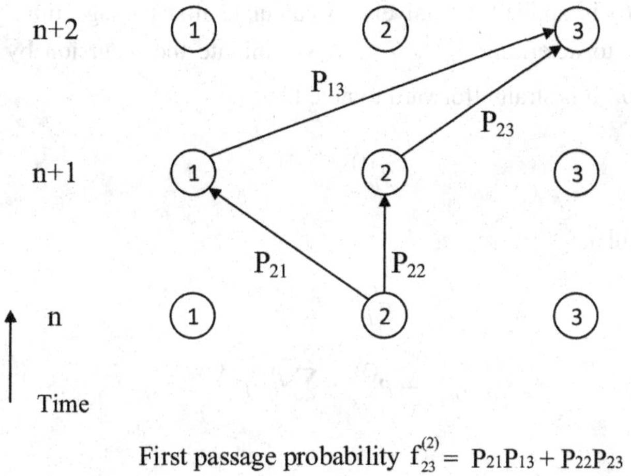

Fig. 3.9 Case of First Passage Probability

$$P_{2,3}^{(2)} = f_{2,3}^{(2)} + p_{2,3}p_{3,3}.$$

It is straightforward to see that

$$p_{ij}^{(n)} = f_{ij}^{(n)} + (1 - \delta_{n1}) \sum_{v=1}^{n-1} f_{ij}^{(v)} p_{jj}^{(n-v)}, \; n \geq 1,$$

where δ_{ij} is the Kronecker delta, i.e. $\delta_{ij} = \begin{cases} 1, & i = j \\ 0, & \text{otherwise} \end{cases}$.

The arguments leading to this expression are as follows: The system is in state j at the nth transition, given that it started from state i at time zero in which case, either

1. the first time that state j is visited is at the nth transition - this leads to the first term, or
2. the first time that state j is visited is at the vth transition $(v < n)$ and then for the remaining $n - v$ transitions the system could visit the j state as many times as possible (without visiting state i in between) but must end up in state j by the last transition.

After rewriting this equation, we obtain

$$f_{ij}^{(n)} = p_{ij}^{(n)} - (1 - \delta_{n1}) \sum_{v=1}^{n-1} f_{ij}^{(v)} p_{jj}^{(n-v)}, \; n \geq 1 \tag{3.6}$$

Equation (3.6) is applied recursively to calculate first passage time probability. For example, to determine $f_{ij}^{(T)}$, $T > 1$, we initiate the recursion by computing $p_{ij}^{(n)} \geq f_{ij}^{(n)}$, $\forall n$. It is straightforward to note that

$$f_{ij}^{(1)} = p_{ij}^{(1)}$$

and then calculate

$$\begin{aligned} f_{ij}^{(2)} &= p_{ij}^{(2)} - f_{ij}^{(1)} p_{jj}^{(1)} \\ f_{ij}^{(3)} &= p_{ij}^{(3)} - \sum_{v=1}^{2} f_{ij}^{(v)} p_{jj}^{(3-v)} \\ \vdots &= \vdots \\ f_{ij}^{(T)} &= p_{ij}^{(T)} - \sum_{v=1}^{T-1} f_{ij}^{(v)} p_{jj}^{(T-v)} \end{aligned}$$

Note that we have to calculate P^n, $n = 2, 3, \ldots, T-1$, during the recursion by this method. However, Neuts (1973) has derived a more efficient procedure for computing the first passage probability and is given as:

$$f_{ij}^{(n+1)} = \sum_{v \in \mathscr{I}} p_{iv} f_{vj}^{(n)} - p_{ij} f_{jj}^{(n)}, \; n \geq 1. \tag{3.7}$$

The arguments leading to this result are as follows:

$$\begin{aligned} f_{ij}^{(2)} &= \Sigma_{v \in \mathscr{N}, v \neq j} p_{iv} f_{vj}^{(1)} = && \Sigma_{v \in \mathscr{N}} p_{iv} f_{vj}^{(1)} - p_{i,j} f_{jj}^{(1)} \\ f_{ij}^{(3)} &= \Sigma_{v \in \mathscr{N}, v \neq j} p_{iv} f_{vj}^{(2)} = && \Sigma_{v \in \mathscr{N}} p_{iv} f_{vj}^{(2)} - p_{i,j} f_{jj}^{(2)} \\ \vdots &= \qquad \vdots = && \vdots \\ f_{ij}^{(n+1)} &= \Sigma_{v \in \mathscr{N}, v \neq j} p_{iv} f_{vj}^{(n)} = && \Sigma_{v \in \mathscr{N}} p_{iv} f_{vj}^{(n)} - p_{i,j} f_{jj}^{(n)}, \; n \geq 1. \end{aligned}$$

If we define the matrix $F^{(n)}$ whose elements are $f_{ij}^{(n)}$, and $F_d^{(n)}$ as the matrix F with only its diagonal elements, i.e. $(F_d^{(n)})_{jj} = f_{jj}^{(n)}$ and $(F_d^{(n)})_{ij} = 0$, $i \neq j$, then the above equation can be written in matrix form as

$$F^{(n+1)} = PF^{(n)} - PF_d^{(n)} = P(F^{(n)} - F_d^{(n)}). \tag{3.8}$$

This is much easier to work with for computational purposes.

3.5.1 Examples

Consider Example 3.4.2.1.1 with the state space $\{0,1,2,3,4\}$ and transition matrix

$$P = \begin{bmatrix} 0.70 & 0.30 & & & \\ 0.42 & 0.46 & 0.12 & & \\ & 0.42 & 0.46 & 0.12 & \\ & & 0.42 & 0.46 & 0.12 \\ & & & 0.42 & 0.58 \end{bmatrix}.$$

Suppose we currently have 2 messages in the buffer, what is the probability that the buffer becomes full (i.e. buffer will reach state 4) for the first time at the fourth transition? We wish to determine $f_{2,4}^{(4)}$, i.e. the probability that the buffer becomes full for the first time at time 4, given that there were 2 in the system at time 0. Also, given that the buffer is now full, what is the probability that the next time that the buffer is full again is at the fifth transition from now?

What we need to calculate is $F^{(4)}$ and select its $(2,4)$ element. By using the formula above we can actually obtain the whole matrix $F^{(4)}$ recursively as follows:

$$F^{(1)} = P = \begin{bmatrix} 0.70 & 0.30 & & & \\ 0.42 & 0.46 & 0.12 & & \\ & 0.42 & 0.46 & 0.12 & \\ & & 0.42 & 0.46 & 0.12 \\ & & & 0.42 & 0.58 \end{bmatrix}, \quad F_d^{(1)} = \begin{bmatrix} 0.70 & & & & \\ & 0.46 & & & \\ & & 0.46 & & \\ & & & 0.46 & \\ & & & & 0.58 \end{bmatrix}$$

We obtain

$$F^{(2)} = \begin{bmatrix} 0.1260 & 0.2100 & 0.0360 & 0.0 & 0.0 \\ 0.1932 & 0.1764 & 0.0552 & 0.0144 & 0.0 \\ 0.1764 & 0.1932 & 0.1008 & 0.0552 & 0.0144 \\ 0 & 0.1764 & 0.1932 & 0.1008 & 0.0552 \\ 0 & 0 & 0.1764 & 0.2436 & 0.0504 \end{bmatrix}.$$

After the fourth recursion we obtain

$$F^{(4)} = \begin{bmatrix} 0.0330 & 0.1029 & 0.0414 & 0.0070 & 0.0005 \\ 0.0701 & 0.0749 & 0.0362 & 0.0117 & 0.0024 \\ 0.1298 & 0.0701 & 0.0302 & 0.0200 & 0.0106 \\ 0.1022 & 0.1298 & 0.0726 & 0.0302 & 0.0200 \\ 0.0311 & 0.1111 & 0.1526 & 0.0819 & 0.0132 \end{bmatrix}.$$

Hence we have $f_{2,4}^{(4)} = 0.0106$.

3.5.2 Some Key Information Provided by First Passage

Define $f_{ij} = \sum_{n=1}^{\infty} f_{ij}^{(n)}$ as the probability of the system ever passing from state i to state j. It is clear that

- $f_{ij} = 1$ implies that state j is reachable from state i.
- $f_{ij} = f_{ji} = 1$ implies that states i and j do communicate
- $f_{ii}^{(n)}$ is the probability that the first recurrence time for state i is n. Letting $f_{ii} = \sum_{n=1}^{\infty} f_{ii}^{(n)}$ then
- $f_{ii} = 1$ for all recurrent states
- $f_{ii} < 1$ for all transient states.

Sometimes we are interested mainly in the first moment of the passage times. We start by obtaining the probability generating function of the first passage time. Let $P_{ij}(z) = \sum_{n=1}^{\infty} p_{ij}^{(n)} z^n$ and $F_{ij}(z) = \sum_{n=1}^{\infty} f_{ij}^{(n)} z^n$, $|z| < 1$ be the probability generating functions (*pgf*) of the probability of transition times and that of the first passage times, respectively. Taking the *pgf* of the first passage equation we obtain:

$$F_{ij}(z) = P_{ij}(z) - F_{ij}(z)P_{jj}(z) = \frac{P_{ij}(z)}{1 + P_{jj}(z)}.$$

The nth factorial moment of the first passage time is then given as $\frac{d^n F_{ij}(z)}{dz^n}|_{z\to 1}$. However, this is not simple to evaluate unless $P_{ij}(z)$ has a nice and simple structure. The first moment can be obtained in an easier manner as follows.

The mean first passage time from state i to state j denoted by M_{ij} is given as

$$M_{ij} = \sum_{n=1}^{\infty} n f_{ij}^{(n)}.$$

Keeping in mind that when the system leaves state i it may go into state j directly with probability p_{ij} in which case it takes one unit of time, or it could go directly to another state $k(\neq j)$ in which case it takes one unit of time with probability p_{ik} plus the mean first passage time from state k to state j. Hence, we can calculate the mean first passage time as follows:

$$M_{ij} = p_{ij} + \sum_{k\neq j}(1 + M_{kj})p_{ik} = 1 + \sum_{k\neq j} p_{ik} M_{jk}. \tag{3.9}$$

Define a matrix $M = (M_{ij})$. Let M_d be a diagonal matrix whose elements are $= M_{ii}$ with $\mathbf{1} = [1\ 1\ \cdots\ 1]^T$ and $\mathbf{e}_1 = [1\ 0\ 0\ \cdots\ 0]^T$, then the above equation can be written in matrix form as

$$M = \mathbf{1} \otimes \mathbf{1}^T + P(M - M_d),$$

where $\otimes$ is the Kronecker product operator, i.e. for two matrices A of order $n \times m$ and B of order $u \times v$ we have $C = A \otimes B$ of dimension $nu \times mv$ with structure

$$C = \begin{bmatrix} A_{11}B & A_{12}B & \cdots & A_{1m}B \\ A_{21}B & A_{22}B & \cdots & A_{2m}B \\ \vdots & \vdots & \cdots & \vdots \\ A_{n1}B & A_{n2}B & \cdots & A_{nm}B \end{bmatrix},$$

where $A_{ij},\ i = 1,2,\cdots,n;\ j = 1,2,\cdots,m$. This equation can in turn be written as a linear equation that can be solved using standard linear algebra techniques as follows.

Let $M_j = [M_{0,j} \ \ M_{1,j} \ \ M_{2,j} \ \ \cdots \ \ M_{N,j}]^T$ for a DTMC with $N+1$ state space numbered $\{0,1,2,\cdots,N\}$. Further let $\mathbf{e}_j$ be the $(j+1)^{st}$ column of an identity matrix of size $N+1$, with $\mathbf{1}$ being a column vector of ones, then we can write

$$P((I - \mathbf{e}_j\mathbf{e}_j^T) - I)M_j = -\mathbf{1}.$$

Further define $\tilde{P}_j = P((I - \mathbf{e}_j\mathbf{e}_j^T) - I)$, $\tilde{P} = \begin{bmatrix} \tilde{P}_0 & & & \\ & \tilde{P}_1 & & \\ & & \ddots & \\ & & & \tilde{P}_N \end{bmatrix}$, and $VecM = \begin{bmatrix} M_0 \\ M_1 \\ \vdots \\ M_N \end{bmatrix}$,

then we have the linear equation

$$VecM = -\tilde{P}^{-1}\mathbf{1}_{(N+1)^2}. \tag{3.10}$$

Note that $\mathbf{1}_S$ represents a column vector of ones of dimension S.

3.5.3 Example

Consider a DTMC with the state space $\{0,1,2\}$ and transition matrix

$$P = \begin{bmatrix} 0.3 & 0.6 & 0.1 \\ 0.2 & 0.5 & 0.3 \\ 0.6 & 0.2 & 0.2 \end{bmatrix}.$$

We get $VecM = \begin{bmatrix} 3.1471 \\ 3.2353 \\ 2.0588 \\ 1.8000 \\ 2.1400 \\ 2.6000 \\ 4.7826 \\ 3.9130 \\ 4.6522 \end{bmatrix}$, which gives

$$M = \begin{bmatrix} 3.1471 & 1.8000 & 4.7826 \\ 3.2353 & 2.1400 & 3.9130 \\ 2.0588 & 2.6000 & 4.6522 \end{bmatrix}.$$

3.5.4 *Mean first recurrence time*

Also, M_{ii} is known as the *mean first recurrence time* for state i.

- For every recurrent state $i, f_{ii} = 1$,
- if $M_{ii} < \infty$, we say state i is positive recurrent.
- if $M_{ii} = \infty$, we say state i is null recurrent.

3.6 Absorbing Markov Chains

A Markov chain which has at least one absorbing state is called an absorbing Markov chain. In an absorbing finite Markov chain, every state is either an absorbing state or a transient state. For example, let the states of an absorbing Markov chain be $\mathscr{N} = \{S_1, S_2, ..., S_N\}$, with $|\mathscr{N}| = N$. Let there be T transient states given as $\mathscr{T} = \{S_1, S_2, ..., S_T\}$, with $|\mathscr{T}| = T$ and $\boldsymbol{N-T}$ absorbing states in this chain given as $\mathscr{A} = \{S_{T+1}, S_{T+2}, ..., S_N\}$, where $\mathscr{N} = \mathscr{T} \cup \mathscr{A}$ and $\mathscr{T} \cap \mathscr{A} = \emptyset$. Then the transition matrix can be written as

$$P = \begin{bmatrix} Q & H \\ 0 & I \end{bmatrix},$$

where Q is a square matrix of dimension T, I is a identity matrix of dimension $N-T$ and H is of dimension $T \times (N-T)$. The matrix Q represents the transitions within the transient states and the matrix H represents the transitions from the transient states into the absorbing states. The matrix Q is sub-stochastic, i.e. the sums of each row is less than or equal to 1, but most important is that at least the sum of one of the rows of Q is strictly less than 1.

As an example consider a 5 state Markov chain with state space $\{0,1,2,3,4\}$ of which two states $\{3,4\}$ are absorbing states. It will have a transition probability matrix of the form

$$P = \begin{bmatrix} q_{00} & q_{01} & q_{02} & h_{03} & h_{04} \\ q_{10} & q_{11} & q_{12} & h_{13} & h_{14} \\ q_{20} & q_{21} & q_{22} & h_{23} & h_{24} \\ 0 & 0 & 0 & 1 & 0 \\ 0 & 0 & 0 & 0 & 1 \end{bmatrix}.$$

An absorbing Markov chain with only one absorbing state is very common and very useful in representing some distributions encountered in real life, e.g. the phase type distributions.

3.6.1 *Characteristics of an absorbing Markov chain*

1. **State of absorption**: It is clear that for an absorbing Markov chain, the system eventually gets absorbed, i.e. the Markov chain terminates eventually. One aspect of interest is usually trying to know into which state the system gets absorbed when there is more than one absorbing state. Let us define a $T \times (N-T)$ matrix B whose elements b_{ij} represent the probability that the system eventually gets absorbed in state $j \in \mathscr{A}$, given that the system started in the transient state $i \in T$. To obtain the equation for B we first obtain the following. The probability that the system is absorbed into state j at the nth transition, given that it started from a transient state i is given by $\sum_{v=1}^{T}(Q^{n-1})_{iv}H_{vj}$, where $(Q^{n-1})_{iv}$ is the $(i,\, v)$ element of Q^{n-1} and H_{vj} is the $(v,\, j)$ element of H. The argument leading to this is that the system remains in the transient states for the first $n-1$ transitions and then enters the absorbing state j at the nth transition. Hence, b_{ij} is obtained as the infinite sum of the series given by

$$b_{ij} = \sum_{n=1}^{\infty}\sum_{v}^{T}(Q^{n-1})_{iv}H_{vj}.$$

This in matrix form is

$$B = (I + Q + Q^2 + Q^3 + \ldots)H$$

It is straightforward to show that

$$B = (I-Q)^{-1}H \tag{3.11}$$

provided the inverse exists. This inverse is known to exist if Q is strictly sub-stochastic and irreducible (see Bhat (1972)). In our case, Q has to be sub-stochastic. The matrix Q is said to be sub-stochastic if at least for one of the rows i we have $\sum_j Q_{ij} < 1,\ \forall i \in \mathscr{T}$ where Q_{ij} is the (i,j)th element of the matrix Q. The only other requirement for the inverse to exist is that it is irreducible, which will depend on other properties of the specific Markov chain being dealt with.

If we let a_i represent a probability of starting from the transient state $i \in \mathscr{T}$ initially, with vector $\boldsymbol{a} = [a_1, a_2, ..., a_T]$, then the probability of getting absorbed into the absorbing state $j' \in \mathscr{A}$ is given by $b_{j'}$ which is obtained from the vector $\boldsymbol{b} = [b_{1'}, b_{2'}, ..., b_{(N-T)'}]$

$$\boldsymbol{b} = \boldsymbol{a}(I-Q)^{-1}H, \tag{3.12}$$

where $j' = T + j$.

2. **Time to absorption:** Let us define a T x $(N-T)$ matrix $C^{(n)}$ whose elements $c_{ij'}^{(n)}$ represent the probability of being absorbed into state $j' \in \mathscr{A}$ at the nth transition, given that the system was in state $i \in \mathscr{T}$ at the start. Then $c_{ij'}^{(n)} = \sum_{v=1}^{T}(Q^{n-1})_{iv}H_{vj'}$ and in matrix form $C^{(n)}$ is given as

$$C^{(n)} = Q^{n-1}H,\ n \geq 1.$$

If we let the vector $\boldsymbol{c}^{(n)} = [c_{1'}^{(n)}, c_{2'}^{(n)}, ..., c_{(N-T)'}^{(n)}]$, where $c_{j'}^{(n)}$ is the probability that the system gets absorbed into state $j' \in A$ at the nth transition then

$$\boldsymbol{c}^{(n)} = \boldsymbol{a}Q^{n-1}H,\ n \geq 1. \tag{3.13}$$

3. **Mean time to absorption**: Let us define a $T \times (N-T)$ matrix $\hat{D}$ whose elements d_{ij} represent the mean time to absorption into state $j \in A$, given that the system started in state $i \in \mathscr{T}$. Then,

$$D = (I + 2Q + 3Q^2 + 4Q^3 + ...)H.$$

After some algebraic manipulations we obtain

$$D = (I-Q)^{-2}H. \tag{3.14}$$

Also, if we let the vector $\boldsymbol{d} = [d_{1'}, d_{2'}, ..., d_{(N-T)'}]$, where $d_{ij'}$ is the mean time to absorption to state j' then

$$\boldsymbol{d} = \boldsymbol{a}(I-Q)^{-2}H \tag{3.15}$$

4. **The case of one absorbing state:** Of particular interest is the case of an absorbing Markov chain with only one absorbing state. In that case, R is a column vector and $(I-Q)\mathbf{1}=H$, where $\mathbf{1}$ is a column vector of 1's. Hence,

$$B=\mathbf{1},\ \ C^{(n)}=Q^{n-1}(I-Q)\mathbf{1} \text{ and } D=(I-Q)^{-1}\mathbf{1}.$$

Later we will find that for the case when $N-T=1$, $c^{(n)}$, $n\geq 1$, is a proper probability mass function. This $c^{(n)}$ is popularly known as the phase type distribution and represented by $(\boldsymbol{a},Q)$.

3.6.1.1 Example:

Consider the 5 state Markov chain with state space $\{0,1,2,3,4\}$ of which two states $\{3,4\}$ are absorbing states. Let its transition probability matrix be of the form

$$P=\begin{bmatrix} 0.1 & 0.2 & 0.1 & 0.4 & 0.2 \\ 0.3 & 0.1 & 0.5 & 0.1 & 0.0 \\ 0.4 & 0.0 & 0.25 & 0.3 & 0.05 \\ 0 & 0 & 0 & 1 & 0 \\ 0 & 0 & 0 & 0 & 1 \end{bmatrix}.$$

Here we have

$$I-Q=\begin{bmatrix} 0.9 & -0.2 & -0.1 \\ -0.3 & 0.9 & -0.5 \\ -0.4 & 0.0 & 0.75 \end{bmatrix},\ \ H=\begin{bmatrix} 0.4 & 0.2 \\ 0.1 & 0.0 \\ 0.3 & 0.05 \end{bmatrix}.$$

Suppose we have $\mathbf{a}=[0.1\ 0.6\ 0.3]$, then we have

$$\mathbf{b}=\mathbf{a}(I-Q)^{-1}R=[0.7683\ 0.2317],$$

$$\mathbf{c}^{(5)}=\mathbf{a}Q^4H=[0.0464\ \ 0.0159],$$

$$\mathbf{d}=\mathbf{a}(I-Q)^{-2}H=[2.1393\ \ 0.7024].$$

Consider another example with the state space $\{0,1,2,3\}$ and transition matrix

$$P=\begin{bmatrix} 0.1 & 0.2 & 0.1 & 0.6 \\ 0.3 & 0.1 & 0.5 & 0.1 \\ 0.4 & 0.0 & 0.25 & 0.35 \\ 0.0 & 0.0 & 0.0 & 1.0 \end{bmatrix}.$$

Here $\mathbf{b} = 1.0$. If we start with $\mathbf{a} = [0.1\ 0.6\ 0.3]$, then we obtain $\mathbf{d} = 2.8417$ and $c^{(n)}$ is obtained as

n	1	2	3	4	5	6	7	8	...
$c^{(n)}$	0.2250	0.3287	0.1909	0.1044	0.0623	0.0367	0.0215	0.0126	...

3.7 Censored Markov Chains

Consider a DTMC with state space $\mathscr{X}$. Let the state space be partitioned into $\mathscr{X}_1$ and $\mathscr{X}_2$, such that

$$\mathscr{X}_1 \cup \mathscr{X}_2 = \mathscr{X}, \text{ and } \mathscr{X}_1 \cap \mathscr{X}_2 = \emptyset.$$

as an example, let $\mathscr{X} = \{1,2,3,4,5,6,7,8,9\}$ and $\mathscr{X}_1 = \{1,4,5,7\}$ and $\mathscr{X}_2 = \{2,3,6,8,9\}$

If we let P be the transition matrix associated with $\mathscr{X}$, then we can write it as

$$P = \begin{bmatrix} P(1,1) & P(1,2) \\ P(2,1) & P(2,2) \end{bmatrix},$$

where $P(i,j)$ is the sub-stochastic matrix representing transitions from $\mathscr{X}_i$ to $\mathscr{X}_j$, $i,j = 1,2$.

Suppose the Markov chain is ergodic and we want to study this system in the long run. If in addition we are only interested in the behaviour of the Markov chain when and only when it is in one of the states of a one subset, e.g. $\mathscr{X}_1$, then we say the Markov chain is censored when it enters states of $\mathscr{X}_2$ and the transition matrix of the new Markov chain, which we call P_1 is given as

$$P_1 = P(1,1) + P(1,2)(I - P(2,2))^{-1}P(2,1).$$

This P_1 is a proper transition matrix representing the behaviour of the Markov chain when it is in $\mathscr{X}_1$ and only. Essentially we study the Markov chain when it is $\mathscr{X}_1$ and when it leaves $\mathscr{X}_1$ we censor it until it returns to $\mathscr{X}_1$. Hence P_1 has two components;

1. The first component captures its behaviour while it is still in $\mathscr{X}_1$, i.e. $P(1,1)$ and
2. the second component captures when it moves into states in $\mathscr{X}_2$ which we do not have interest in, until it returns to the states in $\mathscr{X}_2$, i.e. $P(1,2)(I - P(2,2))^{-1}P(2,1)$.

3.8 Transient Analysis

If we are interested in transient analysis of a Markov chain, i.e. $Pr\{X_n = j|X_0 = i\}$ we may proceed by any of the following three methods (which are algorithmically only feasible for finite state DTMC), among several other methods. It is known that $\boldsymbol{x}^{(n)} = \boldsymbol{x}^{(n-1)}P = \boldsymbol{x}^{(0)}P^n$.

3.8.1 *Direct Algebraic Operations*

3.8.1.1 Naive repeated application of *P*

In this case we simply start with the known values of $\boldsymbol{x}^{(0)}$ and repeatedly apply the relationship

$$\boldsymbol{x}^{(n)} = \boldsymbol{x}^{(n-1)}P,$$

starting from $n = 1$ until the desired time point. For example for a three state DTMC with transition matrix

$$P = \begin{bmatrix} 1/4 & 3/4 & 0 \\ 0 & 1/2 & 1/2 \\ 1/3 & 0 & 2/3 \end{bmatrix},$$

if we are given $\mathbf{x}^{(0)} = [1\ 0\ 0]$ and we want to find the probability distribution of the state of the DTMC at the 4th time epoch we simply apply the recursion four times as

$$\mathbf{x}^{(1)} = \mathbf{x}^{(0)}P,\ \mathbf{x}^{(2)} = \mathbf{x}^{(1)}P,\ \mathbf{x}^{(3)} = \mathbf{x}^{(2)}P,\ \mathbf{x}^{(4)} = \mathbf{x}^{(3)}P,$$

and obtain

$$\mathbf{x}^{(4)} = [0.2122\ \ 0.2695\ \ 0.5182].$$

We will notice that as n gets very large then $\mathbf{x}^{(n+1)} \approx \mathbf{x}^{(n)}$, and in fact the differences get smaller as n increases. This is the limiting behaviour of an ergodic Markov chain and will be presented in detail later.

3.8.1.2 Matrix decomposition approach

A second approach is from observing that the matrix P can be decomposed into the form

$$P = \Lambda D \Lambda^{-1},$$

where D is a diagonal matrix consisting of the eigenvalues of P and Λ is a matrix for which each of its row is the eigenvector corresponding to one of the eigenvalues. Hence we have

$$P^n = \Lambda D^n \Lambda^{-1},$$

which leads to

$$\boldsymbol{x}^{(n)} = \boldsymbol{x}^{(0)} \Lambda D^n \Lambda^{-1}. \tag{3.16}$$

For the example given above we have

$$D = \begin{bmatrix} 0.2083+0.4061i & 0 & 0 \\ 0 & 0.2083-0.4061i & 0 \\ 0 & 0 & 1.0000 \end{bmatrix},$$

$$\Lambda = \begin{bmatrix} 0.7924 & 0.7924 & -0.5774 \\ -0.0440+0.4291i & -0.0440-0.4291i & -0.5774 \\ -0.3228-0.2861i & -0.3228+0.2861i & -0.5774 \end{bmatrix},$$

where the symbol i refers to complex component of the number. This leads to

$$\mathbf{x}^{(4)} = \mathbf{x}^{(0)} \Lambda D^4 \Lambda^{-1} = [0.2122 \ \ 0.2695 \ \ 0.5182].$$

3.8.2 *Transient Analysis Based on the z-Transform*

Consider the relationship $\boldsymbol{x}^{(n)} = \boldsymbol{x}^{(n-1)} P$ and take the z-transform or the probability generating functions of both sides and let $X(z) = \sum_{n=0}^{\infty} \boldsymbol{x}^{(n)} z^n$, $|z| \leq 1$ we obtain

$$X(z) - x^{(0)} = zX(z)P \Rightarrow X(z) = x^{(0)}[I - zP]^{-1}.$$

We know that $A^{-1} = \frac{1}{\det(A)}\mathrm{Adj}(A)$. While this method is based on a classical theoretical approach, it is not simple to use analytically for a DTMC that has more than 2 states. We use an example to show how this method works. Consider the case of $P = \begin{bmatrix} 1/3 & 3/4 \\ 1/2 & 1/2 \end{bmatrix}$. For this example, we have $\det[I - zP]^{-1} = (1-z)\left(1+\frac{1}{4}z\right)$ and $[I - zP]^{-1} = \frac{1}{1-z}\left(1+\frac{1}{4}z\right)\begin{bmatrix} 1-\frac{1}{2}z & \frac{3}{4}z \\ \frac{1}{2}z & 1-\frac{1}{4}z \end{bmatrix}$. Using partial fractions, we have

$$[I - zP]^{-1} = \frac{1}{1-z}\begin{bmatrix} \frac{2}{5} & \frac{3}{4} \\ \frac{2}{5} & \frac{3}{5} \end{bmatrix} + \frac{1}{1+\frac{1}{4}z}\begin{bmatrix} \frac{3}{5} & -\frac{3}{5} \\ -\frac{2}{5} & \frac{2}{5} \end{bmatrix}.$$

Keeping in mind that the inverse of the z-transform $\frac{1}{1-z}$ is the unit step 1 and that of $\frac{1}{1-az}$ is a^n, then we have

$$P^n = \begin{bmatrix} \frac{2}{5} & \frac{3}{5} \\ \frac{2}{5} & \frac{3}{5} \end{bmatrix} + \left(-\frac{1}{4}\right)^n \begin{bmatrix} \frac{3}{5} & -\frac{3}{5} \\ -\frac{2}{5} & \frac{2}{5} \end{bmatrix}.$$

This is the transient solution to this chain. We notice that as $n \to \infty$, the second term on the right hand side vanishes and the first term is the limiting or steady state portion of the solution. In general, we can write the closed form of $X(z)$ as

$$X(z) = \frac{1}{1-z}\boldsymbol{x} + \boldsymbol{x}^{(0)} T^g(z),$$

where $T^{g(z)}$ is the transform part that gives the additional part of the solution to lead to transient solution. When it is removed, we only obtain the steady state solution. For this example

$$T^g(z) = \frac{1}{1+\frac{1}{4}z} \begin{bmatrix} \frac{3}{5} & -\frac{3}{5} \\ -\frac{2}{5} & \frac{2}{5} \end{bmatrix}.$$

For details on this method the reader is referred to Howard (1960)

3.9 Limiting Behaviour of Markov Chains

One very important key behaviour of interest to us with regards to a time-homogeneous DTMC is its limiting behaviour. i.e. after a long period of time. In studying this limiting behaviour of Markov chains, we need to make clear under what conditions we are able to be specific about this behaviour. We identify the difference between the long term behaviour of an ergodic and a non-ergodic DTMC.

3.9.1 Ergodic Chains

Ergodic Markov chains are those chains that are irreducible, aperiodic and positive recurrent. Let P be the transition matrix of an ergodic DTMC. If the probability vector of the initial state of the system is given by $x^{(0)}$, then the probability vector of the state of the system at any time is given by

$$\boldsymbol{x}^{(n)} = \boldsymbol{x}^{(n-1)} P = \boldsymbol{x}^{(0)} P^n.$$

As $n \to \infty$ we have $\boldsymbol{x}^{(n)} \to \boldsymbol{x} \implies \boldsymbol{x}^{(n)} \to \boldsymbol{x}P$, which is known as the invariant (stationary) probability vector of the system.

Similarly, we have

$$P^n \to \begin{bmatrix} \mathbf{y} \\ \mathbf{y} \\ \vdots \\ \mathbf{y} \end{bmatrix} \quad \text{as } n \to \infty$$

where $\mathbf{y}$ is known as the limiting or equilibrium or steady state probability vector of the system.

In the case of the ergodic system, we have

$$\boldsymbol{x} = \mathbf{y}.$$

We return to the Example 2.2.1, where we have

$$P = \begin{bmatrix} 0.70 & 0.30 & & & \\ 0.42 & 0.46 & 0.12 & & \\ & 0.42 & 0.46 & 0.12 & \\ & & 0.42 & 0.46 & 0.12 \\ & & & 0.42 & 0.58 \end{bmatrix}$$

Here $\boldsymbol{x}^{(n+1)} = \boldsymbol{x}^{(n)} = \boldsymbol{x}|_{n\to\infty} = [0.5017, 0.3583, 0.1024, 0.0293, 0.0084]$. Similarly if we find that $P^n \to \begin{bmatrix} \mathbf{y} \\ \mathbf{y} \\ \vdots \\ \mathbf{y} \end{bmatrix}$ as $n \to \infty$, where $\mathbf{y} = [0.5017, 0.3583, 0.1024, 0.0293, 0.0084]$.

3.9.2 Non-Ergodic Chains

For non-ergodic systems

$$\boldsymbol{x} \text{ is not necessarily equal to } \mathbf{y},$$

even when both do exist.

For example, consider the transition matrix of the non-ergodic system given by $P = \begin{bmatrix} 0 & 1 \\ 1 & 0 \end{bmatrix}$. This is a periodic system. For this system, $\boldsymbol{x} = [0.5,\ 0.5]$ and $\boldsymbol{y}$ does not exist because the system alternates between the two states. At any instant in time, $\boldsymbol{y}$ could be $[1,\ 0]$ or $[0,\ 1]$, i.e. we have

$$P^n = \begin{cases} \begin{bmatrix} 0 & 1 \\ 1 & 0 \end{bmatrix}, & \text{if } n \text{ is even} \\ \begin{bmatrix} 1 & 0 \\ 0 & 1 \end{bmatrix}, & \text{if } n \text{ is odd} \end{cases}$$

We consider another example of a non-ergodic system by virtue of not being irreducible. Consider a reducible DTMC described by a transition matrix of the form

$$P = \begin{bmatrix} 0.5 & 0.5 & & \\ 0.6 & 0.4 & & \\ & & 0.2 & 0.8 \\ & & 0.9 & 0.1 \end{bmatrix}.$$

For this system, the solution to $\mathbf{x} = \mathbf{x}P,\ \ \mathbf{x1} = 1$ does not exist, yet we have

$$P^n = \begin{bmatrix} \mathbf{y} \\ \mathbf{y} \\ \mathbf{z} \\ \mathbf{z} \end{bmatrix}, \text{ where}$$

$$\mathbf{y} = [0.5445\ \ 0.4545\ \ 0\ \ 0] \neq \mathbf{z} = [0.0\ \ 0.0\ \ 0.5294\ \ 0.4706].$$

Finally, we consider another example of a non-ergodic DTMC by virtue of not being positive recurrent. This Markov chain is not positive recurrent or even recurrent at all. This is because it drifts to ∞ by the fact that $P_{i,i+1}$ is much greater than $P_{i,i-1}$, hence the probability of the chain returning to state i once it leaves it is less than 1. Suppose we have an infinite number of states $\{0,1,2,3,4,\cdots\}$ and the transition matrix of this is given as

$$P = \begin{bmatrix} 0.3 & 0.7 & & & & \\ 0.1 & 0.1 & 0.8 & & & \\ & 0.1 & 0.1 & 0.8 & & \\ & & 0.1 & 0.1 & 0.8 & \\ & & & \ddots & \ddots & \ddots \end{bmatrix}.$$

This DTMC will neither have an invariant vector nor a steady state vector.

In the rest of the book we focus mainly on ergodic DTMCs.

3.9.3 Mean First Recurrence Time and Steady State Distributions

For an ergodic system we know that x_i is the probability of being in state i at any time, after steady state has been reached. But we also know that M_{ii} is the mean recurrence time for state i. Hence, the probability of being in state i is given as $1/M_{ii}$ based on renewal theory.

3.10 Bivariate Discrete Time Markov Chains

Consider a bivariate process $\{(X_n, Y_n), n = 0, 1, 2, \cdots, \}$ that has a Markov structure. We assume that X_n assumes values in the set $\mathscr{I}$ and Y_n assumes values in the set $\mathscr{M} = \{1, 2, \cdots, M < \infty\}$. An example of this is a system that has two classes of data packets of types 1 and 2, where type 1 has a higher priority than type 2. Suppose we do not allow more than M of type 2 packets to be in the system at any one time, we can represent X_n as the number of type 1 and Y_n as the number of type 2 packets in the system at time n. If we assume that $\{(X_n, Y_n); n = 0, 1, 2, \cdots\}$ is a DTMC we can write the transition matrix of this Markov chain as

$$P = \begin{bmatrix} P_{0,0} & P_{0,1} & P_{0,2} & \cdots \\ P_{1,0} & P_{1,1} & P_{1,2} & \cdots \\ P_{2,0} & P_{2,1} & P_{2,2} & \cdots \\ \vdots & \vdots & \vdots & \vdots \end{bmatrix}, \quad (3.17)$$

where $P_{i,j}, (i,j) = 0, 1, 2, \cdots$ is a block matrix of order $M \times M$. The element $(P_{i,j})_{l,k}$, is $Pr\{X_{n+1} = j, Y_{n+1} = k | X_n = i, Y_n = l\}$, $(i,j) = 0, 1, 2, \cdots; (l,k) = 1, 2, \cdots, M$. Usually we call the set $L_i = \{(i,1), (i,2), \cdots, (i,M)\}$ level i. We also say in state (i,l) that we are in level i and phase l.

Define $x_{i,k}^{(n)} = Pr\{X_n = i, Y_n = k\}$ and write $\boldsymbol{x}^{(n)} = [\boldsymbol{x}_0^{(n)} \ \boldsymbol{x}_1^{(n)} \ \boldsymbol{x}_2^{(n)} \ \cdots]$ with $\boldsymbol{x}_i^{(n)} = [x_{i,1}^{(n)} \ x_{i,2}^{(n)} \ \cdots \ x_{i,M}^{(n)}]$. We have

$$\boldsymbol{x}^{(n+1)} = \boldsymbol{x}^{(n)} P.$$

If the DTMC is positive recurrent then we have $x_{i,k}^{(n)}|_{n\to\infty} = x_{i,k}$, hence

$$\boldsymbol{x} = \boldsymbol{x}P, \ \ \boldsymbol{x}\mathbf{1} = 1.$$

3.11 Problems

1. Consider a discrete time Markov chain with state space $\{1, 2, 3\}$ and transition matrix P given as

$$P = \begin{bmatrix} a_1 & a_2 & a_3 \\ a_3 & a_1 & a_2 \\ a_2 & a_3 & a_1 \end{bmatrix},$$

where $0 < a_j < 1,\ \ j = 1,2,3$ and $a_1 + a_2 + a_3 = 1$. Let $\mathbf{x} = [x_1\ x_2\ x_3]$ be the stationary vector associated with P, i.e.

$$\mathbf{x} = \mathbf{x}P, \quad \mathbf{xe} = 1.$$

a. Find the values of x_j, $\forall j$.
b. What do you observe about the columns of the matrix P? Can you make any generalization?
c. Suppose $a_1 = 0.3$, $a_2 = 0.5$, $a_3 = 0.2$.

   i. If the Markov chain starts from state 2, find the probability that it reaches state 1 before state 3.
   ii. What is the probability that it takes at least three transitions before it reaches state 3 for the first time, given that it starts from state 1?

2. A total of N balls are put in an urn, with i of the balls marked as belonging to person A and $N - i$ belonging to person B. At each step, one ball is chosen at random from the N balls. If it is one for A it is relabelled and assigned to B and then returned to the urn, and vice-versa. Let X_n be the number of balls belonging to A after n steps, and assume that successive random drawings of the balls are independent. It can be shown that $\{X_n,\ n = 0, 1, 2, \cdots\}$ is a discrete time Markov chain. For $N = 4$ display the transition matrix of this Markov chain. Suppose we start with 1 ball in the urn belonging to A find the probability distribution of the number of balls belonging to B after 3 transitions.
3. At a shopping mall passengers looking for taxis arrive according to the Bernoulli process with parameter a and taxis arrive according to the Bernoulli process with parameter b. Passengers wait in a queue until they can get a taxi. There is no upper limit to the number of passengers allowed to wait. However, as soon as the number of waiting taxis is 4 arriving taxis will depart and not join the taxi queue. We assume that only one passenger is served by each taxi. Set up a Markov chain and the associated generator matrix that can be used to study the number of taxis and passengers waiting. Suppose $a = 0.3$ and $b = 0.5$ and consider steady state. Find the

   a. probability that the number of passengers waiting for a taxi is at least 5,
   b. probability that the number of taxis waiting is no more than 2,
   c. average number of passengers waiting,
   d. the average number of taxis waiting.

4. Consider a data communication system with a finite-capacity buffer for temporarily storing messages. The messages arrive according to a Bernoulli process with rate a. The transmission time of each message has geometric distribution with parameter b. The system can transmit only one message at a time. The buffer has room for only K messages excluding the message in transmission (if any). The following protocol for the admission of messages is used. As soon as the buffer is

full, the arrival process of messages is stopped until the number in the buffer has fallen to the resume level K_1, with $0 < K_1 < K$. Set up a DTMC for the number of messages in the system at arbitrary times for the case of $K = 5,\ K_1 = 3$ by displaying the associated generator matrix.

References

Alfa AS, Margolius BH (2008) Two classes of time-inhomogeneous Markov chains: Analysis of the periodic case. Annals of Oper Res 160:121–137

Bhat NU (1972) Elements of Applied Stochastic Proceses. Wiley, New York

Howard R (1960) Dynamic Programming and Markov Processes. MIT Press, Cambridge

Markov AA (1907) Extension of limit theorems of probability theory to sum of variables connected in a chain. The Notes of the Imperial Academy of Sciences of St Petersburg VIII Series, Physio-Mathematical College XXII(9)

Neuts MF (1973) Probability. Allyn and Bacon, Boston

Chapter 4
Numerical Computations with Discrete-Time Markov Chains

4.1 Introduction

For most practical queueing problems solving for the stationary (invariant) vector associated with the Markov chain involves numerical computations. Carrying out the computations efficiently becomes a major factor in using the queueing results. In this chapter we present few well known methods for computing the invariant vectors for both finite and infinite DTMCs with special structures. What we present in this chapter is mainly how to compute the stationary (invariant) vector associated with a DTMC, when such a vector exists. Throughout this chapter we assume that the DTMCs we are dealing with are positive recurrent and point out when that assumption is not necessary.

4.2 Numerical Computations of the Invariant Vectors

In this section, we distinguish between finite and infinite state Markov chains as we present the methods. Our interest is to compute the invariant probability vector $\boldsymbol{x}$ rather than the limiting probability vector $\mathbf{y}$. We know and show later that $\boldsymbol{x}$ exists for every irreducible positive recurrent Markov chain. We therefore focus on such chains only.

We want to determine $\boldsymbol{x}$ for a DTMC which has a transition matrix P, such that

$$\boldsymbol{x} = \boldsymbol{x}P \text{ and } \boldsymbol{x}\mathbf{1} = 1. \tag{4.1}$$

A.S. Alfa, *Applied Discrete-Time Queues*, DOI 10.1007/978-1-4939-3420-1_4

4.3 Finite State Markov Chains

There are mainly two types of methods used for computing the stationary vector of a DTMC; direct methods and iterative methods. Iterative methods are used more frequently for most DTMC which have a large state space – very common in most practical situations. However, the computational effort for iterative methods are dependent on the size and structure of the DTMC and also the starting solution. Direct methods are suitable for smaller DTMC and the number of steps required to get a solution is known ahead of time.

For the iterative methods we work with the equation in the form

$$\boldsymbol{x} = \boldsymbol{x}P, \ \ \boldsymbol{x}\mathbf{1} = 1,$$

whereas for most direct methods we convert the problem to a classical linear algebra problem of the form $AX = b$, by creating a matrix $\tilde{P}$ which is the same as $I - P$, with the last column replaced by a column of ones, i.e. $\mathbf{1}$. For our problem, of N states Markov chain, we have

$$\tilde{P} = (I - P)(I - \mathbf{e}_N\mathbf{e}_N^T) + \mathbf{1}\mathbf{e}_N^T,$$

which then leads to

$$\mathbf{e}_N^T = \boldsymbol{x}\tilde{P} \ \rightarrow \ \tilde{P}^T\boldsymbol{x}^T = \mathbf{e}_N.$$

This is now in the standard form. We use the notation $\mathbf{e}_j$ to represent a column vector which has all zero entries except in the jth column that we have 1 and $\mathbf{e}_j^T$ is its transpose.

For example, consider a three state DTMC $\{0,1,2\}$ with transition matrix P given as

$$P = \begin{bmatrix} 0.3 & 0.6 & 0.1 \\ 0.2 & 0.3 & 0.5 \\ 0.4 & 0.5 & 0.1 \end{bmatrix}.$$

We will use this example in all of our numerical computation methods for invariant vectors.

For the iterative method we work with

$$[x_0 \ x_1 \ x_2] = [x_0 \ x_1 \ x_2] \begin{bmatrix} .3 & .6 & .1 \\ .2 & .3 & .5 \\ .4 & .5 & .1 \end{bmatrix}, \ \ [x_1 \ x_2 \ x_3] \begin{bmatrix} 1.0 \\ 1.0 \\ 1.0 \end{bmatrix} = 1,$$

whereas for the direct method we work with

$$\begin{bmatrix} 0.7 & -0.2 & -0.4 \\ -0.6 & 0.7 & -0.5 \\ 1.0 & 1.0 & 1.0 \end{bmatrix} \begin{bmatrix} x_1 \\ x_2 \\ x_3 \end{bmatrix} = \begin{bmatrix} 0.0 \\ 0.0 \\ 1.0 \end{bmatrix},$$

where

$$\tilde{P} = \begin{bmatrix} 0.7 & -0.6 & 1.0 \\ -0.2 & 0.7 & 1.0 \\ -0.4 & -0.5 & 1.0 \end{bmatrix} \rightarrow \tilde{P}^T = \begin{bmatrix} .7 & -.2 & -.4 \\ -.6 & .7 & -.5 \\ 1 & 1 & 1 \end{bmatrix}.$$

Direct methods do not involve iterations. They simply involve a set of operations which is usually of a known fixed number, depending on the size of the Markov chain. They are efficient for small size Markov chains and usually very time consuming for large Markov chains. One direct method which seems to be able to handle a large size Markov chain is the state reduction method (Grassmann et al (1985)). Theoretically, one may state that it has no round off errors. The most popular of the direct techniques is the Gaussian elimination. For more details of this algorithm as applicable to Markov chains (see Stewart (1994)).

4.3.1 Direct Methods

In what follows we present some of the well known direct methods. We present only a synoptic kind of view of this approach in this book. First we assume that the equations have been transformed to the standard form

$$AX = b. \tag{4.2}$$

Using the above example we have

$$P = \begin{bmatrix} 0.3 & 0.6 & 0.1 \\ 0.2 & 0.3 & 0.5 \\ 0.4 & 0.5 & 0.1 \end{bmatrix},$$

which results in

$$A = \begin{bmatrix} 0.7 & -0.2 & -0.4 \\ -0.6 & 0.7 & -0.5 \\ 1.0 & 1.0 & 1.0 \end{bmatrix}, \; X = \begin{bmatrix} x_1 \\ x_2 \\ x_3 \end{bmatrix} \; b = \begin{bmatrix} 0.0 \\ 0.0 \\ 1.0 \end{bmatrix}.$$

- **Inverse matrix approach**: This is the simplest of all the direct methods. It simply solves for X as $X = A^{-1}b$, assuming that the inverse of A exists. For most of the finite DTMC we will be discussing in this book, the inverse of A would exist. From the above example we have

$$\boldsymbol{x}^T = A^{-1}b = \begin{bmatrix} 0.8955 & -0.1493 & 0.2836 \\ 0.0746 & 0.8209 & 0.4403 \\ -0.9701 & -0.6716 & 0.2761 \end{bmatrix} \begin{bmatrix} 0.0 \\ 0.0 \\ 1.0 \end{bmatrix} = \begin{bmatrix} 0.2836 \\ 0.4403 \\ 0.2761 \end{bmatrix}.$$

- **Gaussian Elimination**: This is another popular method which is more suitable for moderate size matrices than the inverse approach. The matrix A is transformed through row operations into an upper triangular matrix U, such that $U\boldsymbol{x}^T = b = \mathbf{e}_N$ and then the elements of $\boldsymbol{x}$ are obtained from back substitution as

$$x_N = 1/u_{N,N}, \; x_i = -(\sum_{k=i+1}^{N} u_{i,k}x_k)/u_{i,i}, i = 0, 1, 2, \cdots, N-1. \tag{4.3}$$

For this example, we have

$$U = \begin{bmatrix} 0.7 & -0.2 & -0.4 \\ & 0.5286 & -0.8429 \\ & & 3.6216 \end{bmatrix}.$$

Applying the back substitution equation we obtain

$$\boldsymbol{x}^T = \begin{bmatrix} 0.2836 \\ 0.4403 \\ 0.2761 \end{bmatrix}.$$

- **State Reduction Method**:This method is also popularly known as the GTH (Grassmann et al (1985)) method. It is actually based on Gaussian elimination with state space reduction during each iteration and it uses the form of equation $\boldsymbol{x} = \boldsymbol{x}P$, $\boldsymbol{x}\mathbf{1} = 1$, unlike the other direct methods. Consider the equation $x_j = \sum_{i=0}^{N} x_i p_{ij}, j = 0, 1, 2, ..., N$.
The procedure starts by eliminating x_N from the Nth equation to obtain

$$x_N = \sum_{i=0}^{N-1} x_i p_{iN}/(1 - p_{N,N}).$$

Replacing $1 - p_{NN}$ with $\sum_{j=0}^{N-1} p_{N,j}$ and using the property that $x_j = \sum_{i=0}^{N-1} x_i p_{ij}^{(N-1)}$, $j = 0, 1, 2, ..., N-1$, where

$$p_{i,j}^{(N-1)} = p_{i,j} + p_{i,N}p_{N,j}/(1-p_{N,N}) = p_{i,j} + p_{i,N}p_{N,j}/\sum_{j=0}^{N-1} p_{N,j}.$$

This is the principle by which state reduction works. It is based on some form of censoring.

Generally, $x_n = \sum_{i=1}^{n-1} x_i q_{i,n}$, $n = N, N-1, N-2, ..., 2, 1$ where $q_{i,n} = p_{i,n}^{(n)} / \sum_{j=0}^{n-1} p_{n,j}^{(n)}$.

The algorithm can be stated as follows (Grassmann et al (1985)) referred to as the GTH:

GTH Algorithm:

1. For $n = N, N-1, N-2, ..., 2, 1$, do

$$q_{i,n} \leftarrow p_{i,n}/\sum_{j=0}^{n-1} p_{n,j},\ i = 0,1,...,n-1$$

$$p_{i,j} \leftarrow p_{i,j} + q_{i,n}p_{n,j},\ i,j = 0,1,...,n-1$$

$$r_1 \leftarrow 1$$

2. For $j = 2,3,...,N$, $r_j \leftarrow \sum_{i=0}^{j-1} r_i q_{i,j}$
3. For $j = 2,3,...,N$, $x_j \leftarrow r_j / \sum_{i=0}^{N} r_i$, $j = 0,1,...,N$

The advantage with this approach is that it is numerically stable even if the system is ill-conditioned.

4.3.2 *Iterative Methods:*

Iterative methods are usually more efficient and more effective than the direct methods when the size of the matrix is large. The number of iterations required to achieve the desired convergence is not usually known in advance and the resulting distributions have round-off errors associated with them.

Four commonly used iterative methods will be discussed in this section. They are: i)the Power method which is more applicable to transition matrices, ii) the Jacobi method, iii) the Gauss-Seidel method, and iv) the method of successive over-relaxation.

- **The Power Method**: The power method simply involves starting with a probability vector $\boldsymbol{x}^{(0)}$ which satisfies the condition $\boldsymbol{x}^{(0)}\mathbf{1} = 1$ and then applying the relationship $\boldsymbol{x}^{(n+1)} = \boldsymbol{x}^{(n)}P$, $n \geq 0$ until $[\boldsymbol{x}^{(n+1)} - \boldsymbol{x}^{(n)}]_i < \varepsilon$, $\forall i$, then we stop the iteration, where ε is the convergence criterion. After each iteration, the nth iteration, the resulting $\boldsymbol{x}^{(n)}$ is re-normalized before proceeding to the next iteration. Often the number of iterations required depends on both P and the starting vector $\boldsymbol{x}^{(0)}$. It is a very naive approach which is only useful for small size problems, and especially if we also need to know the transient behaviour of the DTMC before it reaches steady state.

The remaining three approaches: The next three approaches will be discussed here in a modified form for the transition matrix.

First let us write $Q = P - I$. Our interest is to solve for $\boldsymbol{x} = [x_1, x_2, ..., x_n]$, where $\boldsymbol{x}Q = 0$ and $\boldsymbol{x}\mathbf{1} = 1$. In order to conform to the standard representations of these methods we write the same equation in a transposed form as $Q^T \boldsymbol{x}^T = 0$.

Let us now separate Q^T into a diagonal component D, an upper triangular component U and a lower triangular component L with ij elements given as $-d_{ii}$, u_{ij} and l_{ij}, respectively. It is assumed that $d_{ii} > 0$. Note that the D, L and U are not the same as the well known LU or LDU decompositions.

For example given a matrix

$$Q^T = \begin{bmatrix} q_{0,0} & q_{1,0} & q_{2,0} \\ q_{0,1} & q_{1,1} & q_{2,1} \\ q_{0,2} & q_{1,2} & q_{2,2} \end{bmatrix},$$

we can write $Q^T = L - D + U$, where

$$L = \begin{bmatrix} 0 & 0 & 0 \\ q_{0,1} & 0 & 0 \\ q_{0,2} & q_{1,2} & 0 \end{bmatrix}, \quad U = \begin{bmatrix} 0 & q_{1,0} & q_{2,0} \\ 0 & 0 & q_{2,1} \\ 0 & 0 & 0 \end{bmatrix},$$

$$D = \begin{bmatrix} -q_{0,0} & 0 & 0 \\ 0 & -q_{1,1} & 0 \\ 0 & 0 & -q_{2,2} \end{bmatrix}.$$

- **The Jacobi Method**: Then the Jacobi iteration is given in the scalar form as

$$x_i^{(k+1)} = \frac{1}{d_{ii}} \{ \sum_{j \neq i} (l_{ij} + u_{ij}) x_j^{(k)} \}, \quad i = 0, 1, 2, ..., N \tag{4.4}$$

or in matrix form, it can be written as

$$(\boldsymbol{x}^{(k+1)})^T = D^{-1}(L + U)(\boldsymbol{x}^{(k)})^T. \tag{4.5}$$

For example, consider the Markov chain given by

$$P = \begin{bmatrix} p_{0,0} & p_{0,1} & p_{0,2} \\ p_{1,0} & p_{1,1} & p_{1,2} \\ p_{2,0} & p_{2,1} & p_{2,2} \end{bmatrix} = \begin{bmatrix} 0.1 & 0.2 & 0.7 \\ 0.3 & 0.1 & 0.6 \\ 0.2 & 0.5 & 0.3 \end{bmatrix}$$

Suppose we have an intermediate solution $\boldsymbol{x}^{(k)}$ for which $\boldsymbol{x}^{(k)}\mathbf{1} = 1$, then we have

$$x_0^{(k+1)} = \frac{1}{1-p_{0,0}}[x_1^{(k)}p_{1,0} + x_2^{(k)}p_{2,0}]$$

$$x_1^{(k+1)} = \frac{1}{1-p_{1,1}}[x_0^{(k)}p_{0,1} + x_2^{(k)}p_{2,1}]$$

$$x_2^{(k+1)} = \frac{1}{1-p_{2,2}}[x_0^{(k)}p_{0,2} + x_1^{(k)}p_{1,2}]$$

- **Gauss-Seidel Method**: The Gauss-Seidel method uses the most up-to-date information on solutions available for computations. It is given by the following iteration in the scalar form:

$$x_i^{(k+1)} = \frac{1}{d_{ii}}\{\sum_{j=0}^{i-1} l_{ij}x_j^{(k+1)} + \sum_{j=i+1}^{n} u_{ij}x_j^{(k)}\} \tag{4.6}$$

and in matrix form as

$$(\boldsymbol{x}^{(k+1)})^T = D^{-1}(L(\boldsymbol{x}^{(k+1)})^T + U(\boldsymbol{x}^{(k)})^T) \tag{4.7}$$

This method seems to be more frequently used than the Jacobi. Using the above example, we have

$$x_0^{(k+1)} = \frac{1}{1-p_{0,0}}[x_1^{(k)}p_{1,0} + x_2^{(k)}p_{2,0}]$$

$$x_1^{(k+1)} = \frac{1}{1-p_{1,1}}[x_0^{(k+1)}p_{0,1} + x_2^{(k)}p_{2,1}]$$

$$x_2^{(k+1)} = \frac{1}{1-p_{2,2}}[x_0^{(k+1)}p_{0,2} + x_1^{(k+1)}p_{1,2}]$$

- **The Successive Over-relaxation Method**: This method introduces an extra parameter ω. The method is a generalization of the Gauss-Seidel. The iteration is implemented as follows:

$$x_i^{(k+1)} = (1-\omega)x_i^{(k)} + \omega\{\frac{1}{d_{ii}}\{\sum_{j=0}^{i-1} l_{ij}x_j^{(k+1)} + \sum_{j=i+1}^{N} u_{ij}x_j^{(k)}\}\}, \tag{4.8}$$

and in matrix form

$$(\boldsymbol{x}^{(k+1)})^T = (1-\omega)(\boldsymbol{x}^{(k)})^T + \omega\{D^{-1}(L(\boldsymbol{x}^{(k+1)})^T + U(\boldsymbol{x}^{(k)})^T)\}. \tag{4.9}$$

Using the above example, we have

$$x_0^{(k+1)} = (1-\omega)x_0^{(k)} + \omega\frac{1}{1-p_{0,0}}[x_1^{(k)}p_{1,0} + x_2^{(k)}p_{2,0}]$$

$$x_1^{(k+1)} = (1-\omega)x_1^{(k)} + \omega\frac{1}{1-p_{1,1}}[x_0^{(k+1)}p_{0,1} + x_2^{(k)}p_{2,1}]$$

$$x_2^{(k+1)} = (1-\omega)x_2^{(k)} + \omega\frac{1}{1-p_{2,2}}[x_0^{(k+1)}p_{0,2} + x_1^{(k+1)}p_{1,2}]$$

The value of ω is usually selected between 0 and 2. When $\omega < 1$, we say we have an under-relaxation and when $\omega = 1$, we have the Gauss-Seidel iteration. The greatest difficulty with this method is how to select the value of ω.
The Jacobi method is very appropriate for parallelization of computing. Note that diagonal dominance of the matrix $I-P$ is required for the Jacobi and in fact also for the Gauss-Seidel methods but it is always satisfied for a Markov chain.

4.4 Bivariate Discrete Time Markov Chains

Consider a bivariate process $\{(X_n,Y_n), n = 0,1,2,\cdots,\}$ that has a Markov structure. We assume that X_n assumes values in the set $\mathscr{I}$ and Y_n assumes values in the set $\mathscr{M} = \{1,2,\cdots,M<\infty\}$. An example of this is a system that has two classes of data packets of types 1 and 2, where type 1 has a higher priority than type 2. Suppose we do not allow more than M of type 2 packets to be in the system at any one time, we can represent X_n as the number of type 1 and Y_n as the number of type 2 packets in the system at time n. If we assume that $\{(X_n,Y_n); n = 0,1,2,\cdots\}$ is a DTMC we can write the transition matrix of this Markov chain as

$$P = \begin{bmatrix} P_{0,0} & P_{0,1} & P_{0,2} & \cdots \\ P_{1,0} & P_{1,1} & P_{1,2} & \cdots \\ P_{2,0} & P_{2,1} & P_{2,2} & \cdots \\ \vdots & \vdots & \vdots & \vdots \end{bmatrix}, \tag{4.10}$$

where $P_{i,j}, (i,j) = 0,1,2,\cdots$ is a block matrix of order $M \times M$. The element $(P_{i,j})_{l,k}$, is $Pr\{X_{n+1} = j, Y_{n+1} = k | X_n = i, Y_n = l\}$, $(i,j) = 0,1,2,\cdots; (l,k) = 1,2,\cdots,M,$. Usually we call the set $L_i = \{(i,1),(i,2),\cdots,(i,M)\}$ level i. We also say in state (i,l) that we are in level i and phase l.

Define $x_{i,k}^{(n)} = Pr\{X_n = i, Y_n = k\}$ and write $\boldsymbol{x}^{(n)} = [\boldsymbol{x}_0^{(n)}\ \boldsymbol{x}_1^{(n)}\ \boldsymbol{x}_2^{(n)}\ \cdots]$ with $\boldsymbol{x}_i^{(n)} = [x_{i,1}^{(n)}\ x_{i,2}^{(n)}\ \cdots\ x_{i,M}^{(n)}]$. We have

$$\boldsymbol{x}^{(n+1)} = \boldsymbol{x}^{(n)}P.$$

If the DTMC is positive recurrent then we have $x_{i,k}^{(n)}|_{n\to\infty} = x_{i,k}$, hence

$$\boldsymbol{x} = \boldsymbol{x}P, \ \ \boldsymbol{x}\mathbf{1} = 1.$$

4.4.1 Computing Stationary Distribution for the Finite Bivariate DTMC

The methods presented in Section 4.3 are easily extended to the bivariate and all multivariate finite DTMC.

To avoid unnecessary notational complexities we present only the case where $P_{i,j}$ is of dimension $M \times M$, $\forall (i,j)$. We can easily modify the result for cases where the dimension of $P_{i,j}$ depends on i and j. Let $\mathbf{1}_M$ be an M column vector of ones and $\mathbf{0}_M$ an $M \times 1$ column vector of zeros. Assume X_n has a state space given as $0,1,2,\cdots,N < \infty$.

Define

$$\tilde{P}_{i,i}^T = P_{i,i}^T - I, \ i = 0,1,2,\cdots,N-1,$$

$$\hat{P}_{i,N}^T = (I_M - \mathbf{e}_M^T \otimes \mathbf{e}_M)P_{i,N}^T + \mathbf{e}_M\mathbf{1}^T, \ i = 0,2,\cdots,N,$$

then we can write the stationary equation, after transposing, as

$$\begin{bmatrix} \tilde{P}_{0,0}^T & P_{1,0}^T & P_{2,0}^T & \cdots & P_{N,0}^T \\ P_{0,1}^T & \tilde{P}_{1,1}^T & P_{2,1}^T & \cdots & P_{N,1}^T \\ \vdots & \vdots & \ddots & \cdots & \vdots \\ P_{0,N-1}^T & \cdots & \ddots & \tilde{P}_{N-1,N-1} & P_{N,N-1} \\ \hat{P}_{0,N}^T & \hat{P}_{1,N}^T & \hat{P}_{2,N}^T & \cdots & \hat{P}_{N,N}^T \end{bmatrix} \begin{bmatrix} \boldsymbol{x}_0 \\ \boldsymbol{x}_1 \\ \vdots \\ \boldsymbol{x}_{N-1} \\ \boldsymbol{x}_N \end{bmatrix} = \begin{bmatrix} \mathbf{0}_M \\ \mathbf{0}_M \\ \vdots \\ \mathbf{0}_M \\ \mathbf{e}_M \end{bmatrix}. \qquad (4.11)$$

This is now of the form similar to

$$A\mathbf{x} = \mathbf{b}$$

and can be solved using the techniques discussed earlier. The techniques are briefly listed here:

- **Direct Methods**
  - **The inverse matrix approach**: This approach can be used for the bivariate DTMC just like in the case of the univariate DTMC. Simply we compute $\boldsymbol{x}$ from

$$X = A^{-1}b. \qquad (4.12)$$

- **The block Gaussian elimination method**: The matrix A is transformed through block row operations into an upper triangular block matrix U, such that $U\boldsymbol{x}^T = b$. The triangular matrix consists of block matrices $U_{i,j}$, for which $U_{i,j} = 0,\ i > j$. The block vectors $(\boldsymbol{x}_i)$ of $\boldsymbol{x}$ are obtained from back substitution as

$$\boldsymbol{x}_N^T = u_{N,N}^{-1}\mathbf{1},\ \ \boldsymbol{x}_i^T = -U_{i,i}^{-1}(\sum_{k=i+1}^{N} U_{i,k}\boldsymbol{x}_k^T), i = 0,1,2,\cdots,N-1. \tag{4.13}$$

- **Block state reduction**: This technique and its scalar counterpart are based on censoring of the Markov chain. Consider the DTMC with $N+1$ levels labelled $\{0,1,2,\cdots,N\}$. For simplicity we will call this the block state space $\{0,1,2,\cdots,N\}$ and let it be partitioned into two subsets E and $\bar{E}$ such that $E \cup \bar{E} = \{0,1,2,\cdots,N\}$ and $E \cap \bar{E} = \emptyset$. For example, we may have $E = \{0,1,2,\cdots,M\}$ and $\bar{E} = \{M+1, M+2,\cdots,N\},\ M \leq N$. Let the transition matrix of the DTMC be written as

$$P = \begin{bmatrix} H & S \\ T & U \end{bmatrix}.$$

We can write the stationary vector of this DTMC as $\boldsymbol{x} = [\boldsymbol{x}_e\ \boldsymbol{x}_{e'}]$ from which we obtain

$$\boldsymbol{x}_e = \boldsymbol{x}_e H + \boldsymbol{x}_{e'} T,\ \ \boldsymbol{x}_{e'} = \boldsymbol{x}_e S + \boldsymbol{x}_{e'} U. \tag{4.14}$$

From this we obtain

$$\boldsymbol{x}_e = \boldsymbol{x}_e(H + S(I-U)^{-1}T). \tag{4.15}$$

The matrix $H + S(I-U)^{-1}T$ is a new transition matrix and it is stochastic. Its stationary distribution is the vector $\boldsymbol{x}_e$ which can be obtained as $\boldsymbol{x}_e = \boldsymbol{x}_e P_e,\ \boldsymbol{x}_e\mathbf{1} = 1$ and then normalized for it to be stochastic. It is immediately clear that we have reduced or restricted this DTMC to be observed only when it is in the states of E. This is a censored DTMC. This idea is used to block-reduce the DTMC and apply the block state reduction to solve for $\boldsymbol{x}$ in the original DTMC.

In block state reduction we first partition the DTMC into two sets with the first set given as $E = \{0,1,2,\cdots,N-1\}$ and the second set $\bar{E} = \{N\}$. We study the states of E only as censored DTMC, and then partition it into $E = \{0,1,2\cdots,N-2\}$ and $\bar{E} = \{N-1\}$, and the process continues until we reduce it to just the set $E = \{0\}$. We obtain the stationary distribution of this reduced DTMC. We then start to expand it to states $E = \{0,1\}$ and continue to expand until we reach the original set of states $\{0,1,2,\cdots,N\}$.

We demonstrate how this procedure works through a small example. Consider a bivariate DTMC that has four levels in its states given by $\{0,1,2,3\}$ and the phases are of finite dimension M. Its transition matrix P has block elements $P_{i,j}$ of order $M \times M$. If we partition the state space into $E = \{0,1,2\}$ and $\bar{E} = \{3\}$, then

$$P_e = \begin{bmatrix} P_{00} & P_{01} & P_{02} \\ P_{1,0} & P_{11} & P_{1,2} \\ P_{2,0} & P_{2,1} & P_{2,2} \end{bmatrix} + \begin{bmatrix} P_{03} \\ P_{13} \\ P_{23} \end{bmatrix} (I - P_{33})^{-1} \begin{bmatrix} P_{30} & P_{31} & P_{32} \end{bmatrix}.$$

When this new set $\{0,1,2\}$ is further partitioned into $E = \{0,1\}$ and $\bar{E} = \{2\}$ we get another transition matrix P_e. Another step further leads to a partitioning into states of the levels $E = \{0\}$ and $\bar{E} = \{1\}$. Writing $\boldsymbol{x} = \boldsymbol{x}P$, we can write the last block equation of this stationary distribution as

$$\boldsymbol{x}_3 = \boldsymbol{x}_0 P_{0,3} + \boldsymbol{x}_1 P_{1,3} + \boldsymbol{x}_2 P_{2,3} + \boldsymbol{x}_3 P_{33}.$$

From this we have

$$\boldsymbol{x}_3 = (\boldsymbol{x}_0 P_{0,3} + \boldsymbol{x}_1 P_{1,3} + \boldsymbol{x}_2 P_{2,3})(I - P_{33})^{-1}.$$

The block equation before the last can be written as

$$\boldsymbol{x}_2 = \boldsymbol{x}_0 P_{0,2} + \boldsymbol{x}_1 P_{1,2} + \boldsymbol{x}_2 P_{2,2} + \boldsymbol{x}_3 P_{32},$$

which can be also be written as

$$\boldsymbol{x}_2 = \boldsymbol{x}_0 (P_{0,2} + P_{0,3} K_{3,2}) + \boldsymbol{x}_1 (P_{1,2} + P_{1,3} K_{3,2}) + \boldsymbol{x}_2 (P_{2,2} + P_{2,3} K_{3,2}),$$

where $K_{3,2} = (I - P_{3,3})^{-1} P_{3,2}$. This in turn leads to

$$\boldsymbol{x}_2 = (\boldsymbol{x}_0 (P_{0,2} + P_{0,3} K_{3,2}) + \boldsymbol{x}_1 (P_{1,2} + P_{1,3} K_{3,2}))(I - (P_{2,2} + P_{2,3} K_{3,2}))^{-1}.$$

This process is continued further to reduce the block equation and obtain $\boldsymbol{x}_1$ in terms of $\boldsymbol{x}_0$ and then solve for $\boldsymbol{x}_0$ after one more reduction. This process is then reversed to expand into obtaining $\boldsymbol{x}_1$ from $\boldsymbol{x}_0$, followed by obtaining $\boldsymbol{x}_2$ using $\boldsymbol{x}_0$ and $\boldsymbol{x}_1$ and finally $\boldsymbol{x}_3$ using $\boldsymbol{x}_0$, $\boldsymbol{x}_1$ and $\boldsymbol{x}_2$. The final result is then normalized. A detailed description of this algorithm together with the steps can be found in Grassmann et al (1985) .

- **Iterative Methods**: The same sets of methods described earlier can be used for bivariate DTMC, i.e. the power method, Jacobi, Gauss-Seidel and the Successive over relaxation. The power method approach is straightforward and will not be elaborated on here.

First let us write $Q = P - I$. In that case, our interest is to solve for $\boldsymbol{x} = [\boldsymbol{x}_0\ \boldsymbol{x}_1\ \boldsymbol{x}_2\ \cdots\ \boldsymbol{x}_n]$, where $\boldsymbol{x}Q = 0$ and $\boldsymbol{x}\,\boldsymbol{1} = 1$. In order to conform to the standard representations of these methods we write the same equation in a transposed form as $Q^T\boldsymbol{x}^T = 0$.

Let us now separate Q^T into a diagonal component D, an upper triangular component U and a lower triangular component L with ij block elements given as $-D_{ij}$, U_{ij} and L_{ij}, respectively. It is assumed that the inverse of D_{ii} exists. Note that the D, L and U are not the same as the well known LU or LDU decompositions.

- **The Jacobi Method**: Then the Jacobi iteration is given in the scalar form as

$$(\boldsymbol{x}^{(k+1)})^T = D^{-1}(L+U)(\boldsymbol{x}^{(k)})^T \tag{4.16}$$

- **Gauss-Seidel Method**: The Gauss-Seidel method uses the most up-to-date information on solutions available for computations. It is given by the following iteration

$$(\boldsymbol{x}^{(k+1)})^T = (D)^{-1}(L(\boldsymbol{x}^{(k+1)})^T + U(\boldsymbol{x}^{(k)})^T). \tag{4.17}$$

- **The Successive Over-relaxation Method**: This method introduces an extra parameter ω. The method is a generalization of the Gauss-Seidel. The iteration is implemented as follows:

$$(\boldsymbol{x}^{(k+1)})^T = (1-\omega)(\boldsymbol{x}^{(k)})^T + \omega\{D^{-1}(L(\boldsymbol{x}^{(k+1)})^T + U(\boldsymbol{x}^{(k)})^T)\}. \tag{4.18}$$

4.5 Special Finite State DTMC

For special types of finite state DTMC which are of the block tridiagonal structures there are specially effective algorithms for obtaining the stationary vectors. Such a DTMC has a matrix of the form

$$P = \begin{bmatrix} A_0 & U_0 & & \\ D_1 & A_1 & U_1 & \\ & \ddots & \ddots & \ddots \\ & & D_N & A_N \end{bmatrix}, \tag{4.19}$$

where the entries $A_i, U_i, D_i, i = 0, 1, \cdots, N$ could be block matrices of finite dimensions. Our interest is to solve for the vector $\boldsymbol{x} = [\boldsymbol{x}_0,\ \boldsymbol{x}_1,\ \cdots,\ \boldsymbol{x}_N]$ where

$$\boldsymbol{x} = \boldsymbol{x}P,\ \boldsymbol{x}\mathbf{1} = 1,$$

and $\boldsymbol{x}_k$, $k = 0, 1, 2, \cdots, N$, is also a vector.

Three methods are presented here due to Gaver et al (1984), Grassmann, Grassmann et al (1985), and Ye and Li (1994). All the three methods are based on direct methods relating to the Gaussian elimination.

- **Level Reduction**: This method is due to Gaver et al (1984). It is based on Gaussian elimination and closely related to the block state reduction which is due to Grassmann et al (1985). They are all based on censored Markov chain idea in that a set of states is restricted – thereby the name level reduction. We consider the level reduced DTMC associated with this Markov chain at the kth stage, we can write

$$P_k = \begin{bmatrix} C_k & U_k & & \\ D_{k+1} & A_{k+1} & U_{k+1} & \\ & \ddots & \ddots & \ddots \\ & & D_N & A_N \end{bmatrix}, \quad 0 \le k \le N-1, \tag{4.20}$$

where

$$C_0 = A_0, \; C_k = A_k + D_k(I - C_{k-1})^{-1} U_{k-1}, \; 1 \le k \le N.$$

The solution is now given as

$$\boldsymbol{x}_N = \boldsymbol{x}_N C_N, \tag{4.21}$$

$$\boldsymbol{x}_k = \boldsymbol{x}_{k+1} D_{k+1} (I - C_k)^{-1}, \; 0 \le k \le N-1, \tag{4.22}$$

and

$$\sum_{k=0}^{N} \boldsymbol{x}_k \mathbf{1} = 1.$$

The first equation is solved using one of the methods proposed earlier, if the blocks of matrices are of dimension more than one.

- **Block State Reduction**: This method which is due to Grassmann et al (1985), is the reverse of the first method. We censor the Markov chain from the bottom as follows

$$P_{N-k} \begin{bmatrix} A_0 & U_0 & & \\ D_1 & A_1 & U_1 & \\ & \ddots & \ddots & \ddots \\ & & D_{N-k} & E_{N-k} \end{bmatrix}, \quad 0 \le k \le N-1, \tag{4.23}$$

where

$$E_N = A_N, \; E_{N-k} = A_{N-k} + U_{N-k}(I - A_{N-k+1})^{-1} D_{N-k+1}, \; 1 \le k \le N.$$

The solution is now given as

$$\boldsymbol{x}_0 = \boldsymbol{x}_0 E_0, \tag{4.24}$$

$$\boldsymbol{x}_{N-k} = \boldsymbol{x}_{N-k-1} U_{N-k-1}(I - E_{N-k})^{-1}, \; 1 \leq k \leq N, \tag{4.25}$$

and

$$\sum_{k=0}^{N} \boldsymbol{x}_k \mathbf{1} = 1.$$

- **Folding Algorithm**: This method is related to the first two in that they all group the DTMC into two classes. The folding algorithm groups them into odd and even whereas the other two group the DTMC into one and others. It then rearranges the DTMC as $\{0,2,4,6,\cdots,N,1,3,5,\cdots,N-1\}$, if N is even, as an example. The transition matrix is then partitioned accordingly leaving it in a good structure. For the details about this algorithm you are referred to Ye and Li (1994).

4.6 Infinite State Markov Chains

Consider a DTMC $\{X_n,\ n = 0,1,2,\cdots\}$, $X_n = 0,1,2,\cdots$. Since $X_n \in \mathscr{I}$ we say its an infinite DTMC. The finite case is when $X_n \in \{0,1,2,\cdots,N < \infty\}$. If the infinite DTMC is positive recurrent we know that it has a stationary distribution $\boldsymbol{x}$ such that

$$\boldsymbol{x} = \boldsymbol{x}P, \;\; \boldsymbol{x}\mathbf{1} = 1.$$

Even if we know that it has a stationary distribution we still face a challenge on how to compute this distribution since all the methods presented for finite DTMC can no longer be used because the state space is infinite.

One may suggest that we truncate the infinite DTMC at an appropriate state $N < \infty$ to a finite DTMC and apply the earlier techniques. A truncation method could involve selecting the smallest N such that $1 - \sum_{j=0}^{N} p_{ij} < \varepsilon$, $\forall i$, where ε is a very small value, say 10^{-12}, for example. Such truncations could lead to problems at times. However for some special infinite DTMCs there are well established methods for determining the stationary distribution without truncation. In addition there are methods for establishing whether such DTMCs are positive recurrent. We will be presenting those classes of infinite DTMCs.

First we discuss the general case of infinite DTMC and present some known results. We then present the three classes of infinite DTMCs, called the GI/M/1 types, M/G/1 types and QBD types, which are numerically tractable and for which the results are well known. Since the results to be developed here are valid for both univariate and multivariate DTMCs, with only the first variable allowed to be unbounded, we will present results for the bivariate case.

4.7 Some Results for Infinite State Markov Chains with Repeating Structure

In this section, we introduce some associated measures and results related to Markov chains with infinite state block-structured transition matrices. Results presented here are for discrete time Markov chains. However, corresponding results for continuous time Markov chains can be obtained in parallel.

Let $\{Z_n = (X_n, Y_n);\ n = 0, 1, 2, ...\}$, with $X_n = 0, 1, 2, 3 \cdots, ;\ Y_n = 1, 2, \cdots, K_{X_n} < \infty$, be the Markov chain, whose transition matrix P is expressed in block matrix form:

$$P = \begin{bmatrix} P_{0,0} & P_{0,1} & P_{0,2} & \cdots & \cdots \\ P_{1,0} & P_{1,1} & P_{1,2} & \cdots & \cdots \\ P_{2,0} & P_{2,1} & P_{2,2} & \cdots & \cdots \\ \vdots & \vdots & \vdots & \vdots & \vdots \end{bmatrix}, \tag{4.26}$$

where $P_{i,j}$ is a matrix of size $K_i \times K_j$ with both $K_i < \infty$ and $K_j < \infty$. In general, P is allowed to be sub-stochastic. Let the state space be S and partitioned accordingly into

$$S = \bigcup_{i=0}^{\infty} L_i, \tag{4.27}$$

with

$$L_i = \{(i,1), (i,2), \ldots, (i,K_i)\}. \tag{4.28}$$

In state (i,r), i is called the level variable and r the state variable. We also use the notation

$$L_{\leq i} = \bigcup_{k=0}^{i} L_k. \tag{4.29}$$

Partitioning the transition matrix P into blocks is not only done because it is convenient for comparison with results in the literature, but also because it is necessary when the Markov chain exhibits some kind of block structure. For the above Markov chain, we define matrices $R_{i,j}$ for $i < j$ and $G_{i,j}$ for $i > j$ as follows. $R_{i,j}$ is a matrix of size $K_i \times K_j$ whose $(r,\ s)$th entry is the expected number of visits to state $(j,\ s)$ before hitting any state in $L_{\leq(j-1)}$, given that the process starts in state $(i,\ r)$. $G_{i,j}$ is a matrix of size $K_i \times K_j$ whose (r,s)th entry is the probability of hitting state $(j,\ s)$ when the process enters $L_{\leq(i-1)}$ for the first time, given that the process starts in state $(i,\ r)$. We call matrices $R_{i,j}$ and $G_{i,j}$ respectively, the matrices of expected number of visits to higher levels before returning to lower levels and the matrices of the first passage probabilities to lower levels.

The significance of $R_{i,j}$ and $G_{i,j}$ in studying Markov chains, especially for ergodic Markov chains, has been pointed out in many research papers, including Zhao et al (1998), Grassmann and Heyman (1990), Grassmann and Heyman (1993), and Heyman (1995)). In these papers, under the ergodic condition, stationary distribution vectors, the mean first passage times and the fundamental matrix are discussed in terms of these matrices. For some recent results (see Zhao et al (1999), and Zhao et al (1998)).

For special type of Markov chains in block form that have the property of repeating rows (or columns), which means that the transition probability matrix partitioned as in Zhao et al (1999) has the following form:

$$P = \begin{bmatrix} P_{0,0} & P_{0,1} & P_{0,2} & P_{0,3} & \cdots & \cdots \\ P_{1,0} & A_0 & A_1 & A_2 & \cdots & \cdots \\ P_{2,0} & A_{-1} & A_0 & A_1 & \cdots & \cdots \\ P_{3,0} & A_{-2} & A_{-1} & A_0 & \cdots & \cdots \\ \vdots & \vdots & \vdots & \vdots & \vdots & \vdots \end{bmatrix} \tag{4.30}$$

where $P_{0,0}$ is a matrix of size $K_0 \times K_0$ and all A_v for $v = 0, \pm 1, \pm 2, \ldots$, are matrices of size $m \times m$ with $m < \infty$. The sizes of other matrices are determined accordingly. $R_{n-i} = R_{i,n}$ and $G_{n-i} = G_{n,i}$ for $i \geq 0$. We only give the following results. The details, i.e. theorems and proofs can be found in Zhao et al (1999) and Zhao et al (1998). We define $sp(A)$ as the spectral radius of a square matrix A, i.e. its largest absolute eigenvalue.

1. Let P be stochastic.
   - Let $R = \sum_{n=1}^{\infty} R_n$. The Markov chain P in (4.30) is positive recurrent if and only if $R < \infty$ and $\lim_{k \to \infty} R^k = 0$.
   - Let $G = \sum_{n=1}^{\infty} G_n$. If $\sum_{k=-\infty}^{\infty} A_k$ is stochastic, then the Markov chain P in (4.30) is recurrent if and only if G is stochastic.
2. Let P be stochastic. If $\sum_{k=0}^{\infty} k P_{0,k} e < \infty$ and if $A = \sum_{k=-\infty}^{\infty} A_k$ is stochastic and irreducible with the stationary distribution π, then
   - P is positive recurrent if and only if $\pi R < \pi$;
   - P is null recurrent if and only if $\pi R = \pi$ and $Ge = e$; and
   - P is transient if and only if $Ge < e$.
3. If $\sum_{k=1}^{\infty} k P_{0,k} e < \infty$ then P is positive recurrent if and only if $sp(R) < 1$;
   - P is transient if and only if $sp(G) < 1$; and
   - P is null recurrent if and only if $sp(R) = sp(G) = 1$.

These are some of the general results. In what follows we focus on specific cases of infinite DTMCs.

4.8 Matrix-Analytical Method for Infinite State Markov Chains

The matrix-analytical method (MAM) is most suited to three classes of Markov chains, viz: i) those with the GI/M/1 structure ii) those with the M/G/1 structure, and those with the QBD structure which actually embodies the combined properties of both the GI/M1 and M/G/1 types of structures. Before we go into the details, let us define the state space which is partially general for our discussions.

Consider a bivariate Markov chain $\{(X_n, Y_n);\ n \geq 0\}$ such that $P_{i,j;v,k} = \Pr\{X_{n+1} = i, Y_{n+1} = j | X_n = v, Y_n = k\};\ i,v \geq 0,\ 1 \leq j,\ k \leq M$ and i and j are referred to as the levels i and phase j, (> 0), respectively. Further define $P^{(m)}_{i,j;v,k} = \Pr\{X_{n+m} = i, Y_{n+m} = j | X_n = v, Y_n = k\}$ and ${}_iP^{(m)}_{i,j;v,k} = \Pr\{X_{n+m} = i, Y_{n+m} = j | X_n = v, Y_n = k\}, X_{n+w} \neq i,\ w = 1,2,\cdots,m$. This probability ${}_iP^{(m)}_{i,j;v,k}$ is the taboo probability that given the DTMC starts from state (i,j) it visits state (v,k) at time m without visiting level i during that period.

4.8.1 The GI/M/1 Type

Let the transition matrix P be given in the block partitioned form as follows:

$$P = \begin{bmatrix} B_{00} & B_{01} & & & & \\ B_{10} & B_{11} & A_0 & & & \\ B_{20} & B_{21} & A_1 & A_0 & & \\ B_{30} & B_{31} & A_2 & A_1 & A_0 & \\ \vdots & \vdots & \vdots & \vdots & \vdots & \ddots \end{bmatrix}. \tag{4.31}$$

The blocks A_{i-j+1} refer to transitions from level i to j for $i \geq 1, j \geq 2, i \geq j-1$. The blocks B_{ij} refer to transitions in the boundary areas from level i to level j for $i \geq 0, j = 0, 1$. This transition matrix is of the GI/M/1 type and very well treated in Neuts (1981). It is called the GI/M/1 type because it has the same structure as the transition matrix of the GI/M/1 queue embedded at arrival points. It is skip-free to the right. This system will be discussed at later stages. It suffices at this point just to accept the name assigned to this DTMC.

We consider only the case for which the matrix P is irreducible and positive recurrent. Neuts (1981) shows that for the matrix P to be positive recurrent, we need the condition that

$$\boldsymbol{\pi} \boldsymbol{\nu} > 1, \tag{4.32}$$

where $\boldsymbol{\pi} = \boldsymbol{\pi} A$, $\boldsymbol{\pi}\mathbf{1} = 1$, $A = \sum_{i=0}^{\infty} A_i$ and $\boldsymbol{\nu} = \sum_{k=0}^{\infty} k A_k \mathbf{1}$.

We define $\boldsymbol{x} = [\boldsymbol{x}_0, \boldsymbol{x}_1, \boldsymbol{x}_2, \ldots]$ as the invariant probability vector associated with P, then for $\boldsymbol{x} = \boldsymbol{x}P$, we have

$$\boldsymbol{x}_0 = \sum_{i=0}^{\infty} \boldsymbol{x}_i B_{i0}, \tag{4.33}$$

$$\boldsymbol{x}_1 = \sum_{i=0}^{\infty} \boldsymbol{x}_i B_{i1}, \tag{4.34}$$

$$\boldsymbol{x}_j = \sum_{i=0}^{\infty} \boldsymbol{x}_{j+v-1} A_v, \quad j \geq 2. \tag{4.35}$$

Let us define a matrix R whose elements $R_{i,j}$ are given as

$$R_{j,w} = \sum_{n=0}^{\infty} {}_iP^{(n)}_{i,j;i+1,w}, \; i \geq 0, 1 \leq j \leq m, 1 \leq w \leq m.$$

The element $R_{j,w}$ is the expected number of visits to state $(i+1, w)$ before returning to level i, given that the DTMC started in state (i,j). If the matrix P satisfies (4.32) then there exists a non-negative matrix R which has all its eigenvalues in the unit disk for which the probabilities $\boldsymbol{x}_{i+1}$ can be obtained recursively as

$$\boldsymbol{x}_{i+1} = \boldsymbol{x}_i R, \text{ or } \boldsymbol{x}_{i+1} = \boldsymbol{x}_1 R^i, \quad i \geq 1. \tag{4.36}$$

This is the matrix analogue of the scalar-geometric solution of the GI/M/1 queue and that is why it is called the matrix-geometric solution. For details and rigorous proof of this results see Neuts (1981). We present a skeletal proof only in this book. Following Neuts (1981) closely, the derivation of these result can be summarized as follows:

$$P^{(n)}_{i+1,j;i+1,j} = {}_iP^{(n)}_{i+1,j;i+1,j} + \sum_{v=1}^{M} \sum_{r=0}^{n} P^{(r)}_{i+1,j;i,v} \times {}_iP^{(n-r)}_{i,v;i+1,j}, \; \forall n \geq 1. \tag{4.37}$$

As shown in Neuts (1981), if we sum this equation from $n = 1$ to $n = N$ and divide by N and then let $N \to \infty$, we have

$$\frac{1}{N} \sum_{n=1}^{N} P^{(n)}_{i+1,j;i+1,j}|_{N\to\infty} \to x_{i+1,j}$$

$$\frac{1}{N} \sum_{n=1}^{N} {}_iP^{(n)}_{i+1,j;i+1,j}|_{N\to\infty} \to 0$$

$$\frac{1}{N} \sum_{n=1}^{N} \sum_{r=0}^{n} P^{(r)}_{i+1,j;i,v}|_{N\to\infty} \to x_{i,v}$$

and

$$\sum_{n=1}^{N} {}_iP^{(n)}_{i,v;i+1,j}|_{N\to\infty} \to R_{v,j}.$$

Hence,

$$\sum_{v=1}^{M} \frac{1}{N} \sum_{n=1}^{N} \sum_{r=0}^{n} P^{(r)}_{i+1,j;i,v}|_{N\to\infty} \to \sum_{v=1}^{M} x_{i,v} R_{v,j}.$$

Neuts (1981) also showed that

$$(R^q)_{j,w} = R^{(q)}_{j,w} = \sum_{n=0}^{\infty} {}_iP^{(n)}_{i,j;i+q,w}, \tag{4.38}$$

and we use this result in what follows.

The matrix R is the minimum non-negative solution to the matrix polynomial equation

$$R = \sum_{k=0}^{\infty} R^k A_k. \tag{4.39}$$

The arguments leading to this are as follows. Consider the equation for $\boldsymbol{x}_j$ given as

$$\boldsymbol{x}_j = \sum_{v=0}^{\infty} \boldsymbol{x}_{j+v-1} A_v, \quad j \geq 2.$$

Now if we replace $\boldsymbol{x}_{j+v}$ with $\boldsymbol{x}_{j-1} R^{v+1}$, $v > 0$, $j > 2$, we obtain

$$\boldsymbol{x}_{j-1} R = \sum_{k=0}^{\infty} \boldsymbol{x}_{j-1} R^k A_k, \quad j \geq 2. \tag{4.40}$$

After some algebraic manipulations, we obtain Equation (4.39).

The matrix R is known as the rate matrix. The elements R_{ij} of the matrix R is the expected number of visits into state $(k+1,j)$, starting from state (k,i), until the first return to state $(k,\cdot)$, $k > 1$. The results quoted in (4.37) and (4.38) are based on taboo probabilities and a detailed presentation of the derivations can be found in Neuts (1981).

The boundary probabilities $(\boldsymbol{x}_0, \boldsymbol{x}_1)$ are determined as

$$\boldsymbol{x}_0 = \sum_{i=0}^{\infty} \boldsymbol{x}_i B_{i0} = \boldsymbol{x}_0 B_{00} + \sum_{i=1}^{\infty} \boldsymbol{x}_1 R^{i-1} B_{i0} \tag{4.41}$$

and

$$\boldsymbol{x}_1 = \sum_{i=0}^{\infty} \boldsymbol{x}_i B_{i1} = \boldsymbol{x}_0 B_{01} + \sum_{i=1}^{\infty} \boldsymbol{x}_1 R^{i-1} B_{i1}. \tag{4.42}$$

When rearranged in matrix form, we obtain

$$(\boldsymbol{x}_0 \ \boldsymbol{x}_1) = (\boldsymbol{x}_0, \boldsymbol{x}_1) B[R], \tag{4.43}$$

where

$$B[R] = \begin{bmatrix} B_{00} & B_{01} \\ \sum_{n=1}^{\infty} R^{n-1} B_{n0} & \sum_{n=1}^{\infty} R^{n-1} B_{n1} \end{bmatrix} \tag{4.44}$$

and then $(\boldsymbol{x}_0, \boldsymbol{x}_1)$ is normalized by

$$\boldsymbol{x}_0 \mathbf{1} + \boldsymbol{x}_1 (I - R)^{-1} \mathbf{1} = 1. \tag{4.45}$$

The argument behind this is that

$$\boldsymbol{x}_0 \mathbf{1} + \boldsymbol{x}_1 \mathbf{1} + \boldsymbol{x}_2 \mathbf{1} + \ldots = \boldsymbol{x}_0 \mathbf{1} + \boldsymbol{x}_1 (I + R + R^2 + \ldots) \mathbf{1} = 1$$

Because of the geometric nature of Equation (4.36), this method is popularly known as the *matrix - geometric method.*

4.8.2 Key Measures

The key measures usually of interest in this DTMC are

- The marginal distribution of the first variable (level) is given as

$$y_k = \boldsymbol{x}_k \mathbf{1} = \boldsymbol{x}_1 R^{k-1} \mathbf{1}, \ k \geq 1. \tag{4.46}$$

- The first moment is given as

$$E[X] = \boldsymbol{x}_1 [I - R]^{-2} \mathbf{1}. \tag{4.47}$$

- Tail behaviour, i.e. $p_k = Pr\{X \geq k\} = \sum_{v=k}^{\infty} \boldsymbol{x}_v \mathbf{1} = \boldsymbol{x}_1 R^{k-1} [I - R]^{-1} \mathbf{1}$. Let $\eta = sp(R)$, i.e. the spectral radius of R (i.e. its maximum absolute eigenvalue), and let $\mathbf{v}$ and $\mathbf{u}$ be the corresponding left and right eigenvectors which are normalized such that $\mathbf{u1} = 1$ and $\mathbf{uv} = 1$. It is known that $R^k = \eta^k \mathbf{vu} + o(\eta^k), \ k \to \infty$. Hence $\boldsymbol{x}_j = \boldsymbol{x}_1 R^{j-1} \mathbf{1} = \kappa \eta^{j-1} + o(\eta^j), \ j \to \infty$, where $\kappa = \boldsymbol{x}_1 \mathbf{v}$. Hence we have

$$p_k = \kappa \eta^{k-1}(1-\eta)^{-1} + o(\eta^k), \;\; k \to \infty. \tag{4.48}$$

The key ingredient to the matrix-geometric method is the matrix R and how to compute it efficiently becomes very important especially when it has a huge dimension and its spectral radius is close to 1.

4.8.2.1 Computing matrix R

There are several methods for computing the matrix R and they are all iterative. Letting $R(v)$ be the computed value of matrix R at the vth iteration, the computation is stopped once $|R(v+1) - R(v)|_{i,j} < \varepsilon$, where ε is a small convergent criterion with values of about 10^{-12}. The simplest but not necessarily most efficient methods are the linear ones, given as follows:

- Given that

  $$R = \sum_{k=0}^{\infty} R^k A_k$$

  we can write

  $$R(v+1) = \sum_{k=0}^{\infty} R^k(v) A_k, \tag{4.49}$$

  with $R(0) := A_0$.
- Given that

  $$R = (\sum_{k=0,k\neq 1}^{\infty} R^k A_k)(I - A_1)^{-1}$$

  we can write

  $$R(v+1) = (\sum_{k=0,k\neq 1}^{\infty} R^k(v+1) A_k)(I - A_1)^{-1}, \tag{4.50}$$

  with $R(0) := A_0$.
- Given that

  $$R = A_0[I - \sum_{k=1}^{\infty} R^{k-1} A_k]^{-1}$$

we can write

$$R(v+1) = A_0[I - \sum_{k=1}^{\infty} R^{k-1}(v)A_k]^{-1} \tag{4.51}$$

with $R(0) := A_0$,

More efficient methods are the cyclic reduction method by Bini and Meini (1996) and the method using the non-linear programming approach by Alfa et al (2002). These methods are more efficient than the linear approach however, they cannot take advantage of the structures of the matrices $A_k,\ k = 0,1,2,\cdots$, when they can be exploited. The method of cyclic reduction will be discussed under QBDs and specifically regarding the computation of matrix G.

- **NON-LINEAR PROGRAMMING APPROACH FOR R:** By definition, R is the minimal non-negative solution to the matrix polynomial equation $R = \sum_{k=0}^{\infty} R^k A_k$. Consider the matrix $A^*(z) = \sum_{k=0}^{\infty} A_k z^k,\ |z| \le 1$, and let $\chi^*(z)$ be the eigenvalue of $A^*(z)$. It is known from Neuts (1981) that the Perron eigenvalue of R is the smallest solution to $z = \chi^*(z)$ and that $\mathbf{u}$ is its left eigenvector. Let X be any non-negative square matrix of order m and a function $f(X) = \sum_{k=0}^{\infty} X^k A_k$. For two square matrices Y and Z of the same dimensions, let $Y \circ Z$ denote their elementwise product. Now we define a function

$$H(X) = \sum_{(i,j)=1}^{m} ([f(X)]_{i,j} - X_{i,j})^2 = \mathbf{1}^T((f(X) - X) \circ (f(X) - X))\mathbf{1}. \tag{4.52}$$

It was shown in Alfa et al (2002) that if the matrix P is positive recurrent then the matrix R is the unique optimal solution to the non-linear programming problem

$$\textit{minimize} \quad H(X) \tag{4.53}$$

$$\textit{subject to} \quad \mathbf{u}X = \eta\mathbf{u}, \tag{4.54}$$

$$X \ge 0. \tag{4.55}$$

Two optimization algorithms were presented by Alfa et al (2002) for solving this problem.

- **THE INVARIANT SUBSPACE METHOD FOR R:** Akar and Sohraby (1997) developed this method for R matrix. The method requires that $A^*(z) = \sum_{k=0}^{\infty} z^k A_k$ be rational. However, whenever we have $A_k = 0,\ k > K < \infty$, this condition is usually met. For most practical situations $A_k = 0,\ k > K < \infty$, hence this method is appropriate in such situations.

4.8.3 *Some Special Structures of the Matrix R often Encountered*

The matrix-geometric method is a very convenient method to work with, and also it has a probabilistic interpretation that makes it more meaningful with certain appeal to application oriented users. One of its weaknesses however, is that as the size of the matrix R gets very large, the computational aspects become cumbersome, especially when its spectral radius, $sp(R)$ is very close to 1. In that respect, it is usually wise to search for special structures of matrix R that can be exploited. Some of the very useful special structures are as follows:

- If A_0 is of rank one, i.e. $A_0 = \omega\beta$, then

$$R = \omega\xi = A_0[I - \sum_{v=1}^{\infty} \eta^{v-1} A_v]^{-1}, \tag{4.56}$$

  where $\eta = \xi\omega$ is the maximal eigenvalue of the matrix R.
  The argument leading to this are based on the iteration

$$R(0) := 0,$$

$$R(k+1) := A_0(I - A_1)^{-1} + \sum_{v=2}^{\infty} R^v(k) A_v (I - A_1)^{-1},\ k \geq 0,$$

  and letting $R(k) = \omega\xi(k)$ we have

$$\xi(1) = \beta(I - A_1)^{-1},$$

$$\xi(k) = \beta(I - A_1)^{-1} + \sum_{v=2}^{\infty} (\xi(k-1)\omega)^{v-1} \xi(k-1) A_v (I - A_1)^{-1},\ k \geq 0.$$

  This leads to the explicit result given above.
- For every row of A_0 that is all zeros, the corresponding rows of R are also zeros.
- When the matrices A_i, $i \geq 0$, are of the sparse block types, it is advantageous to write out the equations of R in smaller blocks in order to reduce the computations.
- Other special structures can be found in Chakravarthy and Alfa (1993), Alfa and Chakravarthy (1994) and Alfa et al (1995).

4.8.4 The M/G/1 Type:

Let the transition matrix P be given in the block partitioned form as follows:

$$P = \begin{bmatrix} B_0 & B_1 & B_2 & B_3 & \cdots \\ A_0 & A_1 & A_2 & A_3 & \cdots \\ & A_0 & A_1 & A_2 & \cdots \\ & & A_0 & A_1 & \cdots \\ & & & \vdots & \vdots \end{bmatrix}. \tag{4.57}$$

The blocks A_{j-i+1} refer to transitions from level i to level j for $j \geq i-1$. The blocks B_j refer to transitions at the boundary area from level 0 to level j. It is skip-free to the left.

The transition matrix is of the M/G/1 type and very well treated in Neuts (1989). Let $A = \sum_{k=0}^{\infty} A_k$ and let $\nu = \sum_{k=1}^{\infty} kA_k\mathbf{1}$ be finite. If A is irreducible, then there exists an invariant vector $\boldsymbol{\pi}$ such that $\boldsymbol{\pi} = \boldsymbol{\pi}A$, and $\boldsymbol{\pi}\mathbf{1} = 1$. The Markov chain P is irreducible if and only if $\rho = \boldsymbol{\pi}\nu < 1$.

If the Markov chain P is irreducible, then there exists a stochastic matrix G which is the minimal non-negative solution to the matrix polynomial equation

$$G = \sum_{k=0}^{\infty} A_k G^k. \tag{4.58}$$

The elements G_{ij} of this matrix refer to the probability that the system will eventually visit state $(v-1, j)$ given that it started from state (v, i), $(v > 1)$. The arguments that lead to Equation (4.58) will be presented in a skeletal form and the rigorous proof can be found in Neuts (1989).

It is clear that starting from level $i+1$ the DTMC can reach level i in one step according to the matrix A_0, or in two steps according to A_1G, i.e. it stays in level $i+1$ in the first step and then goes to level i in the second step; or in three steps according to A_2G^2, i.e. it goes up to level $i+2$ in the first step, comes down two consecutive times in the second and third steps according to $G \times G = G^2$; and this continues. Hence

$$G = A_0 + A_1G + A_2G^2 + \cdots = \sum_{\nu=0}^{\infty} A_\nu G^\nu.$$

It was shown in Neuts (1989) in Section 2.2.1 that

- If P is irreducible, G has no zero rows and it has eigenvalue η of maximum modulus.
- If P is recurrent G is stochastic.

4.8.4.1 Stationary distribution

In order to obtain the stationary distribution for the M/G/1 system we proceed as follows.

Let $\mathbf{g} = \mathbf{g}G$, where $\boldsymbol{gI} = 1$. Further define M such that

$$M = G + \sum_{v=1}^{\infty} A_v \sum_{k=0}^{v-1} G^k M G^{v-k-1} \tag{4.59}$$

and also define $\mathbf{u}$ such that

$$\mathbf{u} = (I - G + \mathbf{1g})[I - A - (\mathbf{1} - \boldsymbol{\beta})\mathbf{g}]^{-1}\mathbf{1}. \tag{4.60}$$

Further define K such that

$$K = B_0 + \sum_{v=1}^{\infty} B_v G^v \tag{4.61}$$

and let $\kappa = \kappa K$, where $\kappa \mathbf{1} = 1$. The (i,j) elements of the matrix K refer to the probability that the system will ever return to state $(0,j)$ given that is started from state $(0,i)$. Note that we say "return" because we are considering the same level 0 in both cases, so we mean a return to level 0.

If we now define the vector $x = [x_0, x_1, x_2, \ldots]$ then we can obtain x_0 as

$$x_0 = \left(\kappa \widetilde{\kappa}\right)^{-1} \kappa \tag{4.62}$$

where

$$\widetilde{\kappa} = \mathbf{1} + \sum_{v=1}^{\infty} B_v \sum_{k=0}^{v-1} G^k \mathbf{u}$$

Once x_0 is obtained the remaining x_n, $n > 1$, are obtained as follows (see Ramaswami (1988)):

$$x_n = \left[x_0 \overline{B}_n + (1 - \delta_{n,1}) \sum_{j=1}^{n-1} x_j \overline{A}_{n-j+1} \right] \left(I - \overline{A}_1\right), \; n \geq 1 \tag{4.63}$$

where

$$\overline{B}_v = \sum_{i=v}^{\infty} B_i G^{i-v} \text{ and } \overline{A}_i = \sum_{i=v}^{\infty} A_i G^{i-v}, \; v \geq 0.$$

The key ingredient to using this method is the matrix G. Just like its GI/M/1 counterpart the effort involved in its computation could be enormous if its size is huge and traffic intensity is close to 1.

4.8.5 Computing Matrix G

There are several methods for computing matrix G and they are all iterative. Letting $G(v)$ be the computed value of matrix G at the vth iteration, the computation is stopped once $|G(v+1)-G(v)|_{i,j} < \varepsilon$, where ε is a small convergent criterion with values of about 10^{-12}. The simplest, but not necessarily most efficient, methods are the linear ones, given as follow:

- Given that

$$G = \sum_{k=0}^{\infty} A_k G^k$$

we can write

$$G(v+1) = \sum_{k=0}^{\infty} A_k G^k(v) A_k, \tag{4.64}$$

with $G(0) := \mathbf{0}$. It has been shown that starting with a $G(0) = I$ sometimes works better.

- Given that

$$G = (I - A_1)^{-1}\Big(\sum_{k=0,k\neq 1}^{\infty} A_k G^k\Big)$$

we can write

$$G(v+1) = (I - A_1)^{-1}\Big(\sum_{k=0,k\neq 1}^{\infty} A_k G^k(v+1)\Big), \tag{4.65}$$

with $G(0) := \mathbf{0}$.

- Given that

$$G = [I - \sum_{k=1}^{\infty} A_k G^{k-1}]^{-1} A_0$$

we can write

$$G(v+1) = [I - \sum_{k=1}^{\infty} A_k G^{k-1}(v)]^{-1} A_0 \tag{4.66}$$

with $G(0) := \mathbf{0}$.

More efficient methods are the cyclic reduction method by Bini and Meini (1996), subspace method by Akar and Sohraby (1997) and the method using the non-linear programming approach by Alfa et al (1998). These methods are more efficient than the linear approach however, they cannot take advantage of the structures of the matrices $A_k,\ k = 0, 1, 2, \cdots$, when they can be exploited.

- **CYCLIC REDUCTION FOR** G**:** The cyclic reduction (CR) method developed by Bini and Meini (1996) capitalizes on the structure of the matrix P in trying to compute G. Note that the matrix equation for G, i.e. $G = \sum_{k=0}^{\infty} A_k G^k$, can be written as $G^v = \sum_{k=0}^{\infty} A_k G^{k+v},\ v \geq 0$, or in matrix form as

$$\begin{bmatrix} I-A_1 & -A_2 & -A_3 & \cdots \\ -A_0 & I-A_1 & -A_2 & \cdots \\ & -A_0 & I-A_1 & \cdots \\ & & -A_0 & \cdots \\ & & & \ddots \end{bmatrix} \begin{bmatrix} G \\ G^2 \\ G^3 \\ G^4 \\ \vdots \end{bmatrix} = \begin{bmatrix} A_0 \\ 0 \\ 0 \\ 0 \\ \vdots \end{bmatrix}. \tag{4.67}$$

By using the odd-even permutation idea, used by Ye and Li (1994) for the folding algorithm together with an idea similar to the Logarithmic Reduction for QBDs (to be presented later), on the block matrices we can write this matrix equation as

$$\begin{bmatrix} I - U_{1,1} & -U_{1,2} \\ -U_{2,1} & I - U_{2,2} \end{bmatrix} \begin{bmatrix} V_0 \\ V_1 \end{bmatrix} = \begin{bmatrix} 0 \\ B \end{bmatrix},$$

where

$$U_{1,1} = U_{2,2} = \begin{bmatrix} A_1 & A_3 & \cdots \\ & A_1 & \cdots \\ & & \ddots \end{bmatrix},$$

$$U_{1,2} = \begin{bmatrix} A_0 & A_2 & \cdots \\ & A_0 & \cdots \\ & & \ddots \end{bmatrix},$$

$$U_{2,1} = \begin{bmatrix} A_2 & A_4 & \cdots \\ A_0 & A_2 & \cdots \\ 0 & \ddots & \ddots \end{bmatrix},\ B = \begin{bmatrix} A_0 \\ 0 \\ \vdots \end{bmatrix},$$

$$V_0 = G^{2k},\ V_1 = G^{2k+1},\ k = 1, 2, \cdots.$$

By applying standard block Gaussian elimination to the above equation we obtain

$$[I - U_{2,2} - U_{2,1}(I - U_{1,1})^{-1} U_{1,2}] V_1 = B. \tag{4.68}$$

By noticing that this equation (4.68) is also of the form

$$\begin{bmatrix} I-A_1^{(1)} & -A_2^{(1)} & -A_3^{(1)} & \cdots \\ -A_0^{(1)} & I-A_1^{(1)} & -A_2^{(1)} & \cdots \\ & -A_0^{(1)} & I-A_1^{(1)} & \cdots \\ & & -A_0^{(1)} & \cdots \\ & & & \ddots \end{bmatrix} \begin{bmatrix} G \\ G^3 \\ G^5 \\ G^7 \\ \vdots \end{bmatrix} = \begin{bmatrix} A_0 \\ 0 \\ 0 \\ 0 \\ \vdots \end{bmatrix},$$

Bini and Meini (1996) showed that after the nth repeated application of this operation we have

$$\begin{bmatrix} I-U_{1,1}^{(n)} & -U_{1,2}^{(n)} \\ -U_{2,1}^{(n)} & I-U_{2,2}^{(n)} \end{bmatrix} \begin{bmatrix} V_0^{(n)} \\ V_1^{(n)} \end{bmatrix} = \begin{bmatrix} 0 \\ B \end{bmatrix},$$

where

$$U_{1,1}^{(n)} = U_{2,2}^{(n)} = \begin{bmatrix} A_1^{(n)} & A_3^{(n)} & \cdots \\ & A_1^{(n)} & \cdots \\ & & \ddots \end{bmatrix},$$

$$U_{1,2}^{(n)} = \begin{bmatrix} A_0^{(n)} & A_2^{(n)} & \cdots \\ & A_0^{(n)} & \cdots \\ & & \ddots \end{bmatrix},$$

$$U_{2,1}^{(n)} = \begin{bmatrix} A_2^{(n)} & A_4^{(n)} & \cdots \\ A_0^{(n)} & A_2^{(n)} & \cdots \\ 0 & \ddots & \ddots \end{bmatrix}, \quad B = \begin{bmatrix} A_0 \\ 0 \\ \vdots \end{bmatrix},$$

$$V_0 = G^{2k.2^n}, \; V_1 = G^{2k.2^n+1}, \; k = 1,2,\cdots, \; n \geq 0.$$

In the end they show that the matrix equation is also of the form

$$[I - U_{2,2}^{(n)} - U_{2,1}^{(n)}(I - U_{1,1}^{(n)})^{-1} U_{1,2}^{(n)}] V_1^{(n)} = B,$$

which results in

$$G = (I - A_1^{(n)})^{-1}(A_0 + \sum_{k=1}^{\infty} A_k^{(n)} G^{k.2^n+1}). \tag{4.69}$$

They further showed that as $n \to \infty$ the second term in the brackets tends to zero, and $(I - A_1^{(n)})^{-1}$ exists, hence

$$G = (I - A_1^{(\infty)})^{-1} A_0. \tag{4.70}$$

Hence G can be obtained from this expression. Of course the operation can only be applied in a finite number of times. So it is a matter of selecting how many operations to carry out given the tolerance desired.

- **THE INVARIANT SUBSPACE METHOD FOR** G**:** Akar and Sohraby (1997) developed this method for G matrix. The method requires that $A^*(z) = \sum_{k=0}^{\infty} z^k A_k$ be rational. However, whenever we have $A_k = 0,\ k > K < \infty$, this condition is usually met. For most practical situations $A_k = 0,\ k > K < \infty$, hence this method is appropriate in such situations.
- **NON-LINEAR PROGRAMMING APPROACH FOR** G**:** The non-linear programming for the matrix G is similar in concept to that used for the matrix R, with some minor differences. Consider an $m \times m$ matrix of non-negative real values such that $g(X) = \sum_{k=0}^{\infty} A_k X^k$, where $g : \Re^m \to \Re^m$. Further define a function $\phi(X)$ such that $\phi : \Re^m \to \Re^1$ and it is order preserving, i.e. $X < Y$ implies $\phi(X) < \phi(Y)$ where both $(X,Y) \in \Re^m$. A good example is $\phi(X) = \mathbf{1}^T X \mathbf{1}$. It was shown by Alfa et al (1998) that the matrix G is the unique and optimal solution to the non-linear programming problem

$$\textit{minimize} \quad \phi(X) \tag{4.71}$$

$$\textit{subject to} \quad g(X) - X \leq 0, \tag{4.72}$$

$$X \geq 0. \tag{4.73}$$

Several algorithms were proposed by Alfa et al (1998) for solving this problem.

4.8.6 Some Special Structures of the Matrix G often Encountered

The matrix analytic method is a very convenient method to work with, and also because it has a probabilistic interpretation that makes it more meaningful with certain appeal to application oriented users. One of its weaknesses is that as the size of the matrix G gets very large and the traffic intensity tends to 1, the computational aspects become cumbersome. In that respect, it is usually wise to search for special structures of matrix G that can be exploited. Some of the very useful special structures are as follows:

- If A_0 is of rank one, i.e. $A_0 = \mathbf{v}\boldsymbol{\alpha}$, with $\boldsymbol{\alpha}\mathbf{1} = 1$, then

$$G = \mathbf{1}\boldsymbol{\alpha}, \tag{4.74}$$

if the Markov chain with transition matrix P is positive recurrent. This is obtained through the following arguments. Note that

$$G(1) := A_0,$$

$$G(k+1) := \sum_{\ell=0}^{\infty} A_\ell G(k)^\ell, \; k \geq 1.$$

If we write $G(k) = \gamma(k)\boldsymbol{\alpha}, \; k \geq 1$, where

$$\gamma(1) = \mathbf{v},$$

$$\gamma(k+1) = \sum_{\ell=0}^{\infty} A_\ell(\boldsymbol{\alpha}\gamma(k-1))^{\ell-1}\gamma(k-1).$$

Hence we have $G = \gamma\boldsymbol{\alpha}$, and since the Markov chain is positive recurrent we have $G\mathbf{1} = \mathbf{1}$, implying that $\gamma = \mathbf{1}$ and hence

$$G = \mathbf{1}\boldsymbol{\alpha}.$$

- For every column of A_0 that is all zeros, the corresponding columns of G are also all zeros.
- When the matrices A_i, $i \geq 0$, are of the sparse block types, it is advantageous to write out the equations of G in smaller blocks in order to reduce the computations.
- Other special structures can be found in Chakravarthy and Alfa (1993), Alfa and Chakravarthy (1994), and Alfa et al (1995).

4.8.7 QBD

The QBD (Quasi-Birth-and-Death) DTMC is the most commonly encountered DTMC in discrete time queues. It embodies the properties of both the GI/M/1 and M/G/1 types of DTMCs. This infinite DTMC will be dealt with in more details because of its importance in queueing theory.

Let the transition matrix P be given in the block partitioned form as follows:

$$P = \begin{bmatrix} B & C & & & \\ E & A_1 & A_0 & & \\ & A_2 & A_1 & A_0 & \\ & & A_2 & A_1 & A_0 \\ & & & \ddots & \ddots & \ddots \end{bmatrix}. \qquad (4.75)$$

Note the change in notations. For the QBD the labelling of the block matrices is based on the GI/M/1 type format.

The QBD only goes a maximum of one level up or down. It is skip-free to the left and to the right. These are very useful properties which are fully exploited in the theory of QBDs. For detailed treatment of this see Latouche and Ramaswami (1999).

We assume that the matrices A_k, $k = 0,1,2$ are of dimension $n \times n$ and matrix B is of dimension $m \times m$, hence C and E are of dimensions $m \times n$ and $n \times m$, respectively. The key matrices that form the ingredients for analyzing a QBD are the R and G matrices, which are given as

$$R = A_0 + RA_1 + R^2A_2 \tag{4.76}$$

and

$$G = A_2 + A_1G + A_0G^2. \tag{4.77}$$

R and G are the minimal non-negative solutions to the first and second equation, respectively. Another matrix that is of importance is the matrix U, which records the probability of visiting level i before level $i-1$, given the DTMC started from level i. The matrix U is given as

$$U = A_1 + A_0(I-U)^{-1}A_2. \tag{4.78}$$

There are known relationships between the three matrices R, G and U, and are given as follows

$$R = A_0(I-U)^{-1}, \tag{4.79}$$

$$G = (I-U)^{-1}A_2, \tag{4.80}$$

$$U = A_1 + A_0G, \tag{4.81}$$

$$U = A_1 + RA_2. \tag{4.82}$$

Using these known relationships we have

$$R = A_0(I - A_1 - A_0G)^{-1}, \tag{4.83}$$

$$G = (I - A_1 - RA_2)^{-1}A_2. \tag{4.84}$$

These last two equations are very useful for obtaining R from G and vice-versa, especially when it is easier to compute one of them when our interest is in the other.

In what follows we present a skeletal development of the relationship between the vector $\boldsymbol{x}_{i+1}$ and $\boldsymbol{x}_i$ through the matrix R using a different approach from the one used for the GI/M/1 system. The method is based on censoring used differently by Latouche and Ramaswami (1999) and Grassmann and Heyman (1993). Consider the matrix P above. It can be partitioned as follows

$$P = \begin{bmatrix} T_L & T_R \\ B_L & B_R \end{bmatrix}. \tag{4.85}$$

Suppose we select n from $0, 1, 2, \cdots, n-1, n, n+1, \cdots$, as the point of partitioning. Then the vector $\boldsymbol{x}$ is also partitioned into $\boldsymbol{x} = [\boldsymbol{x}_T \ \boldsymbol{x}_B]$, with

$$\boldsymbol{x}_T = [\boldsymbol{x}_0, \ \boldsymbol{x}_1, \ \cdots, \ \boldsymbol{x}_n], \ \ \boldsymbol{x}_B = [\boldsymbol{x}_{n+1}, \ \boldsymbol{x}_{n+2}, \ \cdots].$$

The result is that we have

$$T_R = \begin{bmatrix} 0 & 0 & 0 & \cdots \\ \vdots & \vdots & \vdots & \vdots \\ 0 & 0 & 0 & \cdots \\ A_0 & 0 & 0 & \cdots \end{bmatrix}, \tag{4.86}$$

and

$$B_R = \begin{bmatrix} A_1 & A_0 & & \\ A_2 & A_1 & A_0 & \\ & A_2 & A_1 & A_0 \\ & & \ddots & \ddots & \ddots \end{bmatrix}. \tag{4.87}$$

Keeping in mind that $\boldsymbol{x} = \boldsymbol{x}P$, we can write

$$\boldsymbol{x}_T = \boldsymbol{x}_T T_L + \boldsymbol{x}_B B_L \tag{4.88}$$

$$\boldsymbol{x}_B = \boldsymbol{x}_T T_R + \boldsymbol{x}_B B_R. \tag{4.89}$$

Equation (4.89) can be written as

$$\boldsymbol{x}_B = \boldsymbol{x}_T T_R (I - B_R)^{-1}, \tag{4.90}$$

which can be further written as

$$[\boldsymbol{x}_{n+1}, \ \boldsymbol{x}_{n+2}, \ \cdots] = [\boldsymbol{x}_0, \ \boldsymbol{x}_1, \ \cdots, \ \boldsymbol{x}_n] T_R (I - B_R)^{-1}.$$

By the structure of T_R we know that the structure of $T_R(I - B_R)^{-1}$ is as follows

$$T_R(I-B_R)^{-1} = \begin{bmatrix} 0 & 0 & 0 & \cdots \\ \vdots & \vdots & \vdots & \vdots \\ 0 & 0 & 0 & \cdots \\ A_0V_{11} & A_0V_{12} & A_0V_{13} & \cdots \end{bmatrix}, \tag{4.91}$$

where V_{ij} are the block elements of $(I-B_R)^{-1}$. It is clear that

$$\boldsymbol{x}_{n+1} = \boldsymbol{x}_n A_0 V_{11} = \boldsymbol{x}_n R, \tag{4.92}$$

where R is by definition equivalent to A_0V_{11}. This is another way to see the matrix geometric result for the QBD. See Latouche and Ramaswami (1999) for details.

Similar to the GI/M/1 type, if the QBD DTMC is positive recurrent then the matrix R has a spectral radius less than 1 and also the matrix G is stochastic just like the case for the M/G/1 type. Once we know R we can use the result $\boldsymbol{x}_{i+1} = \boldsymbol{x}_i R$, $i \geq 1$. However, we still need to compute R and determine the boundary behaviour of the matrix P. First we discuss the boundary behaviour.

The censored stochastic transition matrix representing the boundary is $B[R]$ and can we can write

$$\boldsymbol{x}_0 = \boldsymbol{x}_0 B + \boldsymbol{x}_1 E,$$

$$\boldsymbol{x}_1 = \boldsymbol{x}_0 C + \boldsymbol{x}_1 (A_1 + RA_2),$$

from which we obtain

$$B[R] = \begin{bmatrix} B & C \\ E & A_1 + RA_2 \end{bmatrix}.$$

Using this matrix we obtain

$$\boldsymbol{x}_1 = \boldsymbol{x}_1 [E(I-B)^{-1}C + A_1 + RA_2],$$

which after getting solved for $\boldsymbol{x}_1$ we can then solve for $\boldsymbol{x}_0$ as $\boldsymbol{x}_0 = \boldsymbol{x}_1 E(I-B)^{-1}$, and then normalized through

$$\boldsymbol{x}_0 \mathbf{1} + \boldsymbol{x}_1 (I-R)^{-1} \mathbf{1} = 1.$$

4.8.8 Computing the matrices R and G

Solving for R and G can be carried out using the techniques for the GI/M/1 type and M/G/1 type respectively. However, very efficient methods have been developed for the QBD, such as

- **THE LOGARITHMIC REDUCTION (LR) METHOD:** This is a method developed by Latouche and Ramaswami (1993) for QBDs and is quadratically convergent. The Cyclic Reduction method discussed earlier is similar in principle to the Logarithmic Reduction method. The LR method is vastly used for analyzing QBDs. Readers are referred to the original paper Latouche and Ramaswami (1993) and the book Latouche and Ramaswami (1999), both by Latouche and Ramaswami, for detail coverage of the method. In this section we present the results and give an outline of the algorithm. Essentially the algorithm works on computing the matrix G by kind of restricting the DTMC to only time epochs when it changes main levels, i.e. it goes up or down one, hence the transitions without a level change are censored out. Once the matrix G is obtained, the matrix R can be easily calculated from the relationship given earlier between R and G.

  Consider the transition matrix P and write the G matrix equation given as

$$G = A_2 + A_1 G + A_0 G^2.$$

  This can be written as $(I - A_1)G = A_2 + A_0 G^2$ or better still as

$$G = (I - A_1)^{-1} A_2 + (I - A_1)^{-1} A_0 G^2 = L + HG^2, \tag{4.93}$$

  with

$$L = (I - A_1)^{-1} A_2, \text{ and } H = (I - A_1)^{-1} A_0. \tag{4.94}$$

  If we restrict the process to only up and down movements while not considering when there is no movement, then the matrix L captures its down movement while the matrix H captures its up movements. Note that G and G^2 record first passage probabilities across one and two levels, respectively. So if we repeat this process we can study first passage probabilities across 4, 8, $\cdots$ levels. So let us proceed and define

$$A_k^{(0)} = A_k, \; k = 0, 1, 2$$

$$H^{(0)} = (I - A_1^{(0)})^{-1} A_0^{(0)}, \; L^{(0)} = (I - A_1^{(0)})^{-1} A_2^{(0)},$$

$$U^{(k)} = H^{(k)} L^{(k)} + L^{(k)} H^{(k)}, \;\; k \geq 0,$$

$$H^{(k+1)} = (I - U^{(k)})^{-1} (H^{(k)})^2, \; L^{(k+1)} = (I - U^{(k)})^{-1} (L^{(k)})^2, \; k \geq 0.$$

  After applying this process repeatedly Latouche and Ramaswami (1993) showed that

$$G = \sum_{k=0}^{\infty} \left(\prod_{i=0}^{k-1} H^{(i)} \right) L^{(k)}. \tag{4.95}$$

Since we can not compute the sum series infinitely, the sum is truncated at some point depending on the predefined tolerance.

- **THE CYCLIC REDUCTION METHOD:** The cyclic reduction method developed by Bini and Meini (1996) for the M/G/1 is another appropriate algorithm for the QBD in the same sense as the LR. We compute the matrix G and use that to compute the matrix R. The algorithm will not be repeated here since it has been presented under the M/G/1 techniques.
- **THE INVARIANT SUBSPACE METHOD:** Akar and Sohraby (1997) developed this method for G and R matrices, and since for the QBDs $A(z)$ are rational this method is appropriate for such analysis.

4.8.9 Some Special Structures of the Matrix R and the matrix G

If A_2 or A_0 is of rank one, then the matrix R can be obtained explicitly as the inverse of another matrix, and G matrix can also be obtained as product of a column vector and a row vector, as follows.

- if $A_2 = \mathbf{v}.\boldsymbol{\alpha}$, then $R = A_0[I - A_1 - A_0 G]^{-1}$, where $G = \mathbf{1}.\boldsymbol{\alpha}$,
- if $A_0 = \omega.\boldsymbol{\beta}$, then $R = A_0[I - A_1 - \eta A_2]^{-1}$, where $\xi = \boldsymbol{\beta}[I - A_1 - \eta A_2]^{-1}$, $\xi\omega = \eta$, $\xi A_2 \mathbf{1} = 1$ and the matrix R satisfies $R^i = \eta^{i-1} R$, $i \geq 1$.
- All the special structures identified for the M/G/1 and GI/M/1 types are also observed for the QBDs.

4.9 Other special QBDs of interest

4.9.1 Level-dependent QBDs:

A level dependent QBD has a transition matrix P given in the block partitioned form as follows:

$$P = \begin{bmatrix} A_{0,0} & A_{0,1} & & & \\ A_{1,0} & A_{1,1} & A_{1,2} & & \\ & A_{2,1} & A_{2,2} & A_{2,3} & \\ & & A_{3,2} & A_{3,3} & A_{3,4} \\ & & & \ddots & \ddots & \ddots \end{bmatrix}, \tag{4.96}$$

We assume that the matrices $A_{k,k},\ k = 0,1,2$ are of dimension $m_k \times m_k$ and the dimensions of the matrices $A_{k,k+1}$ are $m_k \times m_{k+1}$ and of $A_{k+1,k}$ are $m_{k+1} \times m_k$. Because the transition matrix is level dependent, the matrices R and G are now level dependent. We have R_k which records the rate of visiting level k before coming back to level $k-1$, given it started from level $k-1$, and G_k records the probability of first visit from level k to level $k-1$. The matrices are obtained as

$$R_k = A_{k-1,k} + R_k A_{k,k} + R_k R_{k+1} A_{k+1,k}, \tag{4.97}$$

and

$$G_k = A_{k,k-1} + A_{k,k} G_k + A_{k,k+1} G_{k+1} G_k. \tag{4.98}$$

For this level dependent QBD we have matrix-product solution given as

$$\boldsymbol{x}_{k+1} = \boldsymbol{x}_k R_k, \ \ k \geq 0. \tag{4.99}$$

The condition for this QBD to be positive recurrent is given in Latouche and Ramaswami (1999) as follows:

Condition: That

$$\boldsymbol{x}_0 = \boldsymbol{x}_0 (A_{0,0} + A_{0,1} G_1), \tag{4.100}$$

with

$$\boldsymbol{x}_0 \sum_{n=0}^{\infty} \prod_{k=1}^{n} R_k \mathbf{1} = 1. \tag{4.101}$$

The level dependent QBD is usually difficult to implement in an algorithm form in general. However, special cases can be dealt with by using specific features of the DTMC. For more details on level dependent QBDs and algorithms see Latouche and Ramaswami (1999). Some of the special cases will be presented later in the queueing section of this book, especially when we present some classes of vacation models.

4.9.2 *Tree structured QBDS*

Consider a DTMC $\{(X_n, Y_n), n \geq 0\}$ in which X_n assumes values of the nodes of a d-ary tree, and Y_n as the auxiliary variables such as phases. As an example a 2-ary has for level 1 the set $\{1,2\}$, for level 2 the set $\{11,12,21,22\}$ for level 3 the set $\{111,112,121,122,211,212,221,222\}$ and so on. This is shown in Fig. 4.1. If we now allow this DTMC to go up and down not more than one level, we then

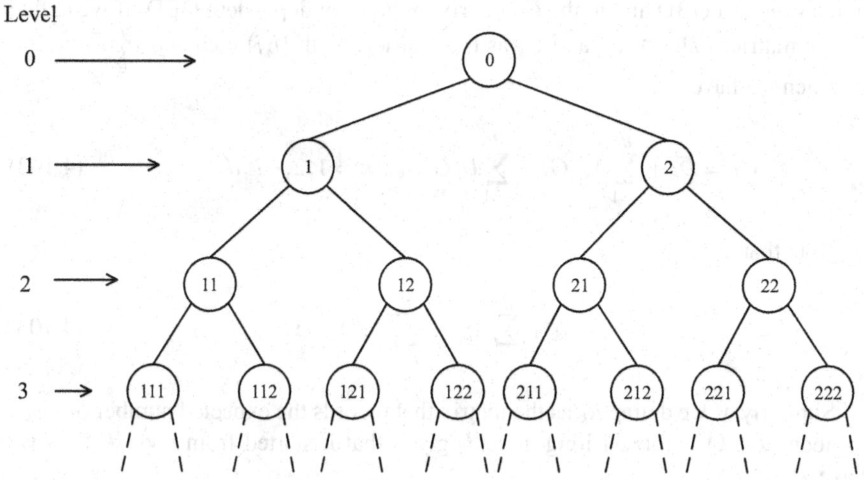

Fig. 4.1 Tree Structured Markov Chain

have a tree structured QBD. Details of such DTMC can be found in Yeung and Alfa (1999). We can also have the GI/M/1 type and M/G/1 type of tree structured DTMC. For further reference see Yeung and Sengupta (1994) and Takine et al (1995), respectively.

A d-ary tree is a tree which has d children at each node, and the root of the tree is labelled $\{0\}$. The rest of the nodes are labelled as strings. Each node, which represents a level say i in a DTMC, has a string of length i with each element of the string consisting of a value in the set $\{1,2,\cdots,d\}$. We use $+$ sign in this section to represent concatenation on the right. Consider the case of $d=2$ and for example if we let $J=1121$ be a node and $k=2$ then $J+k=11212$. We adopt the convention of using upper case letters for strings and lower case letters for integers. Because we are dealing with a QBD case we have that the DTMC, when in node $J+k$ could only move in one step to node J or node $J+k+s$ or remain in node $J+k$, where $k,s=1,2,\cdots,d$. The last come first served single server queues have been analyzed using this tree structured QBD.

For this type of QBD most of the results for the standard QBD still follow with only minor modifications to the notations. Consider the chain in a node $J+k$ and phase i at time n, i.e. the DTMC is in state $(J+k,i)$, where i is an auxiliray variable $i=1,2,\cdots,m$. At time $n+1$ the DTMC could be at state:

- (J,j) with probability $d_k^{i,j}$, $k=1,2,\cdots,d$
- $(J+s,j)$ with probability $a_{k,s}^{i,j}$, $k,s=1,2,\cdots,d$
- $(J+ks,j)$ with probability $u_s^{i,j}$, $k,s=1,2,\cdots,d$

Let G_k be an $m\times m$ matrix which records the probability that the DTMC goes from node $J+k$ to node J for the first time. This is really equivalent to the G matrix in the standard QBD and to the G_k matrix in the level dependent QBD. If we define $m \times m$ matrices D_k, $A_{k,s}$, and U_s as the matrices with (i,j) elements $d_k^{i,j}$, $a_{k,s}^{i,j}$ and $u_s^{i,j}$, then we have

$$G_k = D_k + \sum_{s=1}^{d} A_{k,s} G_s + \sum_{s=1}^{d} U_s G_s G_k, \quad k = 1, 2, \cdots, d. \tag{4.102}$$

Note that

$$(D_k + \sum_{s=1}^{d} A_{k,s} + \sum_{s=1}^{d} U_s)\mathbf{1} = \mathbf{1}. \tag{4.103}$$

Similarly, if we define R_k as the matrix that records the expected number of visits to node $(J+k)$ before visiting node J, given that it started from node J, then we have

$$R_k = U_k + \sum_{s=1}^{d} R_s A_{s,k} + \sum_{s=1}^{d} R_k R_s D_s, \quad k = 1, 2, \cdots, d. \tag{4.104}$$

Because the summations operations in the equations for both the G_k and R_k matrices are finite, we can simply apply the linear approach for the standard R and G matrices to compute these G_k and R_k matrices.

Provided the system is stable we can apply the linear algorithm as follows:

- Set

$$G_k(0) := D_k, \quad R_k(0) := U_k, \quad k = 1, 2, \cdots, d$$

- Compute

$$G_k(n+1) := D_k + \sum_{s=1}^{d} A_{k,s} G_s(n) + \sum_{s=1}^{d} U_s G_s(n) G_k(n), \ k = 1, 2, \cdots, d,$$

$$R_k(n+1) := U_k + \sum_{s=1}^{d} R_s(n) A_{s,k} + \sum_{s=1}^{d} R_k(n) R_s(n) D_s, \ k = 1, 2, \cdots, d.$$

- Stop if

$$(G_k(n+1) - G_k(n))_{i,j} < \varepsilon, \forall i, j, k,$$

and

$$(R_k(n+1) - R_k(n))_{i,j} < \varepsilon, \forall i, j, k,$$

where ε is a preselected tolerance value.

4.9.3 Re-blocking of transition matrices

For most practical situations encountered we find that the resulting transition matrices do not have exactly any of the three structures presented. However, in most cases the structures are close to one of the three and by re-blocking the transition matrix we can achieve one of them

4.9.3.1 The non-skip-free M/G/1 type

As an example, we can end up with a non-skip-free M/G/1 type DTMC with a transition matrix of the following form

$$P = \begin{bmatrix} C_{0,0} & C_{0,1} & C_{0,2} & C_{0,3} & C_{0,4} & \cdots \\ C_{1,0} & C_{1,1} & C_{1,2} & C_{1,3} & C_{1,4} & \cdots \\ \vdots & \vdots & \vdots & \vdots & \vdots & \cdots \\ C_{k-1,0} & C_{k-1,1} & C_{k-1,2} & C_{k-1,3} & C_{k-1,4} & \cdots \\ A_0 & A_1 & A_2 & A_3 & A_4 & \cdots \\ & A_0 & A_1 & A_2 & A_3 & \cdots \\ & & A_0 & A_1 & A_2 & \cdots \\ & & & \ddots & \ddots & \ddots \end{bmatrix}. \quad (4.105)$$

This can be re-blocked further into $k \times k$ super-blocks so that we have

$$P = \begin{bmatrix} C_0 & C_1 & C_2 & C_3 & \cdots \\ H_0 & H_1 & H_2 & H_3 & \cdots \\ & H_0 & H_1 & H_2 & \cdots \\ & & H_0 & H_1 & \cdots \\ & & & \ddots & \ddots & \ddots \end{bmatrix}, \quad (4.106)$$

where

$$C_j = \begin{bmatrix} C_{0,jk} & \cdots & C_{0,(j+1)k-1} \\ \vdots & \cdots & \vdots \\ C_{k-1,jk} & \cdots & C_{k-1,(j+1)k-1} \end{bmatrix}, \; j = 0,1,2,\cdots$$

$$H_j = \begin{bmatrix} A_{jk} & \cdots & A_{(j+1)k-1} \\ \vdots & \cdots & \vdots \\ A_{(j-1)k} & \cdots & A_{jk} \end{bmatrix}, \; j = 1,2,\cdots,$$

and

$$H_0 = \begin{bmatrix} A_0 & A_1 & \cdots & A_{k-1} \\ & A_0 & \cdots & A_{k-2} \\ & & \ddots & \vdots \\ & & & A_0 \end{bmatrix}.$$

This non-skip-free M/G/1 type DTMC has now been converted to an M/G/1 type which is skip-free to the left. Standard M/G/1 results and techniques can now be applied to analyze this DTMC. We point out that there is a class of algorithms specifically designed for this class of problems by Gail et al (1997).

Suppose we have a situation where $C_j = 0$ and $H_{j+w} = 0,\ \forall j > M,\ 0 \le w < \infty$, then we can further re-block this transition matrix to the QBD type as follows. We display the case of $w = 0$ only. Let

$$B = C_0,\ C = [C_1,\ C_2,\ \cdots,\ C_M],\ E = \begin{bmatrix} H_0 \\ 0 \\ \vdots \\ 0 \end{bmatrix},$$

$$U = \begin{bmatrix} 0 & 0 & 0 & \cdots & 0 \\ H_M & 0 & 0 & \cdots & 0 \\ H_{M-1} & H_M & \ddots & \ddots & 0 \\ \vdots & \vdots & \ddots & \vdots & \cdots \\ H_1 & H_2 & \cdots & H_M & 0 \end{bmatrix},\quad D = (\mathbf{e}_1 \otimes \mathbf{e}_M^T) \otimes H_0,$$

$$A = \begin{bmatrix} H_1 & H_2 & H_3 & \cdots & H_M \\ H_0 & H_1 & H_2 & \cdots & H_{M-1} \\ & H_0 & H_1 & \cdots & H_{M-2} \\ & & \ddots & \ddots & \vdots \\ & & & H_0 & H_1 \end{bmatrix},$$

where $\mathbf{e}_j$ is a column vector of zeros with one in location j and $\mathbf{e}_j^T$ is its transpose, then we can write the transition matrix as

$$P = \begin{bmatrix} B & C & & \\ E & A & U & \\ & D & A & U \\ & & D & A & U \end{bmatrix},$$

which is of the QBD type. In this case we can apply the results of QBD to analyze the system.

4.9.3.2 The non-skip-free GI/M/1 type

Similar to the M/G/1 type DTMC we can have a GI/M/1 non-skip-free type of DTMC with transition matrix of the form

$$P = \begin{bmatrix} B_{0,0} & B_{0,1} & \cdots & B_{0,k-1} & A_0 & & & \\ B_{1,0} & B_{1,1} & \cdots & B_{1,k-1} & A_1 & A_0 & & \\ B_{2,0} & B_{2,1} & \cdots & B_{2,k-1} & A_2 & A_1 & A_0 & \\ \vdots & \vdots & \cdots & \vdots & \vdots & \vdots & \vdots & \ddots \end{bmatrix}. \tag{4.107}$$

This can also be re-blocked further into $k \times k$ super-blocks to obtain

$$P = \begin{bmatrix} B_0 & F_0 & & & \\ B_1 & F_1 & F_0 & & \\ B_2 & F_2 & F_1 & F_0 & \\ \vdots & \vdots & \vdots & \vdots & \ddots \end{bmatrix}, \tag{4.108}$$

where

$$B_j = \begin{bmatrix} B_{jk,0} & \cdots & B_{jk,k-1} \\ \vdots & \cdots & \vdots \\ B_{(j+1)k-1,0} & \cdots & B_{(j+1)k-1,k-1} \end{bmatrix}, j = 0,1,2,\cdots,$$

$$F_j = \begin{bmatrix} A_{jk} & \cdots & A_{(j-1)k} \\ \vdots & \cdots & \vdots \\ A_{(j+1)k-1} & \cdots & A_{jk} \end{bmatrix}, j = 1,2,\cdots,$$

$$F_0 = \begin{bmatrix} A_0 & & & \\ A_1 & A_0 & & \\ \vdots & \vdots & \ddots & \\ A_{k-1} & A_{k-2} & \cdots & A_0 \end{bmatrix}.$$

This non-skip-free DTMC has now been converted to a GI/M/1 type which is skip-free to the right. Standard GI/M/1 results can now be applied to analyze this DTMC. Once again we point out that there is a class of algorithms specifically designed for this class of problems by Gail et al (1997).

Suppose we have a situation where $B_j = 0$ and $F_{j+v} = 0,\ \forall j > N,\ 0 \leq v < \infty$, then we can further re-block this transition matrix to the QBD type as follows. We display the case of $v = 0$ only. Let

$$B = B_0,\ C = [F_0,\ 0,\ 0,\ \cdots,\ 0],\ E = \begin{bmatrix} B_1 \\ B_2 \\ \vdots \\ B_N \end{bmatrix},$$

$$A = \begin{bmatrix} F_1 & F_0 & & & \\ F_2 & F_1 & F_0 & & \\ F_3 & F_2 & F_1 & F_0 & \\ \vdots & \vdots & \vdots & \ddots & \ddots \\ F_N & F_{N-1} & F_{N-2} & \cdots & F_1 \end{bmatrix},$$

$$D = \begin{bmatrix} 0 & F_N & F_{N-1} & \cdots & F_2 \\ & & F_N & \cdots & F_3 \\ & & & \ddots & \vdots \\ & & & & F_N \\ & & & & 0 \end{bmatrix},\quad U = (\mathbf{e}_1 \otimes \mathbf{e}_1^T) \otimes F_1,$$

then we can write the transition matrix as

$$P = \begin{bmatrix} B & C & & & \\ E & A & U & & \\ & D & A & U & \\ & & D & A & U \end{bmatrix},$$

which is of the QBD type. In this case we can apply the results of QBD to analyze the system.

In general a banded transition matrix can be reduced to a QBD in structure.

4.9.4 Time-inhomogeneous Discrete Time Markov Chains

Up till now we have focussed mainly on Markov chains that are independent of time, i.e. where $Pr\{X_{n+1} = j|X_n = i\} = p_{i,j},\ \forall n$. Here we deal with cases where $Pr\{X_{n+1} = j|X_n = i\} = p^n_{i,j}$, i.e. the transition probability depends on the time of the transition. We present the case of time-inhomogeneous QBDs here. The case of the GI/M/1 and M/G/1 types have been discussed in details by Alfa and Margolius (2008).

For the time-inhomogeneous QBDs we let the bivariate process be $\{X_n, J_n\}, n \geq 0$ with $X_n \geq 0,\ 1 \leq J_n \leq M < \infty$. We can write the associated transition matrix as

$$P^{(n)} = \begin{bmatrix} A_{0,0}^{(n)} & A_{0,1}^{(n)} & & & \\ A_{1,0}^{(n)} & A_{1,1}^{(n)} & A_{1,2}^{(n)} & & \\ & A_{2,1}^{(n)} & A_{2,2}^{(n)} & A_{2,3}^{(n)} & \\ & & \ddots & \ddots & \ddots \end{bmatrix}, \tag{4.109}$$

where the matrices $A_{i,j}^{(n)}$ records the transition from time n to time $n+1$ with X_n changing from i to j. Hence $(A_{i,j}^{(n)})_{\ell,k} = Pr\{X_{n+1} = j, J_{n+1} = k | X_n = i, J_n = \ell\}$. If we now define $(A_{i,j}^{(n,m)})_{\ell,k} = Pr\{X_{m+1} = j, J_{m+1} = k | X_n = i, J_n = \ell\}, m \geq n$. Note that we have $A^{(n,n)} = A^{(n)}$. Let the corresponding transition matrix be $P^{(n,m)}$. For simplicity we only display the case of $m = n+1$. We can write

$$P^{(n,n+1)} = \begin{bmatrix} A_{0,0}^{(n,n+1)} & A_{0,1}^{(n,n+1)} & A_{0,2}^{(n,n+1)} & & & \\ A_{1,0}^{(n,n+1)} & A_{1,1}^{(n,n+1)} & A_{1,2}^{(n,n+1)} & A_{1,3}^{(n,n+1)} & & \\ A_{2,0}^{(n,n+1)} & A_{2,1}^{(n,n+1)} & A_{2,2}^{(n,n+1)} & A_{2,3}^{(n,n+1)} & A_{2,4}^{(n,n+1)} & \\ & A_{3,1}^{(n,n+1)} & A_{3,2}^{(n,n+1)} & A_{3,3}^{(n,n+1)} & A_{3,4}^{(n,n+1)} & A_{3,5}^{(n,n+1)} \\ & \ddots & \ddots & \ddots & \ddots & \ddots \end{bmatrix}, \tag{4.110}$$

where $A_{i,j}^{(n,n+1)} = \sum_{v=i-1}^{i+1} A_{i,v}^{(n)} A_{v,j}^{(n+1)}$, keeping in mind that $A_{\ell,k}^{(m)} = 0,\ \forall |\ell - k| > 1, m \geq 0$. Also note that $A_{i,j}^{(n,n+1)} = 0,\ \forall |i-j| > 2$. For the general case of $P^{(n,m)}$, the block matrices $A_{i,j}^{(n,m)}$ can be obtained through recursion as

$$A_{i,j}^{(n,m+1)} = \sum_{v=i-m-1}^{i+1} A_{i,v}^{(n,m)} A_{v,j}^{(m+1)}, \tag{4.111}$$

keeping in mind that some of the block matrices inside the summation will have zero values only and $A_{i,j}^{(n,m)} = 0,\ \forall |i-j| > m-n+1$.

If we know the state of the system at time n we can easily determine its state at time $m+1$. For example, let $\boldsymbol{x}_n = [\boldsymbol{x}_0^{(n)}, \boldsymbol{x}_1^{(n)}, \cdots]$ and $\boldsymbol{x}_i^{(n)} = [x_{i,1}^{(n)}, x_{i,2}^{(n)}, \cdots, x_{i,M}^{(n)}]$, then we have

$$\boldsymbol{x}^{(m+1)} = \boldsymbol{x}^{(n)} P^{(n,m)},\ 0 \leq n \leq m. \tag{4.112}$$

Usually of interest to us in this system is the case that we call "periodic", i.e. where we have for some integer $\tau \geq 1$ and thus $P^{(n)} = P^{(n+\tau)}$. In this case we have

$$P^{(n+K\tau,n+(K+1)\tau-1)} = P^{(n,n+\tau-1)},\ \forall K > 0. \tag{4.113}$$

Let $\mathscr{P}^{(n)} = P^{(n,n+\tau-1)}$, and $\mathbf{x}^{(n+K\tau)} = [\boldsymbol{x}^{(n+K\tau)},\ \boldsymbol{x}^{(n+K\tau)+1},\ \cdots,\ \boldsymbol{x}^{(n+(K+1)\tau-1)}]$, then we have

$$\mathbf{x}^{(n+(K+1)\tau)} = \mathbf{x}^{(n+K\tau)} \mathscr{P}^{(n)}. \tag{4.114}$$

Under some stability conditions (which have to be determined for each case) we have $\mathbf{x}^{(n+K\tau)}|_{K\to\infty} = \mathbf{x}^{(n)}$. Hence we have

$$\mathbf{x}^{(n)} = \mathbf{x}^{(n)} \mathscr{P}^{(n)}.$$

4.9.5 Time-inhomogeneous and spatially-homogeneous QBD

Suppose we have a case where

$$P^{(n)} = \begin{bmatrix} B^{(n)} & C^{(n)} & & & \\ E^{(n)} & A_1^{(n)} & A_0^{(n)} & & \\ & A_2^{(n)} & A_1^{(n)} & A_0^{(n)} & \\ & & \ddots & \ddots & \ddots \end{bmatrix}, \tag{4.115}$$

i.e. the DTMC is time-inhomogeneous but spatially-homogeneous. In this case the matrix $\mathscr{P}^{(n)}$ can be re-blocked. For example, if $\tau = 2$ then

$$\mathscr{P}^{(n)} = P^{(n,n+1)} = \begin{bmatrix} A_{0,0}^{(n,n+1)} & A_{0,1}^{(n,n+1)} & A_{0,2}^{(n,n+1)} & & & \\ A_{1,0}^{(n,n+1)} & A_{1,1}^{(n,n+1)} & A_{1,2}^{(n,n+1)} & A_{1,3}^{(n,n+1)} & & \\ A_{2,0}^{(n,n+1)} & A_{2,1}^{(n,n+1)} & A_{2,2}^{(n,n+1)} & A_{2,3}^{(n,n+1)} & A_{2,4}^{(n,n+1)} & \\ & A_{3,1}^{(n,n+1)} & A_{3,2}^{(n,n+1)} & A_{3,3}^{(n,n+1)} & A_{3,4}^{(n,n+1)} & A_{3,5}^{(n,n+1)} \\ & \ddots & \ddots & \ddots & \ddots & \ddots \end{bmatrix}, \tag{4.116}$$

which results in

$$\mathscr{P}^{(n)} = \begin{bmatrix} B_{0,0}^{(n,n+1)} & B_{0,1}^{(n,n+1)} & C_{0,2}^{(n,n+1)} & & & \\ B_{1,0}^{(n,n+1)} & B_{1,1}^{(n,n+1)} & A_1^{(n,n+1)} & A_0^{(n,n+1)} & & \\ B_{2,0}^{(n,n+1)} & A_3^{(n,n+1)} & A_2^{(n,n+1)} & A_1^{(n,n+1)} & A_0^{(n,n+1)} & \\ & A_4^{(n,n+1)} & A_3^{(n,n+1)} & A_2^{(n,n+1)} & A_1^{(n,n+1)} & A_0^{(n,n+1)} \\ & \ddots & \ddots & \ddots & \ddots & \ddots \end{bmatrix}. \tag{4.117}$$

This can be re-blocked to have

$$\mathscr{P}^{(n)} = \begin{bmatrix} \mathscr{B}^{(n)} & \mathscr{C}^{(n)} & & \\ \mathscr{E}^{(n)} & \mathscr{A}_1^{(n)} & \mathscr{A}_0^{(n)} & \\ & \mathscr{A}_2^{(n)} & \mathscr{A}_1^{(n)} & \mathscr{A}_0^{(n)} \\ & & \ddots & \ddots & \ddots \end{bmatrix},$$

where for example

$$\mathscr{A}_1^{(n)} = \begin{bmatrix} A_2^{(n,n+1)} & A_1^{(n,n+1)} \\ A_3^{(n,n+1)} & A_2^{(n,n+1)} \end{bmatrix}.$$

For the general case of $1 \leq \tau < \infty$ we can write and re-block $\mathscr{P}^{(n)}$ into a QBD type transition matrix. Given that $\mathscr{A}^{(n)} = \sum_{k=0}^{2} \mathscr{A}_k^{(n)}$ and let $\boldsymbol{\pi}(n)$ be the stationary distribution associated with it, i.e.

$$\boldsymbol{\pi}(n) = \boldsymbol{\pi}(n)\mathscr{A}^{(n)}, \ \boldsymbol{\pi}(n)\mathbf{1} = 1. \tag{4.118}$$

The stability conditions for standard QBDs apply, i.e. the system is stable iff

$$\boldsymbol{\pi}(n)(\mathscr{A}_1^{(n)} + 2\mathscr{A}_2^{(n)})\mathbf{1} > 1. \tag{4.119}$$

For a stable system we compute the stationary distribution of the transition matrix as

$$\boldsymbol{x}^{(n+j+1)} = \boldsymbol{x}^{(n+j)} R^{(n)}, \tag{4.120}$$

where $R^{(n)}$ is the minimal non-negative solution to the matrix equation

$$R^{(n)} = \mathscr{A}_0^{(n)} + R^{(n)}\mathscr{A}_k^{(n)} + (R^{(n)})^2\mathscr{A}_k^{(n)}. \tag{4.121}$$

In summary, standard matrix-analytic results can be used for the case of time-inhomogeneous DTMCs that are periodic.

4.10 Software Tools for Matrix-Analytic Methods

There are now software tools available to the public for analyzing the M/G/1, GI/M/1 and QBD systems. The website is:

`http://win.ua.ac.be/~vanhoudt/`

The packages use efficient algorithms for the three systems and do provide supporting documents and papers for them also.

4.11 Problems

- Question 1: Consider a discrete time Markov chain with state space $\{1,2,3\}$ and transition matrix P given as

$$P = \begin{bmatrix} a_1 & a_2 & a_3 \\ a_3 & a_1 & a_2 \\ a_2 & a_3 & a_1 \end{bmatrix},$$

where $0 < a_j < 1, \ \ j = 1,2,3$ and $a_1 + a_2 + a_3 = 1$. Let $\mathbf{x} = [x_1 \ x_2 \ x_3]$ be the stationary vector associated with P, i.e.

$$\mathbf{x} = \mathbf{x}P, \ \ \mathbf{xe} = 1.$$

1. Find the values of $x_j, \ \forall j$.
2. What do you observe about the columns of the matrix P? Can you make any generalization?
3. Suppose $a_1 = 0.3, \ a_2 = 0.5, \ a_3 = 0.2$.

   a. If the Markov chain starts from state 2, find the probability that it reaches state 1 before state 3.
   b. What is the probability that it takes at least three transitions before it reaches state 3 for the first time, given that it starts from state 1?

- Question 2: A total of N balls are put in an urn, with i of the balls marked as belonging to person A and $N-i$ belonging to person B. At each step, one ball is chosen at random from the N balls. If it is one for A it is relabelled and assigned to B and then returned to the urn, and vice-versa. Let X_n be the number of balls belonging to A after n steps, and assume that successive random drawings of the balls are independent. It can be shown that $\{X_n, \ n = 0,1,2,\cdots\}$ is a discrete time Markov chain. For $N = 4$ display the transition matrix of this Markov chain. Suppose we start with 1 ball in the urn belonging to A find the probability distribution of the number of balls belonging to B after 3 transitions.
- Question 3: At a shopping mall passengers looking for taxis arrive according to the Bernoulli process with parameter a and taxis arrive according to the Bernoulli process with parameter b. Passengers wait in a queue until they can get a taxi. There is no upper limit to the number of passengers allowed to wait. However, as soon as the number of waiting taxis is 4 arriving taxis will depart and not join the taxi queue. We assume that only one passenger is served by each taxi. Set up a Markov chain and the associated generator matrix that can be used to study the number of taxis and passengers waiting. Suppose $a = 0.3$ and $b = 0.5$ and consider steady state. Find the

1. probability that the number of passengers waiting for a taxi is at least 5,
2. probability that the number of taxis waiting is no more than 2,

3. average number of passengers waiting,
4. the average number of taxis waiting.

- Question 4: Consider a data communication system with a finite-capacity buffer for temporarily storing messages. The messages arrive according to a Bernoulli process with rate a. The transmission time of each message has geometric distribution with parameter b. The system can transmit only one message at a time. The buffer has room for only K messages excluding the message in transmission (if any). The following protocol for the admission of messages is used. As soon as the buffer is full, the arrival process of messages is stopped until the number in the buffer has fallen to the resume level K_1, with $0 < K_1 < K$. Set up a DTMC for the number of messages in the system at arbitrary times for the case of $K = 5$, $K_1 = 3$ by displaying the associated generator matrix.
- Question 5: Consider a system of N computer nodes arranged in a circle, where $N < \infty$. This system uses a token passing ring protocol to get access to the network. Let the nodes be labelled $1, 2, 3, \cdots, N$. When the token is at node i at time n it is passed to the next node in the clockwise direction with probability p or to the next node in the counterclockwise direction with probability $1 - p$. Let X_n be the position of the token at time n. Show that $X_n, n \geq 0$ is a DTMC and write down the transition matrix for X_n. What is the steady state distribution of the location of the token? [Hint: Consider the case of $N = 3$ and then use this information to infer the solution to the question].

References

Akar N, Sohraby K (1997) An invariant subspace approach in M/G/1 and GI/M/1 type Markov chains. Stochastic Models 13:381–416

Alfa AS, Chakravarthy S (1994) A discrete queue with Markovian arrival processes and phase type primary and secondary services. Stochastic Models 10:437–451

Alfa AS, Margolius BH (2008) Two classes of time-inhomogeneous Markov chains: Analysis of the periodic case. Annals of Oper Res 160:121–137

Alfa AS, Dolhun KL, Chakravarthy S (1995) A discrete single server queue with Markovian arrivals and phase type group service. Journal of Applied Mathematics and Stochastic Analysis 8:151–176

Alfa AS, Sengupta B, Takine T (1998) The use of non-linear programming in matrix analytic methods. Stochastic Models 14:351–367

Alfa AS, Sengupta B, Takine T, Xue J (2002) A new algorithm for computing the rate matrix of GI/M/1 type markov chains. Proceedings of the 4th International Conference on matrix Analytic Methods, Adelaide, Australia pp 1–16

Bini D, Meini B (1996) On the solution of a nonlinear matrix equation arising in queueing problems. SIAM Jour Matrix Anal Appl 17:906–926

Chakravarthy S, Alfa AS (1993) A multiserver queue with Markovian arrival process and group services with thresholds. Naval Research Logistics 40:811–827

Gail HR, Hantler SL, Taylor BA (1997) Non-skip-free M/G/1 and GI/M/1 type Markov chains. Adv Appl Prob 29:733–758

Gaver DP, Jacobs PA, Latouche G (1984) Finite birth-and-death models in randomly changing environments. Adv Appl Probab 16:715–731

Grassmann WK, Heyman DP (1990) Equilibrium distribution of block-structured Markov chains with repeating rows. J Appl Prob 27:557–576

Grassmann WK, Heyman DP (1993) Computation of steady-state probabilities for infinite-state Markov chains with repeating rows. ORSA J on Computing 5:292–303

Grassmann WK, Taksar MJ, Heyman DP (1985) Regenerative analysis and steady state distributions for Markov chains. Operations Research 33:1107–1117

Heyman D (1995) Accurate computation of the fundamental matrix of a Markov chain. SIAM Jour Matrix Anl Appl 16:151–159

Latouche G, Ramaswami V (1993) A logarithmic reduction algorithm for quasi-birth-and-death processes. J Appli Probab 30:650–674

Latouche G, Ramaswami V (1999) Introduction to matrix analytic methods in stochastic modeling. In: Pennsylvania, Philadelphia, Pennsylvania

Neuts MF (1981) Matrix-geometric solutions in stochastic models - an algorithmic approach. In: The Johns Hopkins University Press, Baltimore

Neuts MF (1989) Structured Stochastic Matrices of M/G/1 Type and Their Applications. Marcel Decker Inc., New York

Ramaswami V (1988) A stable recursion for the steady state vector in markov chains of M/G/1 type. Stochastic Models 4:183–188

Stewart WJ (1994) Introduction to the numerical solution of Markov chains. Princeton University Press, Princeton, N.J.

Takine T, Sengupta B, Yeung RW (1995) A generalization of the matrix M/G/1 paradigm for markov chains with a tree structure. Stochastic Models 11:411–421

Ye J, Li SQ (1994) Folding algorithm: A computational method for finite QBD processes with level dependent transitions. IEEE Trans Commun 42:625–639

Yeung RW, Alfa AS (1999) The quasi-birth-and-death type Markov chain with a tree structure. Stochastic Models 15:639–659

Yeung RW, Sengupta B (1994) Matrix product-form solutions for Markov chains with a tree structure. Adv Appl Probab 26:965–987

Zhao YQ, Li W, Braun WJ (1998) Infinite block-structured transition matrices and their properties. Adv Appl Prob 30:365–384

Zhao YQ, Li W, Alfa AS (1999) Duality results for block-structured transition matrices. J Appl Prob 36:1045–1057

Chapter 5
Basic Single Node Queueing Models with Infinite Buffers

5.1 Introduction

A single node queue is a system in which a customer comes for service only at one node. The service may involve feedback into the same node. When it finally completes service, it leaves the system. If there is more than one server then they must all be in parallel. It is also possible to have single node queues in which we have several parallel queues and a server (or several servers) moving around from queue to queue to process the items waiting; an example is a polling system to be discussed later. Polling systems are used extensively to model medium access control (MAC) in telecommunications. However, when we have a system where an item arrives for service and after completing the service at a location it moves to another queue in a different location for processing, then this is no longer a single node queue. Such a system could be a network of queues or queues in tandem. Single node queues are special cases of these systems. In Chapter 1, Figs. 1.1 to 1.4 are single node queues and Figs. 1.5 and 1.6 are not.

A single node queue is characterized by the arrival process (A), the service process (B), the number of servers (C) in parallel, the service discipline (D), the available buffer space (E) and the size of the population source (F). Note that the notation (E) is used here to represent the buffer size, whereas in some papers and books it is used to represent the buffer size plus the available number of servers. Using Kendall's notations, a single node queue can be represented as A/B/C/D/E/F. For example, if the arrival process is Geometric, service is Phase type, with 2 servers in parallel, First In First Out (or First Come First Served) service discipline, K buffer spaces, and infinite size of the population source, then the queue is represented as Geo/PH/2/FIFO/K/∞ or Geo/PH/2/FCFS/K/∞, since FIFO and FCFS can be used interchangeably. Usually only the first three descriptors are used when the remaining are common occurrence or their omission will not cause any ambiguity. For example, Geo/PH/2 implies that we have geometric inter-arrival times, Phase type services, and two servers in parallel; and implied is that the FIFO service discipline, waiting space is infinite and size of population source is infinite.

A.S. Alfa, *Applied Discrete-Time Queues*, DOI 10.1007/978-1-4939-3420-1_5

When we need to include a finite buffer space the we use the notation A/B/C/E. For example Geo/Geo/3/K implies geometric inter-arrival times, geometric service times, three identical serves in parallel and a buffer of size K. For more complex single node queues such as polling systems, feedback systems, etc. we may simply add some texts to give further descriptions.

For all the queues which we plan to study in this chapter, we will assume that the inter-arrival times and service times are independent of one another. Also, unless otherwise stated, we will assume that the types of service disciplines we are dealing with most of this chapter and the next one when studying the waiting times are FIFO.

A major emphasis of this book is to present a queuing model analysis via the use of Markov chain techniques, without necessarily overlooking the other popular and effective methods. We therefore use the matrix method very frequently to present this. The aim in setting up every queuing model is to reduce it to a Markov chain analysis, otherwise it would be difficult to solve. Use of recursion is also paramount in presenting the solution technique. The main performance measures that we will focus on are: the number of items in the system, the queue length, and the waiting times. We will present busy period in some instances and the other measures such as workload, departure times and age will only be considered in a few cases of interest. Specifically, workload and age process will be considered for Geo/Geo/1 and PH/PH/1 systems, and departure process for Geo/Geo/1/K and PH/PH/1/K systems.

Considerable effort will be focused in studying the Geo/Geo/1 and the PH/PH/1 systems. The latter is the matrix analogue of the former and a good knowledge of these two classes of systems will go a long way to helping in understanding the general single node queueing systems. Generally the structure of the PH/PH/1 system appears frequently in the book and its results can be extended to complex single node cases with little creative use of the Markov chain idea. Understanding Geo/Geo/1 makes understanding PH/PH/1 simple and besides the former is a special case of the latter.

5.2 Birth-and-Death Process

Birth-and-death (BD) processes are processes in which a single birth can occur at anytime and the rate of birth depends on the number in the system. Similarly, a single death can occur at anytime and the death rate depends on the number in the system. It will be noticed later that a birth-and-death process can only increase the population or decrease it by, at most, one unit at a time, hence the transition matrix of its Markov chain is of the tri-diagonal type. The birth-and-death process is very fundamental in queuing theory. The discrete time BD process has not received much attention in the literature. This is partly because it does not have the simple structure that makes it very general for a class of queues as its continuous time counterpart.

In the discrete time version, we assume that the birth-process is state dependent Bernoulli process and the death process follows a state dependent geometric distribution. In what follows, we develop the model more explicitly. But as a start let us state the problem formally as a Markov chain. Let X_n be the number of units in the system, including the one being processed, at time n. Note that X_n assumes the values in the state space $\mathscr{X} = \{0, 1, 2, 3, \cdots\}$. Let $D(X_n)$ be the number of service completions at time n when there are X_n in the system, where $D(X_n) = 0, 1$ and let $A(X_n)$ be the number of arrivals at time n, when there are X_n in the system, where $A(X_n) = 0, 1$. It is immediately clear that $D(0) = 0$. The stochastic process $\{X_n, n = 0, 1, 2, \cdots\}, X_n \in \mathscr{I}$, is a DTMC, with the relationship

$$X_{n+1} = (X_n - D(X_n))^+ + A(X_n), \tag{5.1}$$

where $(z)^+ = max\{0, z\}$.

The DTMC representing this BD process is now discussed in the next section.

5.3 Discrete time B-D Process

Let a_i = Probability that a birth occurs when there are $i \geq 0$ customers in the system with $\bar{a}_i = 1 - a_i$. Also, let b_i = Probability that a death occurs when there are $i \geq 1$ customers in the systems, with $\bar{b}_i = 1 - b_i$. We let $b_0 = 0$.

If we now define a Markov chain with state space $\{X_n, n \geq 0\}$, where X_n is the number of customers in the system, then the transition matrix P of this chain is given as

$$P = \begin{bmatrix} \bar{a}_0 & a_0 & & & \\ \bar{a}_1 b_1 & \bar{a}_1 \bar{b}_1 + a_1 b_1 & a_1 \bar{b}_1 & & \\ & \bar{a}_2 b_2 & \bar{a}_2 \bar{b}_2 + a_2 b_2 & a_2 \bar{b}_2 & \\ & & \ddots & \ddots & \ddots \end{bmatrix} \tag{5.2}$$

If we define $x_i^{(n)} = Pr\{X_n = i\}$ as the probability that there are i customers in the system at time n, and letting $\boldsymbol{x}^{(n)} = \left[x_0^{(n)}, x_1^{(n)}, \ldots\right]$ then we have

$$\boldsymbol{x}^{(n+1)} = \boldsymbol{x}^{(n)} P, \text{ or } \boldsymbol{x}^{(n+1)} = \boldsymbol{x}^{(0)} P^{n+1}.$$

If P is irreducible and positive recurrent, then there exists an invariant vector, which is also equivalent to the limiting distribution, $\boldsymbol{x} = \boldsymbol{x}^{(n)}|_{n \to \infty}$ and given as

$$\boldsymbol{x}P = \boldsymbol{x}, \quad \mathbf{1} = 1. \tag{5.3}$$

If we let $\boldsymbol{x} = [x_0, x_1, x_2, \ldots]$, then in an expanded form we can write

$$x_0 = x_0 \bar{a}_0 + x_1 \bar{a}_1 b_1 \tag{5.4}$$

$$x_1 = x_0 a_0 + x_1 \left(\bar{a}_1 \bar{b}_1 + a_1 b_1\right) + x_2 \bar{a}_2 b_2 \tag{5.5}$$

$$x_i = x_{i-1} a_{i-1} \bar{b}_{i-1} + x_i \left(\bar{a}_i \bar{b}_i + a_i b_i\right) + x_{i+1} \bar{a}_{i+1} b_{i+1}, \; i \geq 2 \tag{5.6}$$

This results in

$$x_1 = a_0 (\bar{a}_1 b_1)^{-1} x_0 \tag{5.7}$$

from the first equation, and

$$x_i = x_0 \prod_{v=0}^{i-1} [a_v \bar{b}_v [\bar{a}_{v+1} b_{v+1}]^{-1}], \; i \geq 1 \tag{5.8}$$

from the other equations, where $b_0 = 0$.

Using the normalizing condition that $\boldsymbol{x}\mathbf{1} = 1$, we obtain

$$x_0 = \left[1 + \sum_{i=1}^{\infty} \prod_{v=0}^{i-1} [a_v \bar{b}_v [\bar{a}_{v+1} b_{v+1}]^{-1}] \right]^{-1}. \tag{5.9}$$

Note that for stability, we require that $x_0 = \sum_{i=1}^{\infty} \prod_{v=0}^{i-1} (a_v \bar{b}_v) [\bar{a}_{v+1} b_{v+1}]^{-1} < \infty$.

Example of a discrete B-D process: Consider a single server queue with infinite buffer space. Arrivals and service completions are both governed by the Bernoulli process dependent on the number in the system. At any time, when there are N items in the system, there is a probability $\frac{\alpha}{N+1}$ that an item arrives, where $0 < \alpha < 1$. The probability of no service completion is given as $\frac{1-\beta}{N}$, when $N > 0$ and $0 < \beta < 1$. Hence

$$a_i = \frac{\alpha}{i+1}, \; i \geq 0,$$

$$b_i = \frac{i - (1-\beta)}{i}, \; i > 0.$$

The BD process is a special case of the level dependent QBD discussed in Section 4.9.1. Here we have scalar entries in the transition matrix instead of block matrices. It is interesting to note that even for this scalar case we do not have simple general solutions. We have to deal with each case as it arises. An example of a case that we can handle is where

$$a_i = a, \text{ and } b_i = b, \; \forall i \geq K < \infty.$$

In this case we have

$$x_i = \prod_{v=0}^{i-1}[a_v\bar{b}_v[\bar{a}_{v+1}b_{v+1}]^{-1}], \; i < K, \tag{5.10}$$

and

$$x_{i+1} = x_i a\bar{b}(b\bar{a})^{-1}, \; i \geq K. \tag{5.11}$$

The normalization equation is applied appropriately.

Later we show that the Geo/Geo/1 and the Geo/Geo/1/K are special cases of this BD process. In the continuous time case it is possible to show that the M/M/s is a special case of the continuous time BD process. However this is not the case for the discrete time BD; the Geo/Geo/s can not be shown to be a special case of the discrete BD process. We discuss this later in the next chapter.

In what follows, we present special cases of the discrete birth-and-death process at steady state.

5.4 Geo/Geo/1 Queues

We consider the case when $a_i = a, \; \forall i \geq 0$ and $b_i = b, \; \forall i \geq 1$. We define the transforms $b^*(z) = 1 - b + zb$ and $a^*(z) = 1 - a + za$. In this case, the transition matrix P becomes

$$P = \begin{bmatrix} \bar{a} & a & & & \\ \bar{a}b & \bar{a}\bar{b} + ab & a\bar{b} & & \\ & \bar{a}b & \bar{a}\bar{b} + ab & a\bar{b} & \\ & & \ddots & \ddots & \ddots \end{bmatrix}. \tag{5.12}$$

We can use several methods to analyze this Markov chain. First, we present the standard algebraic approach, the z-transform approach and then use the matrix-geometric approach.

An example application of this queueing model in telecommunication is where we have packets arriving at a single server (a simple type of router) according to the Bernoulli process. Each packet has a random size with a geometric distribution, hence the time it takes to process each packet follows a geometric distribution. Because the packets arrive according to the Bernoulli process then their inter-arrival times follow the geometric distribution. Hence we have a Geo/Geo/1 queueing system.

5.4.1 Algebraic Approach

Applying the relationship $\mathbf{x} = \mathbf{x}P$, we obtain the state equations

$$x_0 = x_0\bar{a} + x_1\bar{a}b \tag{5.13}$$

$$x_1 = x_0 a + x_1\left(\bar{a}\bar{b} + ab\right) + x_2\bar{a}b \tag{5.14}$$

$$x_i = x_{i-1}a\bar{b} + x_i\left(\bar{a}\bar{b} + ab\right) + x_{i+1}\bar{a}b,\ i \geq 2 \tag{5.15}$$

If we let $\theta = (\bar{a}b)^{-1}(a\bar{b})$, then we have

$$x_i = (\theta^i/\bar{b})x_0,\ i \geq 1. \tag{5.16}$$

Using the normalizing condition, we obtain

$$x_0 = \frac{b-a}{b}. \tag{5.17}$$

Note that for stability, we require that $a < b$.
Hence, for a stable system, we have

$$x_i = (b\bar{b})^{-1}(b-a)\theta^i,\ i \geq 1. \tag{5.18}$$

5.4.2 Transform Approach

We can use the z-transform approach to obtain the same results as follows:
Let

$$X^*(z) = \sum_{i=0}^{\infty} x_i z^i,\quad |z| \leq 1.$$

Multiplying each sides of the state equations by z^i, $i \geq 0$, we have

$$x_0 = x_0\bar{a} + x_1\bar{a}b \tag{5.19}$$

$$zx_1 = zx_0 a + zx_1\left(\bar{a}\bar{b} + ab\right) + zx_2\bar{a}b \tag{5.20}$$

$$z^i x_i = z^i x_{i-1}a\bar{b} + z^i x_i\left(\bar{a}\bar{b} + ab\right) + z^i x_{i+1}\bar{a}b,\ i \geq 2. \tag{5.21}$$

Taking the sum over i from 0 to ∞ we obtain

$$X^*(z) = X^*(z)[\bar{a}\bar{b} + ab + z^{-1}\bar{a}b + za\bar{b}] + x_0[\bar{a} + za - \bar{a}\bar{b} - ab - z^{-1}\bar{a}b - za\bar{b}]. \tag{5.22}$$

After routine algebraic operations we have

$$X^*(z) = x_0 \frac{u}{v}, \tag{5.23}$$

where

$$u = z(\bar{a} - \bar{a}\bar{b} - ab) - \bar{a}b + z^2(a - a\bar{b})$$

$$= b(za + \bar{a})(z-1),$$

and

$$v = z(1 - \bar{a}\bar{b} - ab) - \bar{a}b - z^2 a\bar{b}$$

$$= z(\bar{a}b + a\bar{b}) - \bar{a}b - z^2 a\bar{b}$$

$$= (\bar{a}b - za\bar{b})(z-1).$$

This leads to the results that

$$X^*(z) = b(za + \bar{a})[\bar{a}b - za\bar{b}]^{-1} x_0. \tag{5.24}$$

It is known that $X(z)|_{z\to 1} = 1$. Using this and the above equations we obtain

$$x_0 = \frac{b-a}{b}, \tag{5.25}$$

a result obtained earlier using algebraic approach.

We can write $X^*(z) = f_0 + f_1 z^1 + f_2 z^2 + f_3 z^3 + \cdots$. Based on the algebra of z transform we know that $x_i = f_i$, so we can obtain the probability distribution of the number in the system by finding the coefficients of z^i in the polynomial.

5.4.3 Matrix-geometric Approach

Using the matrix-geometric approach first we write

$$A_0 = a\bar{b},\ A_1 = \bar{a}\bar{b} + ab \text{ and } A_2 = \bar{a}b,\ A_v = 0,\ v \geq 3.$$

The matrix R is a scalar in this case, we therefore write it as r and it is a solution to the quadratic equation

$$r = a\bar{b} + r(\bar{a}\bar{b} + ab) + r^2 \bar{a}b \tag{5.26}$$

The solutions to this equation are

$$r = 1 \text{ and } r = (\bar{a}b)^{-1}(a\bar{b}) = \theta.$$

The r which we want is the smaller of the two which is $r = \theta$ since $\theta < 1$ for stability. It then follows that

$$x_{i+1} = x_i r = x_i \theta \tag{5.27}$$

We can normalize the vector $\boldsymbol{x}$ by using the matrix $B[R]$. After normalization, we obtain the same results as above, i.e.

$$x_0 = \frac{b-a}{b}.$$

Note that here our $\theta < 1$, if the system is stable. This states

$$(a\bar{b})(b\bar{a})^{-1} < 1 \quad \rightarrow \quad \frac{a}{b} < 1.$$

Example:

Consider an example in which packets arrive according to the Bernoulli process with probability $a = 0.2$ that one arrives in a discrete time slot. Let service be geometric with completion probability of $b = 0.6$ in a single time slot for a packet that is receiving service. We assume we are dealing with a single server. In this case we have

$$\theta = r = \frac{0.2 \times 0.4}{0.6 \times 0.8} = \frac{1}{6},$$

so the system is stable.

The probability of finding an empty system $x_0 = \frac{0.6-0.2}{0.6} = \frac{2}{3}$. Based on these calculations we have

$$x_0 = \frac{2}{3}, \; x_k = \frac{1}{0.6}\left(\frac{1}{6}\right)^k, \; k \geq 1.$$

5.4.4 Performance Measures

One of the most important measures for queueing systems is the number of packets waiting in the system for service, i.e. those in the queue(s) plus the one(s) in service. The distribution of this measure was presented in the previous section. In this section we present other measures that are very commonly used such as mean values, waiting times and delays and duration of busy period.

5.4.4.1 Mean number in system:

The mean number in the system $E[X] = \sum_{i=1}^{\infty} ix_i$ is obtained as

$$E[X] = (b\bar{b})^{-1}(b-a)\left[\theta + 2\theta^2 + 3\theta^3 + \ldots\right]$$
$$= (b\bar{b})^{-1}(b-a)\theta\left[1 + 2\theta + 3\theta^2 + \ldots\right]$$
$$= (b\bar{b})^{-1}(b-a)\theta\frac{d(1-\theta)^{-1}}{d\theta}$$

This results in

$$E[X] = \frac{a\bar{a}}{b-a} \tag{5.28}$$

This same result can be obtained by

$$E[X] = \frac{dX(z)}{dz}|_{z\to 1},$$

where $X(z)$ is the $z-$transform obtained in Eq.(5.24).

Continuing with the example of the previous section, with $a = .2$ and $b = .6$, we have $E[X] = \frac{.2\times.8}{.6-.2} = 0.4$.

5.4.4.2 Mean number in queue:

Let the number in the queue be Y, then the mean queue length $E[Y] = \sum_{i=2}^{\infty}(i-1)x_i$ is obtained as

$$E[Y] = E[X] - \frac{a}{b} \tag{5.29}$$

Again using the example of the previous section, with $a = .2$ and $b = .6$, we have $E[Y] = .4 - \frac{.2}{.6} = 0.0667$.

5.4.4.3 Waiting time in the queue:

We assume throughout this chapter that we are dealing with FCFS system, unless we say differently. Let W_q be the waiting time in the queue for a customer and let $w_i^{(q)} = \Pr\{W_q = i\}$, with $W_q^*(z) = \sum_{i=0}^{\infty} z^i w_i^{(q)}$, then

$$w_0^{(q)} = x_0 = \frac{b-a}{b} \tag{5.30}$$

and

$$w_i^{(q)} = \sum_{v=1}^{i} x_v \binom{i-1}{v-1} b^v (1-b)^{i-v}, \; i \geq 1 \tag{5.31}$$

The arguments leading to this equation are as follows: if we have v items in the system, an arriving item will wait i units of time if in the first $i-1$ time units exactly $v-1$ services are completed and the service completion of the vth item (last one ahead) occurs at time i. This is summed over v from 1 to i.

Since we know that $x_v = (b\bar{b})^{-1}(b-a)\theta^v, \; v \geq 1$, it is straightforward to show that

$$w_i^{(q)} = (b\bar{b})^{-1}(b-a)\sum_{v=1}^{i} \theta^v \binom{i-1}{v-1} b^v (1-b)^{i-v} = \frac{(b-a)}{1-b}\theta(b\theta+1-b)^{i-1}. \tag{5.32}$$

We obtain the z-transform of W_q as follows. If we replace z in $X^*(z)$ with $B^*(z) = bz(1-(1-b)z)^{-1}$, which is the $z-$transform of the geometric service time distribution of each packet, we obtain

$$X^*(B^*(z)) = g_0 + g_1 z + g_2 z^2 + g_3 z^3 + \cdots, \tag{5.33}$$

implying that g_i, the coefficients of z^i in $X(B(z))$ is w_i^q. Hence we have

$$W_q^*(z) = X^*(B^*(z)) = x_0 \frac{bz + \bar{a}(1-z)}{\bar{a}(1-z) + z(b-a)}. \tag{5.34}$$

The mean time in the queue $E[W_q]$ is given as

$$E[W_q] = \frac{a\bar{a}}{b(b-a)}. \tag{5.35}$$

Once again, using the previous example we have

$$w_0^{(q)} = \frac{2}{3}, \; w_1^{(q)} = \frac{1}{6}, \; w_2^{(q)} = \frac{1}{12}, \; \text{and so on.}$$

The mean value of waiting time in the queue $E[W_q] = \frac{2}{3}$.

We can also use matrix-analytic approach to study W_q. Consider an absorbing DTMC $\{V_n, \; n \geq 0\}$ with state space $\{0, 1, 2, \cdots\}$, where each number in the state space represents the number of items in the queue ahead of an arriving item. We let this DTMC represent a situation in which an arbitrary item only sees items ahead of it until it gets served. Since we are dealing with a FCFS system the number ahead of this item can not grow in number. The transition matrix of this DTMC, $P(w_q)$, can be written as

$$P(w_q) = \begin{bmatrix} 1 & & & \\ b & \bar{b} & & \\ & b & \bar{b} & \\ & & b & \bar{b} \\ & & & \ddots & \ddots \end{bmatrix}. \tag{5.36}$$

Define a vector $\mathbf{y}^{(0)} = \boldsymbol{x}$, where $\boldsymbol{x}$ is given by $\boldsymbol{x} = \boldsymbol{x}P$, $\boldsymbol{x}\mathbf{1} = 1$, and consider the recursion

$$\mathbf{y}^{(i+1)} = \mathbf{y}^{(i)} P(w_q), \; i \geq 0.$$

Then we have

$$Pr\{W_q \leq i\} = y_0^{(i)}, \; i \geq 0.$$

5.4.4.4 Waiting time in system:

Let W be the waiting time in the system with $w_i = Pr\{W = i\}$, and let the corresponding z-transform be $W^*(z) = \sum_{j=1}^{\infty} z^j w_j$. By simple probability it is clear that W is the convolution sum of W_q and service times, hence we have

$$W^*(z) = W_q^*(z)B^*(z) = x_0 \frac{(bz + \bar{a}(1-z))bz}{(\bar{a}(1-z) + z(b-a))(1-(1-b)z)}. \tag{5.37}$$

and the mean waiting time in the system $E[W]$ is given as

$$E[W] = E[W_q] + \frac{1}{b}.$$

Using the same numerical example with $a = .2$ and $b = .6$ we have $E[W] = \frac{2}{3} + \frac{1}{.6} = 2.3333$.

The distribution of the waiting time in the system can be obtained very easily by straightforward probabilistic arguments as follows: An item's waiting time in the system will be i if it has waited in the queue for $j < i$ units of time and its service time is $i - j$, i.e. the first $i - j - 1$ time units were without service completion and service completion occurred at the $(i-j)^{th}$ time unit. By summing this over j we obtain the waiting time in the system distribution, i.e.

$$w_i = \sum_{j=0}^{i-1} w_j^{(q)} (1-b)^{i-j-1} b, \; i \geq 1. \tag{5.38}$$

For example:

$$w_3 = w_0^{(q)}(1-b)^{3-0-1}b + w_1^q(1-b)^{3-1-1}b + w_2^q(1-b)^{3-2-1}b$$

$$= \frac{2}{3}(0.4)^2(0.6) + \frac{1}{6}(0.4)(.6) + \frac{1}{12}(0.6) \quad = 0.154.$$

5.4.4.5 Duration of a Busy Period:

The busy period is initiated when a customer arrives to an empty queue and it lasts up to the time the system becomes empty for the first time. Let us define B as the duration of a busy period. This can be studied in at least two ways; one using direct probabilistic arguments and the other is matrix-analytic approach using an absorbing Markov chain.

- **Probabilistic arguments:** For the probabilistic approach, two methods are presented.
  Method 1: Let $S(k)$ be the service time of k items and define $s_j^{(k)} = Pr\{S(k) = j\}$ and $A(n)$ be the number of arrivals in any interval $[m, m+n]$ and define $a_j^{(v)} = Pr\{A(j) = v\}$. If we define $c_i = Pr\{B = i\}$, we then have:

$$a_j^{(v)} = \binom{j}{v} a^{(v)}(1-a)^{(j-v)},\ 0 \le j \le v \ge 1. \tag{5.39}$$

$$s_j^{(1)} = b(1-b)^{j-1},\ s_j^{(j)} = b^j, \tag{5.40}$$

$$s_i^{(j)} = bs_{i-1}^{(j-1)} + (1-b)s_{i-1}^{(j)},\ 1 \le j \le i \ge 1. \tag{5.41}$$

  The busy period can last i units of time if any of the following occur: i) the first arrival requires i units of service and no arrivals occur during this period, and ii) this probability is $s_i^1 a_i^0$, or the first arrival requires $i-k$ service time, however during its service some more items, say v arrive and alls these v items require a total of k time units for service and no arrivals occur during the k time units. This now leads to the result:

$$c_i = s_i^{(1)} a_i^{(0)} + \sum_{k=1}^{i-1} s_{i-k}^{(1)} \sum_{v=1}^{k} a_{i-k}^{(v)} s_k^{(v)} a_k^{(0)},\ i \ge 1. \tag{5.42}$$

  Method 2:We present another probabilistic approach due to Klimko and Neuts (1973) and Heimann and Neuts (1973) as follows. A busy period, in the case of single arrivals, is initiated by a single item. By definition c_i is the probability that a busy period initiated by an item lasts for i units of time. Let us assume that the first item's service lasts k units of time with probability s_k^1. Let the number of arrivals during this first item's service be v with probability a_k^v. These v items may be considered as initial items of v independent busy periods. Keeping in mind that busy periods in this case are independent identically distributed random variables, we know that the busy period generated by v items is a v convolution sum of the independent busy periods. Let $c_j^{(v)}$ be the probability that v independent busy periods last j units of time, then we have

$$c_i = \sum_{k=1}^{i} s_k^{(1)} \sum_{v=0}^{k} a_k^{(v)} c_{i-k}^{(v)}, \quad i \geq 1. \tag{5.43}$$

If we define the z-transform of the busy period as $c^*(z) = \sum_{i=1}^{\infty} z^i c_i, \quad |z| \leq 1$, we have

$$c^*(z) = \sum_{k=1}^{\infty} z^k s_k^{(1)} (1 - a + ac^*(z))^k$$

$$= \sum_{k=1}^{\infty} (1-b)^{k-1} bz^k (1 - a + ac^*(z))^k$$

$$= bg^*(z)[1 + (1-b)g^*(z) + (1-b)^2(g^*(z))^2 + z^3(g^*(z))^3 + \cdots], \tag{5.44}$$

where $g^*(z) = z(1 - a + ac^*(z))$.

After routine algebraic operations we obtain

$$c^*(z) = \frac{bg^*(z)}{1 - (1-b)g^*(z)}, \quad |z| \leq 1. \tag{5.45}$$

Whereas we have presented a more analytic result due to Klimko and Neuts (1973) and Heimann and Neuts (1973), the first method is easier for computational purposes. For further discussions on these results the reader is referred to Hunter (1983)

- **Matrix-analytic approach:** The idea behind this approach is that we consider the DTMC for the Geo/Geo/1 system but modify it at the boundary such that when the system goes into the 0 state it gets absorbed. The DTMC is initiated from state 1. The transition matrix P^* given as:

$$P^* = \begin{bmatrix} 1 & & & \\ \bar{a}b & \bar{a}\bar{b} + ab & a\bar{b} & \\ & \bar{a}b & \bar{a}\bar{b} + ab & a\bar{b} \\ & & \ddots & \ddots & \ddots \end{bmatrix} \tag{5.46}$$

Let the vector $\mathbf{y}^{(n)} = \left[y_0^{(n)}, y_1^{(n)}, \ldots\right]$ be the state vector of the system at time n, with $\mathbf{y}^{(0)} = [0,\ 1,\ 0,\ 0,\ \cdots]$. Now consider the following recursion

$$\mathbf{y}^{(n+1)} = \mathbf{y}^{(n)} P^*. \tag{5.47}$$

The element $y_0^{(n)}$ is the probability that the busy period last for n or less time units, i.e.

$$y_0^{(n)} = Pr\{B \leq n\}. \tag{5.48}$$

The rectangular iteration algorithm due to Shi et al (1996) can be used for computing $\mathbf{y}_0^{(n)}$ as follows. Let P_m^* be the $(m+1) \times (m+2)$ top left block of matrix P^* and let $\mathbf{y}^{(m)} = [y_0^{(m)},\ y_1^{(m)},\ \cdots, y_{m+1}^{(m)}]$, then we have

$$\mathbf{y}^{(m+1)} = \mathbf{y}^{(m)} P_m^*,\ m \geq 0.$$

5.4.4.6 Number Served During a Busy Period:

Another very important measure is the number of packets served during a busy period. Here we adapt the approach used by Kleinrock (1975) for the continuous time analogue to our discrete time case.

Let N_b be the number of packets served during a busy period, with $n_b^{(k)} = Pr\{N_b = k\},\ k \geq 1$. Further let $N_b^*(z) = \sum_{k=1}^{\infty} z^k n_b^{(k)},\ |z| \leq 1$. Since the arrival process is Bernoulli with parameter a and service is geometric with parameter b, we know that if the duration of service of a packet is j then the probability that i packets arrive during that duration will be given by $\binom{j}{i} a^i (1-a)^{j-i}$. Let u_i be the probability that i packets arrive during the service of one packet then we have $u_i = \sum_{j=max(i,1)}^{\infty} \binom{j}{i} a^i (1-a)^{j-i} (1-b)^{j-1} b$.

Letting $U(^*z) = \sum_{i=0}^{\infty} z^i u_i$, we have

$$U^*(z) = [1-(1-b)(1-a+az)]^{-1}(1-a+az)b,$$

which can be written as

$$U^*(z) = ba^*(z)[1-(1-b)a^*(z)]^{-1}, \tag{5.49}$$

where $a^*(z) = 1 - a + az$, is the z-transform of the Bernoulli arrival process.

A busy period in this Geo/Geo/1 system is initiated by one packet that arrives after an idle period. Then the number of packets that arrive during this first packet's service each generate sub-busy periods of their own, and since these sub-busy periods are independent and identically distributed, their convolution sum makes up the busy period. Hence if the number of packets that arrive during the service of the first packet is defined as A_b and letting $A_b = j$ and further letting $V_1, V_2, \cdots, V_j$ be the number of packets that arrive during the service of each of the packets $1, 2, \cdots, j$, then the number served during the busy period is $1 + V_1 + V_2 + \cdots + V_j$. Hence we have

$$N_b^*(z) = z \sum_{j=0}^{\infty} Pr\{A_b = j\}(N_b^*(z))^j = zU^*(N_b^*(z)),$$

which results in

$$N_b^*(z) = zba^*(N_b^*(z))[1-(1-b)a^*(N_b^*(z))]^{-1}.$$

5.4.4.7 Age Process

The age process has to do with how long the leading packet in the system has been in the system. Let Y_n be how long the leading packet, which is currently receiving service, has been in the system. It is immediately clear that $Y_n = 0$ implies there are no packets in the system. Consider the stochastic process $\{Y_n, n \geq 0\}$, where $Y_n = 0, 1, 2, \cdots$. If the arrival and service processes of packets are according to the Geo/Geo/1 definition, then Y_n is a DTMC with the transition matrix P given and

$$P = \begin{bmatrix} \bar{a} & a & & & \\ b\bar{a} & ba & \bar{b} & & \\ b(\bar{a})^2 & ba\bar{a} & ba & \bar{b} & \\ b(\bar{a})^3 & ba(\bar{a})^2 & ba\bar{a} & ba & \bar{b} \\ \vdots & \vdots & \vdots & \vdots & \vdots & \ddots \end{bmatrix}. \tag{5.50}$$

This is a GI/M/1 type DTMC and the system can be analyzed as such.

5.4.4.8 Workload:

Workload is the amount of work that has accumulated for a server to carry out at any time. Let V_n be the workload at time n. The process $\{V_n, n \geq 0\}$, $V_n = 0, 1, 2, \cdots$ is a DTMC, and its transition matrix P can be written as

$$P = \begin{bmatrix} \bar{a} & a\bar{b} & ab\bar{b} & ab(\bar{b})^2 & ab(\bar{b})^3 & \cdots \\ \bar{a} & a\bar{b} & ab\bar{b} & ab(\bar{b})^2 & ab(\bar{b})^3 & \cdots \\ & \bar{a} & a\bar{b} & ab\bar{b} & ab(\bar{b})^2 & \cdots \\ & & \bar{a} & a\bar{b} & ab\bar{b} & \cdots \\ & & & \ddots & \ddots & \ddots \end{bmatrix}. \quad (5.51)$$

This is an M/G/1 type of DTMC, with a special structure. Standard methods can be used to analyze this.

5.4.5 *Discrete time equivalent of Little's Law:*

Little's law is a very popular result in queueing theory. In continuous time it is the well known result that $L = \lambda W$, where L is the average number in the system, λ is the arrival rate and W is the mean waiting time in the system. The variables L, λ, and W are used in this previous sentence only to present the result in a well-known format. They will not be used with the same meaning after this.

Let W be the waiting time in the system as experienced by a packet and X be the number in the system as left behind by a departing packet. Let $w_k = Pr\{W = k\}$, $k \geq 0$, and $x_j = Pr\{X = j\}$, $j \geq 0$. We let the arrival be geometric with parameter a, as before. Let A_k be the number of arrivals during a time interval of length $k \geq 1$. It is clear that

$$Pr\{A_k = v\} = \binom{k}{v} a^v (1-a)^{k-v}, \; 0 \leq v \leq k.$$

Define

$$X^*(z) = \sum_{j=0}^{\infty} z^j x_j, \;\; w^*(z) = \sum_{j=1}^{\infty} z^k w_k, \; |z| \leq 1.$$

Then we have

$$X^*(z) = \sum_{k=1}^{\infty} \sum_{v=0}^{k} \binom{k}{v} a^v (1-a)^{k-v} w_k z^v$$

$$= \sum_{k=1}^{\infty} w_k \sum_{v=0}^{k} \binom{k}{v} (az)^v (1-a)^{k-v}$$

$$= \sum_{k=1}^{\infty} (az + 1 - a)^k w_k. \quad (5.52)$$

After routine algebraic operations we obtain

$$X^*(z) = w^*(az + 1 - a). \tag{5.53}$$

By taking the differentials of both sides and setting $z \to 1$ we have

$$E[X] = aE[W], \tag{5.54}$$

which is equivalent to $L = \lambda W$ in the case of Little's law in continuous time. This is actually very straightforward for the Geo/Geo/1. We find that this same idea is extendable to the Geo/G/1 system as well.

5.5 Geo/Geo/1 System with Start-up Times

Consider a regular Geo/Geo/1 system with arrival parameter a and service parameter b. Suppose when the system starts up from empty the service time parameter of the first item is now b^*, where $b^* \leq b$ and we have $\bar{b}^* = 1 - b^*$. The state space of this DTMC is $\{0 \cup (X_n, J_n), X_n = 1, 2, 3, \cdots; J_n = 0, 1, \}$ where $J_n = 0$ if the item in service is the first one in the busy period and 1 otherwise. The associated transition matrix P can be written as

$$P = \begin{bmatrix} B & C & & & \\ E & A_1 & A_0 & & \\ & A_2 & A_1 & A_0 & \\ & & \ddots & \ddots & \ddots \end{bmatrix},$$

where

$$B = \bar{a},\ C = [a,\ 0],\ E = \begin{bmatrix} \bar{a}b^* \\ \bar{a}b \end{bmatrix},$$

$$A_0 = \begin{bmatrix} a\bar{b}^* & 0 \\ 0 & a\bar{b} \end{bmatrix},\ A_1 = \begin{bmatrix} \bar{a}\bar{b}^* + ab^* & 0 \\ 0 & \bar{a}\bar{b} + ab \end{bmatrix},\ A_2 = \begin{bmatrix} \bar{a}b^* & 0 \\ 0 & \bar{a}b \end{bmatrix}.$$

From here it is straightforward to apply MAM to perform the analysis.

5.6 Geo/G/1 Queues

So far, most of the queues we have studied in discrete time have resulted in univariate DTMCs and both the inter-arrival and service times are geometric which has the *lack of memory property (lom)*. One of the challenges faced by queueing theorists in the 1950*s* and early 1960*s* was that most queueing systems observed in real life did not have geometric (or exponential) service times and hence the *lom* property assumption for service times was not valid. However, introducing more general service times which do not have the *lom* are more difficult to model for a queueing system because then we need to include a supplementary variable in order to capture the Markovian property. By introducing a supplementary variable we end up with a bivariate DTMC, which in theory should be manageable, given the results already discussed in Chapter 3, and which were already available in those days. However, because a general distribution does not necessarily have a finite support then the range of the supplementary variable will be unbounded. This led to a different problem. Until Kendall (1953) introduced the method of imbedded Markov chains before this problem became practically feasible to analyze and obtain numerical results. In what follows we present both the method of supplementary variables and the imbedded Markov chain approach for the Geo/G/1 system. The imbedded Markov chain observes the system at points of events, e.g. points of arrivals or points of departures. This way we are able to obtain a Markov chain with usually only a single index which has infinite range. With this approach even though information in between the events is not known directly they can be computed or recovered using renewal theory. The Geo/G/1 queue is studied as an imbedded Markov chain. For this specific model, the system is observed at points of departures.

Consider a single server queue with geometric inter-arrival times with probability of an arrival at any point in time given as a and $\bar{a} = 1 - a$. This system has a general service time S with probability mass function, *pmf*, of $s_j = \Pr\{S = j,\ j \geq 1\}$, having a mean of $E[S] = \mu^{-1}$ and $E[S^2]$ is given as the second moment of the service times. We define the following $z-$transforms

$$A^*(z) = 1 - a + za, \text{ and } S^*(z) = \sum_{j=1}^{\infty} z^j s_j.$$

5.6.1 Supplementary Variable Technique

At time n let X_n be the number of packets in the system and J_n the remaining service time of the packet in service. It is immediately clear that $\{(X_n, J_n), n \geq 0\}$ is a bivariate DTMC, keeping in mind that J_n has no value when $X_n = 0$. The transition matrix of this DTMC is much more difficult to visualize because then both variables are unbounded. Hence we are interested in $x_{i,j}^{(n)} = Pr\{X_n = i, J_n = j\}$, $i \geq 1, j \geq 1$, and $x_0^{(n)}$, where $x_0^{(n)} = Pr\{X_n = 0\}$. First we write out the related Chapman-Kolmogorov equations as follows:

$$x_0^{(n+1)} = \bar{a}(x_0^{(n)} + x_{1,1}^{(n)}) \tag{5.55}$$

$$x_{1,j}^{(n+1)} = as_j x_0^{(n)} + as_j x_{1,1}^{(n)} + \bar{a}x_{1,j+1}^{(n)} + \bar{a}s_j x_{2,1}^{(n)},\ j \geq 1 \tag{5.56}$$

$$x_{i,j}^{(n+1)} = as_j x_{i-1,1}^{(n)} + as_j x_{i,1}^{(n)} + \bar{a}x_{i-1,j+1}^{(n)} + \bar{a}s_j x_{i+1,1}^{(n)},\ i \geq 1,\ j \geq 1. \tag{5.57}$$

Next we present two approaches for handling the set of equations in order to obtain the stationary distribution of the system. The two approaches are z-transform and matrix-analytic approaches.

5.6.1.1 Transform Approach

The associated transition matrix can be written as follows: Let the z-transform of the process (X_n, J_n) be

$$X_n^*(z_1,z_2) = x_0 + \sum_{i=1}^{\infty}\sum_{j=1}^{\infty} z_1^i z_2^j x_{i,j}^{(n)},\quad |z_1| \leq 1,\ |z_2| \leq 1.$$

Minh (1978) and Dafermos and Neuts (1971) have derived the transforms of this system under time-inhomogeneous conditions. Here we focus on stationary distributions only. In that effect we assume the system is stable and consider the case where

$$x_{i,j} = x_{i,j}^n|_{n\to\infty},\quad x_0 = x_0^{(n)}|_{n\to\infty},$$

hence we have

$$X^*(z_1,z_2) = X_n^*(z_1,z_2)|_{n\to\infty}.$$

For this stationary system it was shown by Chaudhry (2000) that

$$X^*(z_1,z_2) = \frac{z_2 a z_1(1-z_1)x_0\{S(z_1) - S^*(1-a+az_1)\}}{[S^*(1-a-az_1) - z_1]\{z_2 - (1-a+az_1)\}}. \tag{5.58}$$

By setting $z_2 = 1$ in $X^*(z_1,z_2)$ we can get the $z-$transform of the marginal distribution, i.e. $y_i = \sum_{j=1}^{\infty} x_{i,j},\ j \geq 1$, where $y_0 = x_0$. Knowing that $x_0 = 1-\rho,\ \rho = a\mu$, setting $z_2 = 1$ and replacing $z = z_1$ in $X^*(z_1,z_2)$, we have $Y(z) = \sum_{i=0}^{\infty} z^i y_i$ which is given as

$$Y^*(z) = \frac{(1-\rho)(1-z)S^*(1-a+za)}{S^*(1-a+za) - z}. \tag{5.59}$$

5.6.1.2 Matrix-analytic approach

In order to take full advantage of MAM we need to write the service times distribution as a PH distribution, more of an IPH distribution, based on the elapsed time approach. Let the IPH representation be $(\boldsymbol{\beta}, S)$, with $\mathbf{s} = \mathbf{1} - S\mathbf{1}$, where

$$\boldsymbol{\beta} = [1,\ 0,\ 0,\ \cdots],\ \text{ and } S_{i,j} = \begin{cases} \tilde{s}_i & j = i+1, \\ 0 & \text{otherwise}, \end{cases}$$

where $\tilde{s}_i = \frac{u_i}{u_{i-1}}$, $u_i = 1 - \sum_{v=1}^{i} s_v$, $i \geq 1$, $u_0 = 1$. Based on the Chapman-Kolmogorov equations presented above we can write the transition matrix of the DTMC as

$$P = \begin{bmatrix} B & C & & & \\ E & A_1 & A_0 & & \\ & A_2 & A_1 & A_0 & \\ & & A_2 & A_1 & A_0 \\ & & & \ddots & \ddots & \ddots \end{bmatrix},$$

where

$$B = \bar{a},\ C = a\boldsymbol{\beta} = [a,\ 0,\ 0,\ \cdots],\ E = \bar{a}\mathbf{s} = \bar{a}[1-\tilde{s}_1,\ 1-\tilde{s}_2,\ 1-\tilde{s}_3,\ \cdots]^T$$

$$A_0 = aS = \begin{bmatrix} 0 & a\tilde{s}_1 & 0 & \cdots \\ 0 & 0 & a\tilde{s}_2 & \cdots \\ & 0 & 0 & \ddots \\ & & \ddots & \end{bmatrix},$$

$$A_1 = a\mathbf{s}\boldsymbol{\beta} + \bar{a}S = \begin{bmatrix} a(1-\tilde{s}_1) & \bar{a}\tilde{s}_1 & 0 & 0 & \cdots \\ a(1-\tilde{s}_2) & 0 & \bar{a}\tilde{s}_2 & 0 & \cdots \\ a(1-\tilde{s}_3) & 0 & 0 & \bar{a}\tilde{s}_3 & \cdots \\ \vdots & \vdots & \vdots & \vdots & \ddots \end{bmatrix},$$

$$A_2 = \bar{a}\mathbf{s}\boldsymbol{\beta} = \begin{bmatrix} \bar{a}(1-\tilde{s}_1) & 0 & 0 & \cdots \\ \bar{a}(1-\tilde{s}_2) & 0 & 0 & \cdots \\ \bar{a}(1-\tilde{s}_3) & 0 & 0 & \cdots \\ \vdots & \vdots & \vdots & \vdots \end{bmatrix}.$$

The G matrix for a stable version of this DTMC is given by the solution to the matrix quadratic equation

$$G = A_2 + A_1 G + A_0 G^2, \tag{5.60}$$

and this solution is of the form

$$G = \begin{bmatrix} 1 & 0 & 0 & 0 & \cdots \\ 1 & 0 & 0 & 0 & \cdots \\ 1 & 0 & 0 & 0 & \cdots \\ \vdots & \vdots & \vdots & \vdots & \cdots & \vdots \end{bmatrix}. \tag{5.61}$$

By virtue of the fact that the matrix A_2 is rank one it follows that the corresponding R matrix is given as

$$R = aS[I - a\mathbf{s}\boldsymbol{\beta} - \bar{a}S - aS\mathbf{1}\mathbf{e}_1^T]^{-1}. \tag{5.62}$$

Whereas this result looks elegant and can give us the joint distribution of the number of packets in the system and the remaining inter-arrival times, we have to truncate R at some point in order to carry out the computations. However, if the general service time has a finite support, i.e. $s_j = 0,\ j < n_s < \infty$, then we can just apply the matrix-geometric method directly because all the block matrices B, C, E and $A_k,\ k = 0, 1, 2$ will all be finite. This situation will be presented later when dealing with the GI/G/1 system which is a generalization of the Geo/G/1 system.

If however, for the case where the service distribution support is not finite, and information about the number in the system as seen by an arriving packet is all we need then we use the imbedded Markov chain approach.

Note that we could have used the remaining time approach to represent the IPH for service time. However, we would not have been able to obtain a special structure for the matrix G as we did, or even for matrix R, for that matter.

5.6.2 Imbedded Markov Chain Approach

Let X_n be the number of customers left behind by the departing nth customer, and A_n be the number of arrivals during the service of the nth customer. Then we have

$$X_{n+1} = (X_n - 1)^+ + A_{n+1}, \tag{5.63}$$

where $(U-V)^+ = \max\{0,\ U-V\}$.

This model is classified as *the late arrival with delayed access system* by Hunter (1983).

Let $a_v = \Pr\{A_n = v,\ v \geq 0\},\ \forall n$. Then a_v is given as

$$a_v = \sum_{k=v}^{\infty} s_k \binom{k}{v} a^v (1-a)^{k-v},\ v \geq 0. \tag{5.64}$$

Let P be the transition matrix describing this Markov chain, then it is given as

$$P = \begin{bmatrix} a_0 & a_1 & a_2 & \cdots \\ a_0 & a_1 & a_2 & \cdots \\ & a_0 & a_1 & \cdots \\ & & a_0 & \cdots \\ & & & \ddots \end{bmatrix} \tag{5.65}$$

For steady state to exist, we need that $\rho = \sum_{j=1}^{\infty} j a_j < 1$. If we assume that steady state exists, then we have the invariant vector $\boldsymbol{x} = [x_0, \; x_1, \; \ldots]$ as

$$\boldsymbol{x} = \boldsymbol{x}P, \; \boldsymbol{x}\mathbf{1} = 1.$$

This can be written as

$$x_i = x_0 a_i + \sum_{j=1}^{i+1} x_j a_{i-j+1}, \; i \geq 0. \tag{5.66}$$

Now define the probability generating functions

$$X^*(z) = \sum_{i=0}^{\infty} x_i z^i, \text{ and } A^*(z) = \sum_{i=0}^{\infty} a_i z^i = S^*(1 - a + za), \; |z| \leq 1,$$

we obtain

$$X^*(z) = \frac{x_0(1-z)A^*(z)}{A^*(z) - z}. \tag{5.67}$$

Using the fact that $X^*(1) = 1$, $A^*(1) = 1$ and $(A^*)'(1) = a/\mu$, we find that

$$x_0 = 1 - \rho \tag{5.68}$$

and

$$X^*(z) = \frac{(1-\rho)(1-z)A^*(z)}{A^*(z) - z} = \frac{(1-\rho)(1-z)S^*(1-a+za)}{S^*(1-a+za) - z}. \tag{5.69}$$

The mean number in the system $E[X]$ is given as

$$E[X] = \rho + \frac{a^2 E[S^2]}{2(1-\rho)}, \tag{5.70}$$

where $\rho = a/\mu$ and $E[S^2]$ is the second moment of the service times.

If we let Y be the number in the queue, then we have

$$E[Y] = \frac{a^2 E[S^2]}{2(1-\rho)}. \tag{5.71}$$

5.6.2.1 Distribution of number in the system

The probability of the number of packets left behind in the system by a departing packet $x_i,\ i = 0,1,2,\cdots$, can be obtained by finding the coefficients of z^i in the polynomial associated with $X(z)$, but this could be onerous. The simplest approach is to apply the recursion by Ramaswami (1988) as follows. If the system is stable, then the G matrix (a scalar in this case) associated with the M/G/1-type system is a scalar and has the value of 1. Since x_0 is known to be $1-\rho$, then we can successfully apply the recursion to obtain this probability distribution.

Let $\tilde{a}_v = \sum_{i=v}^{\infty} a_i,\ v \geq 0$, then

$$x_i = (1-\tilde{a}_1)^{-1}[x_0\tilde{a}_i + (1-\delta_{i,1})\sum_{j=1}^{i-1} x_j\tilde{a}_{i+1-j}],\ i \geq 1, \tag{5.72}$$

where δ_{ij} is the Kronecker delta, i.e. $\delta_{ij} = \begin{cases} 1, & i=j \\ 0, & \text{otherwise} \end{cases}$.

The recursion is initiated by $x_0 = 1-\rho$.

5.6.2.2 Waiting Time:

Let W be the waiting time in the system, with $w_k = Pr\{W=k\}$ and $W^*(z) = \sum_{k=0}^{\infty} z^k w_k$. We know that the number of items left behind by a departing item is the number of arrivals during its waiting time in the system, hence

$$x_i = \sum_{k=i}^{\infty} \binom{k}{i} a^i (1-a)^{k-i} w_k,\ i \geq 0. \tag{5.73}$$

By taking the $z-$ transform of both sides we end up with

$$X^*(z) = \sum_{k=0}^{\infty} (1-a+az)^k w_k = W^*(1-a+za). \tag{5.74}$$

Hence

$$W^*(1-a+za) = \frac{(1-\rho)(1-z)S^*(1-a+za)}{S^*(1-a+za)-z}. \tag{5.75}$$

5.6.2.3 Workload:

Let $V_n, n \geq 0$ be the workload in the system at time n, with $V_n = 0, 1, 2, \cdots$. It is straightforward to show that V_n is a DTMC and its transition matrix is an M/G/1 type with the following structure

$$P = \begin{bmatrix} c_0 & c_1 & c_2 & c_3 & \cdots \\ c_0 & c_1 & c_2 & c_3 & \cdots \\ & c_0 & c_1 & c_2 & \cdots \\ & & c_0 & c_1 & \cdots \\ & & & \ddots & \ddots \end{bmatrix}, \tag{5.76}$$

with

$$c_0 = 1 - a, \; c_j = ab_j, \; j \geq 1.$$

If the system is stable, i.e. if $1 > \sum_{j=1}^{\infty} jc_j$, then we have a stationary distribution $\mathbf{v} = [v_0, \; v_1, \; v_2, \; \cdots]$ such that

$$\mathbf{v} = \mathbf{v}P, \;\; \mathbf{v1} = 1.$$

This leads to the equation

$$v_i = c_i v_0 + \sum_{j=1}^{i+1} v_j c_{i-j+1}, \;\; i \geq 0. \tag{5.77}$$

It is immediately clear that we can set this up in such a manner that allows us to use Ramaswami's recursion to obtain the stationary distribution at arbitrary times.

By letting

$$C^*(z) = \sum_{i=0}^{\infty} z^i c_i, \;\; V^*(z) = \sum_{i=0}^{\infty} z^i v_i, \; |z| \leq 1,$$

we have

$$V^*(z) = \frac{v_0[(z-1)C^*(z)]}{z - C^*(z)}, \;\; |z| \leq 1. \tag{5.78}$$

Since the resulting equations are exactly of the same form as the ones for the number in the system we simply apply the techniques used earlier for similar structure problems to obtain key measures.

5.6.2.4 Age Process:

Here we study the age process, but first we re-write the service time in the form of elapsed time. Consider the service time distribution $\{b_1, b_2, b_3, \cdots\}$. It can be written as an infinite PH (IPH) distribution represented as $(\boldsymbol{\beta}, S)$ where

$$\boldsymbol{\beta} = [1,\ 0,\ 0,\ \cdots],\ \text{and}\ S_{i,j} = \begin{cases} \tilde{b}_i, j = i+1 \\ 0, \text{otherwise}, \end{cases}$$

where

$$\tilde{b}_i = \frac{e_i}{e_{i-1}},\ e_i = 1 - \sum_{v=1}^{i} b_v,\ e_0 = 1.$$

By now we know that

$$b_k = \boldsymbol{\beta} S^{k-1}\mathbf{s},\ k \geq 1.$$

At time $n \geq 0$, let Y_n be the age of the leading packet in the system, i.e. the one currently receiving service, and let J_n be its elapsed time of service. The bivariate process $\{(Y_n, J_n), n \geq 0\}$ is a DTMC of the GI/M/1 type with transition matrix P written as

$$P = \begin{bmatrix} C_0 & C_1 & & & & \\ B_1 & A_1 & A_0 & & & \\ B_2 & A_2 & A_1 & A_0 & & \\ B_3 & A_3 & A_2 & A_1 & A_0 & \\ \vdots & \vdots & \vdots & \vdots & \vdots & \ddots \end{bmatrix}, \tag{5.79}$$

where,

$$C_0 = \bar{a},\ C_1 = a\boldsymbol{\beta},\ B_k = \bar{a}^k\mathbf{s},\ A_0 = S,\ A_k = a\bar{a}^{k-1}(\mathbf{s}\boldsymbol{\beta}),\ k \geq 1.$$

With this set up the standard GI/M/1 results can be applied to the age process to obtain necessary measures. However, because we have used the IPH caution has to be exercised because now we are dealing with infinite block matrices. If the service time distribution has a finite support then it is straightforward to analyze the system. When it is not a finite support one needs to refer to the works of Tweedie (1982) to understand the conditions of stability.

5.6.2.5 Busy Period:

Let $b_i(j)$ be the probability that the busy period initiated by j customers lasts i units of time. It was shown by Klimko and Neuts (1973) and Heimann and Neuts (1973) that

$$b_i(1) = \sum_{\ell=1}^{i} s_\ell(1) \sum_{k=0}^{\ell} a_k^{(\ell)} b_{i-\ell}(k), \tag{5.80}$$

where $a_k^{(l)}$ is the probability that k arrivals occurs in l units of time, and $s_i(k)$ is the probability that the total service times of k customers lasts exactly i units of time.

5.6.3 *Mean-Value approach*

The Mean-value Approach as the name implies is based on mean values, and the results are not easily obtained for higher moments using the same approach. We first write out the random variable equation as follows

$$X_{n+1} = (X_n - 1)^+ + A_n,$$

where A_n is the number of arrivals in between the service completion times of the nth and $(n+1)^{st}$ item, and X_n is the number left behind by a departing n^{th} item. By definition of the arrival process, $A_n = A, \ \forall n$, i.e. the arrivals are independent and identically distributed. We write $E[Y]$ as the expectation of a random variable Y, then based on our random variable equation we have

$$E[X_{n+1}] = E[(X_n - 1)^+] + E[A].$$

Going back to the random variable equation we can write it as

$$X_{n+1} = X_n - U(X_n) + A,$$

where $U(V) = \begin{cases} 1, & V > 0, \\ 0, & V = 0 \end{cases}.$

$$E[X_{n+1}] = E_{[X_n]} - E[U(X_n)] + E[A].$$

For a stable system we know that $(L_{D+1} - L_D)|_{n\to\infty} = 0$, hence

$$0 = -E[U(X)] + E[A],$$

where $X = X_n|_{n\to\infty}$ hence

$$E[A] = E[U(X)].$$

Now we obtain an expression for $E[A]$.

$$E[A] = \sum_{i=0}^{\infty}\sum_{n=i}^{\infty} i\binom{n}{i} a^i(1-a)^{n-i}s_n$$

$$= a\sum_{i=1}^{\infty}\sum_{n=i}^{\infty} n\binom{n-1}{i-1} a^{i-1}(1-a)^{n-i}s_n$$

$$= a\sum_{n=1}^{\infty} ns_n \sum_{i=1}^{n}\binom{n-1}{i-1} a^{i-1}(1-a)^{n-i}$$

$$= a\sum_{n=1}^{\infty} ns_n = aE[S] = a\mu^{-1}.$$

Let us return to the random variable equation and square it as follows

$$(X_{n+1})^2 = (X_n - U(X_n) + A)^2 = (X_n)^2 - 2X_nU(X_n) + 2AX_n - 2AU(X_n) + (U(X_n))^2 + A^2.$$

Once again if we consider a stable system, then we have

$$0 = 2(AE[X] - E[XU(X)] - AE[U(X)]) + E[U(X)^2] + E[A^2].$$

Because $(U(X))^2 = U(X)$ and $E[U(X)] = E[A] = \frac{a}{\mu} = \rho$, and by definition $E[XU(X)] = E[X]$, we have

$$2E[XU(X)] = 2E[X] = 2E[X].$$

Hence we have

$$2E[X] = E[A] + E[A^2] + 2AE[X] - 2(E[A])^2 = \rho + E[A^2] + 2\rho E[X] - 2(\rho)^2.$$

After some algebraic operations this leads to

$$E[X] = \rho + \frac{a^2E[S^2]}{2(1-\rho)},$$

the same result obtained for the mean number in the system at departure using the embedded Markov chain approach in the previous section.

5.7 GI/Geo/1 Queues

The GI/Geo/1 model is essentially the "dual" of the Geo/G/1. Suppose the arrivals are of the general independent type with inter-arrival times $\mathscr{G}$ having a *pmf* of $g_n = \Pr\{\mathscr{G} = n,\ n \geq 1\}$ whose mean is λ^{-1} and the service time is of the geometric distribution with parameters s and $\bar{s}$ and with mean service time s^{-1}. Let $S^*(z) = \sum_{i=1}^{\infty} z^i g_i,\ \ |z| \leq 1$. Because the inter-arrival time is general and the service time is geometric, we study this also using supplementary variables and the imbedded Markov chain which is imbedded at the points of arrival.

5.7.1 *Supplementary Variable Technique*

Define X_n and K_n as the number of packets in the system and the remaining inter-arrival time at time $n \geq 0$. Further let $x_{i,j}^{(n)} = Pr\{X_n = i, K_n = j\}$. Then we have

$$x_{0,j}^{(n+1)} = x_{0,j+1}^{(n)} + x_{1,j+1}^{(n)} s,\ \ j \geq 1, \tag{5.81}$$

$$x_{1,j}^{(n+1)} = x_{0,1}^{(n)} g_j + x_{1,1}^{(n)} s g_j + x_{1,j+1}^{(n)} \bar{s} + x_{2,j+1}^{(n)} s,\ \ j \geq 1. \tag{5.82}$$

$$x_{i,j}^{(n+1)} = x_{i-1,1}^{(n)} \bar{s} g_j + x_{i,1}^{(n)} s g_j + x_{i,j+1}^{(n)} \bar{s} + x_{i+1,j+1}^{(n)} s,\ \ j \geq 2. \tag{5.83}$$

We consider a stable system and study $x_{i,j} = x_{i,j}^{(n)}|_{n \to \infty}$.

5.7.1.1 Matrix-analytic approach

We have a bivariate DTMC with the transition matrix P written as

$$P = \begin{bmatrix} B & C & & & \\ A_2 & A_1 & A_0 & & \\ & A_2 & A_1 & A_0 & \\ & & \ddots & \ddots & \ddots \end{bmatrix},$$

with

$$B = \begin{bmatrix} 0 & 0 & 0 & \cdots \\ 1 & 0 & 0 & \cdots \\ 0 & 1 & 0 & \cdots \\ & & \ddots & \end{bmatrix},\ \ A_1 = \begin{bmatrix} sg_1 & sg_2 & sg_3 & \cdots \\ \bar{s} & 0 & 0 & \cdots \\ 0 & \bar{s} & 0 & \cdots \\ & & \ddots & \end{bmatrix},$$

$$C = \begin{bmatrix} g_1 & g_2 & g_3 & \cdots \\ 0 & 0 & 0 & \cdots \\ 0 & 0 & 0 & \cdots \\ \vdots & \vdots & \vdots & \cdots \end{bmatrix},$$

$$A_2 = sB, \;\; A_0 = \bar{s}C.$$

We can now easily apply the matrix-geometric results for the QBD to this system, given all the conditions required by the infinite blocks of P are satisfied. With

$$R = A_0 + RA_1 + R^2 A_2,$$

and the known properties of R we can see that it has the following structure

$$R = \begin{bmatrix} r_1 & r_2 & r_3 & \cdots \\ 0 & 0 & 0 & \cdots \\ 0 & 0 & 0 & \cdots \\ \vdots & \vdots & \vdots & \cdots \end{bmatrix},$$

where

$$r_k = \bar{s}g_k + r_1 s g_k + s r_1 r_{k+1}, \;\; k \geq 1. \tag{5.84}$$

This can be solved iteratively as

$$r_k(n+1) := \bar{s}g_k + r_1(n) s g_k + s r_1(n) r_{k+1}(n), \;\; n = 0, 1, 2, \cdots,$$

where $r_i(0) := 0$.

Whereas this approach looks elegant and can be used to obtain the joint distribution of the number of packets in the system and the remaining inter-arrival times, we have to truncate R at some point in order to carry out the computations. However, if the general inter-arrival time has a finite support, i.e. $a_j = 0, \; j < K_a = n_t < \infty$, then we can just apply the matrix-geometric method directly because all the block matrices B, C, E and $A_k, \; k = 0, 1, 2$ will all be finite. The case of finite support, i.e. $n_t < \infty$ is given as follows.

If $n_t < \infty$, then we have

$$r_k = \bar{s}g_k + r_1 s g_k + s r_1 r_{k+1}, \;\; 1 \leq k < n_t, \tag{5.85}$$

$$r_{n_t} = \bar{s}g_{n_t} + r_1 s g_{n_t}. \tag{5.86}$$

Applying the last equation and working backwards we have

$$\begin{aligned} r_{n_t-1} &= \bar{s}g_{n_t-1} + r_1 s g_{n_t-1} + s r_1 r_{n_t} \\ &= \bar{s}g_{n_t-1} + r_1 s g_{n_t-1} + s r_1 \bar{s} g_{n_t} + s^2 (r_1)^2 g_{n_t}. \end{aligned} \tag{5.87}$$

By repeated application we find that r_1 is a polynomial equation of the form

$$r_1 = a_0 + a_1 r_1 + a_2 (r_1)^2 + a_3 (r_1)^3 + \cdots + a_{n_t} (r_1)^{n_t}, \tag{5.88}$$

where a_j, $j = 0, 1, \cdots, n_t$ are known constants. For example when $n_t = 2$ we have

$$a_0 = \bar{s} g_1,\ a_1 = s g_1 + \bar{s} g_2, \text{ and } a_2 = s^2 g_2.$$

We solve this polynomial for the minimum r_1 which satisfies $0 < r_1 < 1$. Once r_1 is known we can then solve for r_{n_t} and then proceed to obtain all the other r_j, $j = n_t - 1, n_t - 2, \cdots, 3, 2$. This case will be presented later when dealing with the general case of GI/G/1 system.

If however, for the case where the inter-arrival time does not have a finite support, and we need information about the number in the system as seen by a departing packet then we use the imbedded Markov chain approach.

5.7.2 Imbedded Markov Chain Approach

Let Y_n be the number of packets in the system at the arrival time of the nth packet, and B_n the number of packets served during the inter-arrival time of the nth packet. Then we have

$$Y_{n+1} = Y_n + 1 - B_n,\ Y_n \geq 0,\ B_n \leq Y_n + 1. \tag{5.89}$$

This system is based on the *early arrival system* as discussed by Hunter Hunter (1983).

Further, let us define b_j as the probability that j packets are served during the $(n+1)$th inter-arrival time if the system is found empty by the $(n+1)$th arriving packet and a_j is the corresponding probability if this packet arrives to find the system not empty with j packets served during its inter-arrival time. Then we have

$$a_j = \sum_{k=j}^{\infty} g_k \binom{k}{j} s^j (1-s)^{k-j},\ j \geq 0 \tag{5.90}$$

$$b_j = 1 - \sum_{i=0}^{j} a_i,\ j \geq 0 \tag{5.91}$$

The transition matrix describing this Markov chain is given by P as follows

$$P = \begin{bmatrix} b_0 & a_0 & & & \\ b_1 & a_1 & a_0 & & \\ b_2 & a_2 & a_1 & a_0 & \\ \vdots & \vdots & \vdots & & \ddots \end{bmatrix}. \tag{5.92}$$

Let $\boldsymbol{x} = [x_0,\ x_1,\ \ldots]$ be the invariant vector associated with this Markov chain. It is simple to use either the matrix-geometric approach or the transform approach to obtain this vector.

5.7.2.1 Matrix-geometric approach:

It is straightforward to see that the matrix P here has the same structure as the GI/M/1-type presented in the previous chapter. The GI/Geo/1 queue is the discrete time analogue of the GI/M/1 system. The descriptor GI/M/1-type got its name from the classical GI/M/1 queue transition probability matrix structure. We note that this GI/Geo/1 queue transition matrix has scalar entries in P and as such the corresponding R is a scalar r.

Provided that the system is stable, i.e. $\rho = \left[\sum_{j=1}^{\infty} jb_j\right]^{-1} < 1$, then there exists r which is the minimal non-negative solution to the polynomial equation

$$r = \sum_{i=0}^{\infty} a_i r^i. \tag{5.93}$$

The resulting solution gives us the geometric form of the solution for $\boldsymbol{x}$ as follows:

$$x_{i+1} = x_i r,\ i \geq 0. \tag{5.94}$$

After solving for the boundary value $x_0 = 1 - \rho = 1 - r$ through $x_0 = \sum_{j=0}^{\infty} x_j b_j$ and applying the normalization equation $\sum_{j=0}^{\infty} x_j = 1$, we obtain

$$x_i = (1-r)r^i,\ i \geq 0. \tag{5.95}$$

The results in (5.93) and (5.95) are actually the scalar-geometric solution analogue of the matrix-geometric solution presented in the previous chapter.

5.7.2.2 Transform approach

Consider the matrix P, we can write the steady state equations for this system as

$$x_0 = \sum_{j=0}^{\infty} x_j b_j = \sum_{j=0}^{\infty} x_j (1 - \sum_{v=0}^{j} a_v), \tag{5.96}$$

$$x_i = \sum_{j=1}^{\infty} x_{i+j-1} a_j,\ \ i \geq 1. \tag{5.97}$$

Adopting the approach in Gross et al (2008) we can solve the above set of difference equations by using the operator approach. Consider the operator F and write

$$x_{i+1} = Fx_i, \; i \geq 1,$$

then substituting this into the above set of equations for $x_i, \; i \geq 1$, i.e.

$$x_i - [\sum_{j=1}^{\infty} x_{i+j-1} a_j] = 0,$$

we have

$$F - a_0 - Fa_1 - F^2 a_2 - F^3 a_3 - \cdots = 0. \tag{5.98}$$

This leads to the result that $F = \sum_{j=0}^{\infty} F^j a_j$. The characteristic equation for this difference equation is

$$z = \sum_{j=0}^{\infty} z^j a_j. \tag{5.99}$$

However, a_j is the probability that the number of service completions during an inter-arrival time is j, so by substituting $sz + 1 - s$ into $\sum_{j=0}^{\infty} z^j a_j$ we get

$$z = G^*(sz + 1 - s). \tag{5.100}$$

It can be shown that a solution z^* that satisfies this equation is the same as r which we obtained earlier using the matrix-geometric approach. Hence

$$r = G^*(sr + 1 - s). \tag{5.101}$$

Once again provided the system is stable we only have one positive value that satisfies this equation.

It can be shown that

$$x_i = Kr^i, \;\; i \geq 0, \tag{5.102}$$

and when this is normalized we find that $K = 1 - r$.

A simple way to obtain r is by iterating the following

$$r_{k+1} = G^*(r_k s + (1 - s)), \text{ with } 0 \leq r_0 < 1 \tag{5.103}$$

until $|r_{k+1} - r_k| < \varepsilon$, where ε is a very small positive value, e.g. 10^{-12}.

Waiting time:

- Waiting time in the queue: Let W_q be the waiting time in the queue for a packet, with $w_j^{(q)} = Pr\{W_q = j\}$, $j \geq 0$. Since this is a FCFS system, then an arriving packet only has to wait for the packets ahead of it. Let $W_q^*(z) = \sum_{j=0}^{\infty} z^j w_j^{(q)}$. It is clear that the $z-$ transform of the probability x_i is $X^*(z) = \sum_{i=0}^{\infty} z^i x_i = (1-r)(1-rz)^{-1}$. Since this is the transform of the number of packets seen by an arriving packet then to obtain $W_q^*(z)$ we simply replace z in $X^*(z)$ with $sz(1-(1-s)z)^{-1}$ leading to

$$W_q^*(z) = (1-r)(1-rsz(1-(1-s)z)^{-1})^{-1}.$$

Hence

$$W_q^*(z) = \frac{(1-r)(1-(1-s)z)}{1-(1-s)z-rsz}. \quad (5.104)$$

- Waiting time in the system: The $z-$ transform of the waiting time in the system is simply a convolution sum of W_q and service times. Let $W^*(z)$ be the $z-$ transform of the waiting time in the system then we have

$$W^*(z) = (1-r)sz(1-rsz(1-(1-s)z)^{-1})^{-1}(1-(1-s)z)^{-1}.$$

Hence

$$W^*(z) = \frac{sz(1-r)}{1-(1-s)z-rsz}. \quad (5.105)$$

5.8 Geo/PH/1 Queues

Consider a single server queue with geometric inter-arrival times with parameter a and PH service with representation $(\boldsymbol{\beta}, S)$ of order n_s. At time n, let X_n and J_n be the number of packets in the system and the phase of service of the packet receiving service. Consider this system as a DTMC with the state space $\{(0) \cup (X_n, J_n), n \geq 0\}, X_n \geq 1, J_n = 1, 2, \cdots, n_s$. This is a QBD with the transition matrix P given as

$$P = \begin{bmatrix} B & C & & & \\ E & A_1 & A_0 & & \\ & A_2 & A_1 & A_0 & \\ & & \ddots & \ddots & \ddots \end{bmatrix},$$

$$B = 1-a,\ E = (1-a)\mathbf{s},\ C = a\boldsymbol{\beta},\ A_0 = aS,$$

$$A_1 = a\mathbf{s}\boldsymbol{\beta} + (1-a)S,\ A_2 = (1-a)\mathbf{s}\boldsymbol{\beta}.$$

We can go ahead and use the results for the QBD presented in Chapter 4 to study this system. However, this system has an added feature that we can exploit. The matrix A_2 is rank one, i.e. $(1-a)\mathbf{s}\boldsymbol{\beta}$. Even though we need to compute the matrix R we try and obtain it through matrix G, as follows. We know from Chapter 4 that $G=(I-U)^{-1}A_2$. Hence we have

$$G=(I-U)^{-1}(1-a)\mathbf{s}\boldsymbol{\beta}=\mathbf{d}\boldsymbol{\beta}, \tag{5.106}$$

where $\mathbf{d}=(I-U)^{-1}(1-a)\mathbf{s}$.

Since G is stochastic, if the DTMC is stable, by Theorem 8.5.1 in Latouche and Ramaswami (1999), we have

$$G\mathbf{1}=\mathbf{d}\boldsymbol{\beta}\mathbf{1}=\mathbf{1} \quad \rightarrow G=\mathbf{1}\boldsymbol{\beta}. \tag{5.107}$$

Using this result and substituting into $R=A_0(I-A_1-A_0G)^{-1}$ we obtain

$$R=A_0[I-A_1-A_0\mathbf{1}\boldsymbol{\beta}]^{-1}, \tag{5.108}$$

an explicit solution for matrix R from an explicit solution for G.

The stationary distribution of this system $x_{i,j}=Pr\{X_n=i,J_n=j\}|_{n\rightarrow\infty}$ and $x_0=Pr\{X_n=0\}|_{n\rightarrow\infty}$ leading to $\boldsymbol{x}=[x_0,\ \boldsymbol{x}_1,\ \boldsymbol{x}_2,\ \cdots,\ ]$, where $\boldsymbol{x}_i=[x_{i,1},\ x_{i,2},\ \cdots,\ x_{i,n_s}]$, we have

$$\boldsymbol{x}=\boldsymbol{x}P,\ \ \boldsymbol{x}\mathbf{1}=1,$$

and

$$\boldsymbol{x}_{i+1}=\boldsymbol{x}_iR,\ i\geq 1,$$

with the boundary equations

$$x_0=x_0(1-a)+\boldsymbol{x}_1E,\ \boldsymbol{x}_1=x_0C+\boldsymbol{x}_1(A_1+RA_2),$$

and normalization equation

$$x_0+\boldsymbol{x}_1(I-R)^{-1}\mathbf{1}=1,$$

we obtain the stationary distribution for this system.

The mean number in the system $E[X]$ is given as

$$E[X]=(\boldsymbol{x}_1+2\boldsymbol{x}_2+3\boldsymbol{x}_3+\cdots)\mathbf{1}=\boldsymbol{x}_1(I+2R+3R^2+\cdots)\mathbf{1},$$

which gives

$$E[X]=\boldsymbol{x}_1(I-R)^{-2}\mathbf{1}. \tag{5.109}$$

5.8.1 Waiting Times

We consider a FCFS system of service. Let W_q be the waiting time in the queue, with $w_i^{(q)} = Pr\{W_q = i\}$, $i \geq 0$ and let $W_q^*(z) = \sum_{j=0}^{\infty} z^j w_j^{(q)}$, $|z| \leq 1$, then we have

$$W_q^*(z) = x_0 + \sum_{i=1}^{\infty} \boldsymbol{x}_i (z\boldsymbol{\beta}(I - zS)^{-1}\mathbf{s})^i,$$

which gives

$$W_q^*(z) = x_0 + \boldsymbol{x}_1 (z\boldsymbol{\beta}(I - zS)^{-1}\mathbf{s})[I - Rz\boldsymbol{\beta}(I - zS)^{-1}\mathbf{s})]^{-1}. \tag{5.110}$$

Alternatively we may compute w_j^q as follows.

Let $B_k^{(k)} = (\mathbf{s}\boldsymbol{\beta})^k$, $k \geq 1$, $B_j^{(1)} = S^{j-1}(\mathbf{s}\boldsymbol{\beta})$, $j \geq 1$, then we have

$$B_j^{(k)} = SB_{j-1}^{(k)} + (\mathbf{s}\boldsymbol{\beta})B_{j-1}^{(k-1)}, \; k \geq j \geq 1. \tag{5.111}$$

We then write

$$w_0^{(q)} = x_0, \tag{5.112}$$

$$w_j^{(q)} = \sum_{v=1}^{j} \boldsymbol{x}_v B_j^{(v)} \mathbf{1}, \; j \geq 1. \tag{5.113}$$

On the other hand we may apply the idea of absorbing DTMC to analyze the waiting time, as follow. Consider a Markov chain $\{(\tilde{X}_n, J_n)\}$ where $\tilde{X}_n$ is the number of packets ahead of a target packet at time n. Since this is a FCFS system, if we let $\tilde{X}_n = 0$ be an absorbing state, then the time to absorption in this DTMC is the waiting time in the queue for the target packet. The transition matrix for the DTMC is

$$P_w = \begin{bmatrix} 1 & & & \\ \mathbf{s} & S & & \\ & \mathbf{s}\boldsymbol{\beta} & S & \\ & & \mathbf{s}\boldsymbol{\beta} & S \end{bmatrix}.$$

The stationary vector of the number in the system now becomes the initial vector of this DTMC, i.e. let $\mathbf{y}(0) = \boldsymbol{x}$, where $\mathbf{y}(0) = [y_0(0), \mathbf{y}_1(0), \mathbf{y}_2(0), \cdots]$, then we can apply the recursion

$$\mathbf{y}(k+1) = \mathbf{y}(k)P_w, \; k \geq 0.$$

From this we have

$$Pr\{W_q \le j\} = W_j^{(q)} = y_0(j),\, j \ge 1, \tag{5.114}$$

$$W_0^{(q)} = y_0(0). \tag{5.115}$$

Waiting time in the system can be obtained directly as a convolution sum of W_q and service time.

5.9 PH/Geo/1 Queues

Consider a single server queue with PH inter-arrival times with parameter $(\boldsymbol{\alpha}, T)$ of order n_t and geometric service with parameter b representing probability of a service completion in a time epoch when a packet is receiving service. At time n_t, let X_n and J_n be the number of packets in the system and the phase of the inter-arrival time. Consider this system as a DTMC with the state space $\{(X_n, J_n), n \ge 0\}, X_n \ge 0, J_n = 1, 2, \cdots, n_t$. This is a QBD with the transition matrix P given as

$$P = \begin{bmatrix} B & C & & & \\ E & A_1 & A_0 & & \\ & A_2 & A_1 & A_0 & \\ & & \ddots & \ddots & \ddots \end{bmatrix},$$

$$B = T,\; E = bT,\; C = \mathbf{t}\boldsymbol{\alpha},\; A_0 = (1-b)(\mathbf{t}\boldsymbol{\alpha}),$$

$$A_1 = b(\mathbf{t}\boldsymbol{\alpha}) + (1-b)T,\; A_2 = bT.$$

We can go ahead and use the results for the QBD to study this system. However, this system also has an added feature that we can exploit. Because the matrix A_0 is rank one, i.e. $(1-b)\mathbf{t}\boldsymbol{\alpha}$, the matrix R can be written explicitly as

$$R = A_0(I - A_1 - \eta A_2)^{-1}, \tag{5.116}$$

where $\eta = sp(R)$, i.e. the spectral radius of matrix R. This is based on Theorem 8.5.2 in Latouche and Ramaswami (1999). Even though this is an explicit expression we still need to obtain η. So in a sense, it is not quite an explicit expression. The term η can be obtained by solving for z in the scalar equation

$$z = \boldsymbol{\alpha}(I - A_1 - zA_2)^{-1}\mathbf{1}. \tag{5.117}$$

A detailed discussion of this can be found in Latouche and Ramaswami (1999).

Let $x_{i,j} = Pr\{X_n = i, J_n = j\}|_{n\to\infty}$ be the stationary distribution of this system. Further define $\boldsymbol{x} = [\boldsymbol{x}_0, \boldsymbol{x}_1, \boldsymbol{x}_2, \cdots,]$, where $\boldsymbol{x}_i = [x_{i,1}, x_{i,2}, \cdots, x_{i,n_t}]$, we have

$$\boldsymbol{x} = \boldsymbol{x}P, \ \boldsymbol{x}\mathbf{1} = 1,$$

and

$$\boldsymbol{x}_{i+1} = \boldsymbol{x}_i R, \ i \geq 1,$$

with the boundary equations

$$\boldsymbol{x}_0 = \boldsymbol{x}_0 B + \boldsymbol{x}_1 E, \ \boldsymbol{x}_1 = \boldsymbol{x}_0 C + \boldsymbol{x}_1 (A_1 + RA_2),$$

and normalization equation

$$[\boldsymbol{x}_0 + \boldsymbol{x}_1 (I - R)^{-1}]\mathbf{1} = 1,$$

we obtain the stationary distribution for this system.

The mean number in the system $E[X] = \boldsymbol{x}_1 (I-R)^{-2}\mathbf{1}$.

5.9.1 Waiting Times

To study the waiting time we first need to obtain the distribution of the number in the system as seen by an arriving packet. For a packet arriving at time n let $\tilde{X}_n$ be the number of packets it finds in the system and let J_n be the phase of the next packet arrival. Define $y_{i,j} = Pr\{\tilde{X}_n = i, J_n = j\}|_{n\to\infty}, \ i \geq 0, 1 \leq j \leq n_t$. Let $\mathbf{y}_i = [y_{i,1}, y_{i,2}, \cdots, y_{i,n_t}]$, then we have

$$\mathbf{y}_0 = \sigma[\boldsymbol{x}_0(\mathbf{t}\boldsymbol{\alpha}) + \boldsymbol{x}_1(\mathbf{t}\boldsymbol{\alpha})b], \tag{5.118}$$

$$\mathbf{y}_i = \sigma[\boldsymbol{x}_i(\mathbf{t}\boldsymbol{\alpha})\bar{b} + \boldsymbol{x}_{i+1}(\mathbf{t}\boldsymbol{\alpha})b], \ i \geq 1. \tag{5.119}$$

The paramter σ is a normalizing constant and is obtained as follows. Since we have $\sum_{i=0}^{\infty} \mathbf{y}_i \mathbf{1} = 1$, then we have

$$\sigma \sum_{i=0}^{\infty} \boldsymbol{x}_i (\mathbf{t}\boldsymbol{\alpha})\mathbf{1} = 1. \tag{5.120}$$

By the definition of $\boldsymbol{x}_i$ it is clear that $\mathbf{c} = \sum_{i=0}^{\infty} \boldsymbol{x}_i$ is a vector representing the phase of the PH renewal process describing the arrival process, i.e. it is equivalent to $\mathbf{c} = \mathbf{c}(T + \mathbf{t}\boldsymbol{\alpha})$ with $\mathbf{c}\mathbf{1} = 1$. Hence

$$\sigma \sum_{i=0}^{\infty} \boldsymbol{x}_i(\mathbf{t}\boldsymbol{\alpha})\mathbf{1} = \sigma \mathbf{c}(\mathbf{t}\boldsymbol{\alpha})\mathbf{1} = \sigma(\mathbf{ct}) = 1. \tag{5.121}$$

Hence we have

$$\sigma = \lambda^{-1}, \tag{5.122}$$

where λ is the mean arrival rate of the packets, with $\lambda^{-1} = \boldsymbol{\alpha}(I-T)^{-1}\mathbf{1}$ which was shown in Alfa Alfa (2004) that $\mathbf{ct} = \lambda$.

Let the waiting time in the queue be W_q with $w_i^{(q)} = Pr\{W_q = i\},\ i \geq 0$. Then we have

$$w_0^{(q)} = \mathbf{y}_0\mathbf{1}, \tag{5.123}$$

and

$$w_i^{(q)} = \sum_{v=1}^{i} \mathbf{y}_v \mathbf{1} \binom{i-1}{v-1} b^v (1-b)^{i-v}, \quad i \geq 1. \tag{5.124}$$

5.10 PH/PH/1 Queues

Both the Geo/PH/1 and PH/Geo/1 queues are special cases of the PH/PH/1 queue. However, the PH/PH/1 is richer than either one of them in that it has more applications in many fields, especially telecommunications. In addition, it can be analyzed in several different ways depending on the structures of the underlying PH distributions.

Consider a single server system with phase type arrivals characterized by $(\boldsymbol{\alpha},\ T)$ of dimension n_t and service times of phase type distribution characterized by $(\boldsymbol{\beta},\ S)$ of dimension n_s. The mean arrival rate $\lambda^{-1} = \alpha(I-T)^{-1}\mathbf{1}$ and the mean service rate $\mu^{-1} = \beta(I-S)^{-1}\mathbf{1}$. We study the DTMC $\{(X_n, J_n, K_n)\}, n \geq 0$, where at time n, X_n, J_n and K_n represent the number of items in the system, the phase of service for the item in service, and the phase of arrival, respectively. The state space can be arranged as $\{(0,k) \cup (i,j,k),\ i \geq 1, j = 1,2,\cdots,n_s; k = 1,2,\cdots,n_t\}$. The transition matrix for this DTMC is a QBD written as

$$P = \begin{bmatrix} B & C & & & \\ E & A_1 & A_0 & & \\ & A_2 & A_1 & A_0 & \\ & & A_2 & A_1 & A_0 \\ & & & \ddots & \ddots & \ddots \end{bmatrix}, \tag{5.125}$$

where the matrices $B, C, E, A_v, v = 0,1,2$ have dimensions $n_t \times n_t,\ n_t \times n_s n_t,\ n_s n_t \times n_t,\ n_s n_t \times n_s n_t$, respectively. These matrices are given as

$$B = T,\ E = T \otimes \mathbf{s},\ C = (\mathbf{t}\boldsymbol{\alpha}) \otimes \boldsymbol{\beta}$$
$$A_0 = (\mathbf{t}\boldsymbol{\alpha}) \otimes S,\ A_1 = (\mathbf{t}\boldsymbol{\alpha}) \otimes (\mathbf{s}\boldsymbol{\beta}) + T \otimes S \text{ and } A_2 = T \otimes (\mathbf{s}\boldsymbol{\beta}).$$

As a reminder the operator $\otimes$ is the Kronecker product such that for two matrices U of dimension $n_a \times n_b$ and V of dimension $n_c \times n_d$ we obtain a matrix $W = U \otimes V$ of dimension $n_a n_c \times n_b n_d$ as

$$W = \begin{bmatrix} U_{1,1}V & U_{1,2}V & \cdots & U_{1,n_b}V \\ \vdots & \vdots & \vdots & \vdots \\ U_{n_a,1}V & U_{n,2}V & \cdots & U_{n_a,n_b}V \end{bmatrix}.$$

Observation: We notice that if we replace $(\boldsymbol{\alpha}, T)$ with $(1, \bar{a})$ and $(\boldsymbol{\beta}, S)$ with $(1, \bar{b})$, we actually have the Markov chain of the Geo/Geo/1 system. It is easy to see that both the GI/Geo/1 and Geo/G/1 are also special cases of the PH/PH/1 if we let the general inter-arrival times for GI/Geo/1 and the general service times for the Geo/G/1 to have finite supports. In summary, the PH/PH/1 system is a generalization of several well-known single server queues, in discrete time. We will also show later that the GI/G/1 is also a special case of the PH/PH/1 system. So the PH/PH/1 system is a very important discrete-time single server queueing model.

We now want to solve for $\boldsymbol{x}$ where

$$\boldsymbol{x} = \boldsymbol{x}P,\ \boldsymbol{x}\mathbf{1} = 1,$$

and

$$\boldsymbol{x} = [\boldsymbol{x}_0,\ \boldsymbol{x}_1,\ \cdots,\ ],\ \boldsymbol{x}_i = [\boldsymbol{x}_{i,1},\ \boldsymbol{x}_{i,2},\ \cdots,\ \boldsymbol{x}_{i,n_s}],\ i \geq 1,$$

$$\boldsymbol{x}_{i,j} = [x_{i,j,1},\ x_{i,j,2},\ \cdots,\ x_{i,j,n_t}],\ \boldsymbol{x}_0 = [x_{0,1},\ x_{0,2},\ \cdots,\ x_{0,n_t}].$$

We know that there is an R matrix which is the minimal nonnegative solution to the matrix quadratic equation

$$R = A_0 + RA_1 + R^2 A_2.$$

After R is known then we have

$$\boldsymbol{x}_{i+1} = \boldsymbol{x}_i R,\ i \geq 1.$$

For most practical problems we need to compute R using one of the efficient techniques discussed in Chapter 4. However, we could try and exploit a structure of the matrices, if they have useful structures. The key ones that often arise in the case of the PH/PH/1 queue is that of some zero rows for A_0 or some zero columns for A_2. As we know, for any zero rows in A_0 the corresponding rows in R are zero. Similarly for any zero columns in A_2 the corresponding columns in G are zero.

The next step is to compute the boundary values $[\boldsymbol{x}_0, \boldsymbol{x}_1]$, which are based on

$$[\boldsymbol{x}_0, \boldsymbol{x}_1] = [\boldsymbol{x}_0, \boldsymbol{x}_1] \begin{bmatrix} B & C \\ E & A_1 + RA_2 \end{bmatrix}, \; [\boldsymbol{x}_0, \boldsymbol{x}_1]\mathbf{1} = 1.$$

This is then normalized as

$$\boldsymbol{x}_0\mathbf{1} + \boldsymbol{x}_1(I-R)^{-1}\mathbf{1} = 1.$$

Several desired performance measures can then be computed accordingly.

5.10.1 Examples

5.10.1.1 A Numerical Example

Consider the case where

$$\boldsymbol{\alpha} = [1,\, 0,\, 0],\; T = \begin{bmatrix} .2 & .8 & 0 \\ 0 & .3 & .7 \\ 0 & 0 & 0.1 \end{bmatrix},\; \boldsymbol{\beta} = [.25,\; .75],\; S = \begin{bmatrix} .8 & 0 \\ 0 & .4 \end{bmatrix}.$$

First we check the stability condition. We have

$$\lambda^{-1} = \boldsymbol{\alpha}(I-T)^{-1}\mathbf{1} = 1/.2639,\; \mu^{-1} = \boldsymbol{\beta}(I-S)^{-1}\mathbf{1} = 1/.4,$$

We obtain $\lambda/\mu = 0.65975 < 1$. Hence the system is stable.

We calculate

$$A_0 = (\mathbf{t}\boldsymbol{\alpha}) \otimes S = \begin{bmatrix} 0 & 0 & 0 & 0 & 0 & 0 \\ 0 & 0 & 0 & 0 & 0 & 0 \\ 0 & 0 & 0 & 0 & 0 & 0 \\ 0 & 0 & 0 & 0 & 0 & 0 \\ .72 & 0 & 0 & 0 & 0 & 0 \\ 0 & .36 & 0 & 0 & 0 & 0 \end{bmatrix}$$

$$A_1 = T \otimes S + (\mathbf{t}\alpha) \otimes \mathbf{s}\beta = \begin{bmatrix} .16 & 0 & .64 & 0 & 0 & 0 \\ 0 & .08 & 0 & .32 & 0 & 0 \\ 0 & 0 & .24 & 0 & .56 & 0 \\ 0 & 0 & 0 & .12 & 0 & .28 \\ .045 & .135 & 0 & 0 & .08 & 0 \\ .135 & .405 & 0 & 0 & 0 & .04 \end{bmatrix}$$

$$A_2 = T \otimes (\mathbf{s}\beta) = \begin{bmatrix} .01 & .03 & .04 & .12 & 0 & 0 \\ .03 & .09 & .12 & .36 & 0 & 0 \\ 0 & 0 & .015 & .045 & .035 & .105 \\ 0 & 0 & .045 & .135 & .105 & .315 \\ 0 & 0 & 0 & 0 & .005 & .015 \\ 0 & 0 & 0 & 0 & .015 & .045 \end{bmatrix}.$$

For this example we have

$$R = \begin{bmatrix} 0 & 0 & 0 & 0 & 0 & 0 \\ 0 & 0 & 0 & 0 & 0 & 0 \\ 0 & 0 & 0 & 0 & 0 & 0 \\ 0 & 0 & 0 & 0 & 0 & 0 \\ .9416 & .2313 & .8829 & .3173 & .5874 & .2363 \\ .0103 & .4195 & .0141 & .1666 & .0105 & .0541 \end{bmatrix}.$$

If we write the matrix blocks A_v, $v = 0,1,2$ in smaller blocks such that $A_{i,j}^{(v)}$, $i,j = 1,2,3$ is the $(i,j)^{th}$ block of the matrix A_v we find that

$$A_0 = \begin{bmatrix} 0 & 0 & 0 \\ 0 & 0 & 0 \\ A_{3,1}^{(0)} & 0 & 0 \end{bmatrix}, A_1 = \begin{bmatrix} A_{1,1}^{(1)} & A_{1,2}^{(1)} & 0 \\ 0 & A_{2,2}^{(1)} & A_{2,3}^{(1)} \\ A_{3,1}^{(1)} & 0 & A_{3,3}^{(1)} \end{bmatrix}, A_2 = \begin{bmatrix} A_{1,1}^{(2)} & A_{1,2}^{(2)} & 0 \\ 0 & A_{2,2}^{(2)} & A_{2,3}^{(2)} \\ 0 & 0 & A_{3,3}^{(2)} \end{bmatrix}.$$

With the structure A_0 we know that our R matrix is of the form $R = \begin{bmatrix} 0 & 0 & 0 \\ 0 & 0 & 0 \\ R_{3,1} & R_{3,2} & R_{3,3} \end{bmatrix}$. This is one example where applying the linear approach for computing R may be beneficial because we only need to compute the three blocks of R, and this could reduce computational efforts compared to the quadratic algorithms in some instances. In this particular case we have

$$R_{3,1} = A_{3,1}^{(0)} + R_{3,1}A_{1,1}^{(1)} + R_{3,3}A_{3,1}^{(1)} + R_{3,3}R_{3,1}A_{1,1}^{(2)}, \tag{5.126}$$

$$R_{3,2} = R_{3,1}A_{1,2}^{(1)} + R_{3,2}A_{2,2}^{(1)} + R_{3,3}R_{3,1}A_{1,2}^{(2)} + R_{3,3}R_{3,2}A_{2,2}^{(2)}, \tag{5.127}$$

$$R_{3,3} = R_{3,2}A_{2,3}^{(1)} + R_{3,3}A_{3,3}^{(1)} + R_{3,3}R_{3,2}A_{2,3}^{(2)} + R_{3,3}R_{3,3}A_{3,3}^{(2)}. \tag{5.128}$$

Writing these equations simply as $R_{3,i} = f_i(R, A_0, A_1, A_2), i = 1,2,3,$ we can now write an iterative process given as

$$R_{3,i}(k+1) = f_i(R(k), A_0, A_1, A_2). \tag{5.129}$$

What we now do is set $R_{3,i}(0) := 0$ and then apply an iterative process until $|R_{3,i}(k+1) - R_{3,i}(k)|_{u,v} < \varepsilon,\ (u,v) = 1,2,3,\ \forall i,$ where ε is a very small number, usually 10^{-12} is acceptable.

Sometimes we find that by re-arranging the state space we can find useful structures.

5.10.1.2 Another Example

Consider a system with Negative binomial arrival process which can be represented by a phase type distribution (α, T) with

$$\alpha = [1,\ 0,\ 0],\ \ T = \begin{bmatrix} 1-a & a & 0 \\ 0 & 1-a & a \\ 0 & 0 & 1-a \end{bmatrix},$$

and Negative binomial service process which is represented by (β, S) with

$$\beta = [1,\ 0],\ \ S = \begin{bmatrix} 1-b & b \\ 0 & 1-b \end{bmatrix}.$$

Here we can see that there are different advantages with arranging our DTMC as i) $\{(X_n, J_n, K_n)\}, n \geq 0$, or as ii) $\{(X_n, K_n, J_n)\}, n \geq 0$. In the first arrangement we have

$$A_0 = S \otimes (\mathbf{t}\alpha),\ \text{ and }\ A_2 = (\mathbf{s}\beta) \otimes T.$$

With this arrangement, our A_2 has only one non-zero column and one zero-column. In fact, if our Negative binomial for service was of higher dimension $n_s > 2$ we would have $n_s - 1$ zero columns. In this case we are better to study the G matrix because this matrix has equivalent zero columns. We can then obtain matrix R from matrix G by using the formula of Equation (4.83).

If however, we use the second arrangement then we have

$$A_0 = (\mathbf{t}\alpha) \otimes S,\ \text{ and }\ A_2 = T \otimes (\mathbf{s}\beta).$$

With this arrangement, our A_0 has two zero rows. Again had our Negative binomial for arrival process been of order $n_t > 3$ we would have $n_t - 1$ zero rows. This would result in corresponding $n_t - 1$ zero rows of the matrix R. So here we are better to study the matrix R directly.

The standard methods presented are now used to obtain the performance measures of this system after matrix R is obtained.

In some cases we may not have zero rows for R or zero columns for G, but the matrices A_v, $v = 0,1,2$ may be very sparse, thereby making it easier to work in smaller blocks of matrices. This sometimes assists in reducing computational efforts required for computing R and G.

The mean waiting time in the system μ_W is given by Little's Law as

$$\mu_L = \lambda \mu_W. \tag{5.130}$$

5.10.2 Waiting Time Distribution

The waiting time distribution can be studied in at least three different ways. All the three methods will be presented here. It goes without saying that any of the methods may also be used for all the single node queues that have been presented so far since they are all special cases of this PH/PH/1 system.

In order to obtain the waiting time distribution for a packet, irrespective of the method used, we first have to determine the state of the system as seen by an arriving packet. Let us define $\mathbf{y}$ as the steady state vector describing the state of the system as seen by an arriving customer, with $\mathbf{y} = [\mathbf{y}_0, \mathbf{y}_1, \ldots]$. Then we have

$$\mathbf{y}_0 = \lambda^{-1} [x_0(\mathbf{t}\boldsymbol{\alpha}) + x_1(\mathbf{t}\boldsymbol{\alpha} \otimes \mathbf{s})], \tag{5.131}$$

$$\mathbf{y}_i = \lambda^{-1} [x_i(\mathbf{t}\boldsymbol{\alpha}) \otimes S) + x_{i+1}(\mathbf{t}\boldsymbol{\alpha} \otimes \mathbf{s}\boldsymbol{\beta})],\ i \geq 1. \tag{5.132}$$

By letting $F = \lambda^{-1}[(\mathbf{t}\boldsymbol{\alpha}) \otimes S + R((\mathbf{t}\boldsymbol{\alpha}) \otimes (\mathbf{s}\boldsymbol{\beta}))]$, we can write

$$\mathbf{y}_i = \boldsymbol{x}_1 R^{i-1} F,\ i \geq 1.$$

Let W_q be the waiting time in the queue, with $w_j^{(q)} = Pr\{W_q = j\}$ and $\tilde{W}_j^{(q)} = Pr\{W_q \geq j\}$.

<u>**Method One:**</u>

Further using $B_r^{(k)}$ as defined in Section 5.8.1 we have

$$w_0^{(q)} = \mathbf{y}_0 \mathbf{1} \tag{5.133}$$

$$w_i^{(q)} = \sum_{k=1}^{i} y_k (\mathbf{1} \otimes I) B_i^{(k)} \mathbf{1},\ i \geq 1. \tag{5.134}$$

Method Two:

Another approach that can be used for studying the waiting time is as follows. An arriving packet will wait for at least k units of time if at its arrival it finds $i \geq 1$ packets ahead of it in the system and the server takes no less than $k \geq 1$ units of time to serve these i packets, i.e. no more than $i-1$ service completions in the interval $(0, k-1)$. Let $U(i,n)$ be the probability that the number of packets served in the interval $(0,n)$ is i. Then we have

$$\tilde{W}_k^{(q)} = \sum_{i=1}^{\infty} \mathbf{y}_i (\mathbf{1}_{n_t} \otimes I_{n_s}) [\sum_{j=0}^{i-1} U(j, k-1)] \mathbf{1}$$

$$= \sum_{i=1}^{\infty} \boldsymbol{x}_1 R^{i-1} F(\mathbf{1}_{n_t} \otimes I_{n_s}) [\sum_{j=0}^{i-1} U(j, k-1)] \mathbf{1},$$

which results in

$$\tilde{W}_k^{(q)} = \boldsymbol{x}_1 (I-R)^{i-1} [\sum_{j=0}^{k-1} R^j F(\mathbf{1}_{n_t} \otimes I_{n_s}) U(j, k-1)] \mathbf{1}. \tag{5.135}$$

Method Three:

A third approach uses the absorbing Markov chain idea. First we let H be the phase of arrival after the target packet has arrived, and define $W_{i,j}^{(q)} = Pr\{W_q \leq i, H = j\},\ i \geq 0,\ 1 \leq j \leq n_t$. Define a transition matrix $\hat{P}$ of an absorbing Markov chain as

$$\hat{P} = \begin{bmatrix} I & & & & \\ I \otimes \mathbf{s} & I \otimes S & & & \\ & I \otimes (\mathbf{s}\boldsymbol{\beta}) & I \otimes S & & \\ & & I \otimes (\mathbf{s}\boldsymbol{\beta}) & I \otimes S & \\ & & & \ddots & \ddots \end{bmatrix}. \tag{5.136}$$

Define $\mathbf{z}^{(0)} = \mathbf{y}$, with $\mathbf{z}_i^{(0)} = \mathbf{y}_i$. Further write $\mathbf{W}^{(q)} = [W_1^{(q)},\ W_2^{(q)},\ \cdots]$ and $W(q)_i = [W_{i,1}^{(q)},\ W_{i,2}^{(q)},\ \cdots,\ W_{i,n_t}^{(q)}]$, then we have

$$\mathbf{z}^{(i)} = \mathbf{z}^{(i-1)} \hat{P} = \mathbf{z}^{(0)} \hat{P}^i,\ i \geq 1, \tag{5.137}$$

with

$$W_i^{(q)} = \mathbf{z}_0^{(i)}, \tag{5.138}$$

which is the waiting time probability vector that also captures the phase of arrival.

The busy period of the PH/PH/1 queue can be obtained as a special case of the busy period of the MAP/PH/1 queue as shown in Frigui and Alfa (1995).

5.10.2.1 Workload

Consider the DTMC $\{(V_n, K_n), n \geq 0; V_n \geq 0\}, K_n = 1, 2, \cdots, n_t$, where at time n, V_n is the workload in the system and K_n is the phase of the inter-arrival time. Let the probability transition matrix of this DTMC be P. Then P is given as

$$P = \begin{bmatrix} C_0 & C_1 & C_2 & C_3 & \cdots \\ C_0 & C_1 & C_2 & C_3 & \cdots \\ & C_0 & C_1 & C_2 & \cdots \\ & & C_0 & C_1 & \cdots \\ & & & \ddots & \ddots \end{bmatrix},$$

where

$$C_0 = T \otimes I, C_k = (\mathbf{t}\beta) \otimes S^{k-1}\mathbf{s}, \ k \geq 1.$$

This workload model has the same structure as the M/G/1 type Markov chain. In addition the boundary has a very special structure similar to that of the workload for the Geo/G/1 system as expected. So we can go ahead and apply the M/G/1 type results here and also try and capitalize on the associated special structure at the boundary.

5.10.2.2 Age Process

Here we present the age process model which studies the age of the leading packet in a FCFS system. Consider the DTMC represented by $(Y_n, J_n, K_n), n \geq 0; Y_n \geq 0$, $K_n = 1, 2, \cdots, n_t; J_n = 1, 2, \cdots, n_s$. Here Y_n is the age at time n, and K_n and J_n are the phases of arrivals and service, respectively. The associated probability transition matrix P has the following form

$$P = \begin{bmatrix} C_0 & C_1 & & & \\ B_1 & A_1 & A_0 & & \\ B_2 & A_2 & A_1 & A_0 & \\ B_3 & A_3 & A_2 & A_1 & A_0 \\ \vdots & \vdots & \vdots & \vdots & \vdots & \ddots \end{bmatrix},$$

where

$$C_0 = T, \ C_1 = (\mathbf{t}\alpha) \otimes \beta, \ B_k = \mathbf{s}T^k, k \geq 1,$$

$$A_0 = S \otimes I, \ A_j = (\mathbf{s}\beta) \otimes T^{j-1}(\mathbf{t}\alpha), \ j \geq 1.$$

It is immediately clear that this has the GI/M/1 structure and the results from that class of Markov chains can be used to analyze this system. If the system is stable then from the steady state results of the DTMC one can obtain several performance measures such as the waiting time distribution and the distribution of the number in the system. For a detailed treatment of such systems see van Houdt and Blondia (2002).

5.10.3 PH/PH/1 Queues at points of events

We study this system at epochs of events. Epochs of events refers to the epochs of an arrival, a departure, or of joint arrival and departure. The idea was first studied by Latouche and Ramaswami (1997) for the continuous case of the PH/PH/1. By studying it at points of events we end up cutting down on the state space requirements, and hence the computational effort required at the iteration stage, even though the front-end work does increase substantially.

Let $\tau_j,\ j \geq 0$ be a collection of the epochs of events. Consider the process $\{(N_j, I_j, J_j), j \geq 0\}$, where $N_j = N(\tau_j+)$ is the number in the system just after the jth epoch, I_j the indicator equal to $+$ if τ_j is an arrival, $-$ if τ_j is a departure, and $*$ if τ_j is a joint arrival and departure. It is immediately clear that if τ_j is an arrival then J_j is the phase of service in progress at time τ_j and similarly if τ_j is a departure epoch then J_j is the phase of the inter-arrival time. Of course, when τ_j is an epoch of joint arrival and departure then both service (in case of a customer in the system) and arrival processes are both initialized. The process $\{(N_j, I_j, J_j), j \geq 0\}$ is an embedded Markov chain with its transition matrix P given as

$$P = \begin{bmatrix} \mathbf{0} & C & & & \\ E & A_1 & A_0 & & \\ & A_2 & A_1 & A_0 & \\ & & A_2 & A_1 & A_0 \\ & & & \ddots & \ddots & \ddots \end{bmatrix}, \tag{5.139}$$

where

$$A_0 = \begin{bmatrix} A_0^{++} & \mathbf{0} & \mathbf{0} \\ A_0^{*+} & \mathbf{0} & \mathbf{0} \\ A_0^{-+} & \mathbf{0} & \mathbf{0} \end{bmatrix}, \quad A_2 = \begin{bmatrix} \mathbf{0} & \mathbf{0} & A_2^{+-} \\ \mathbf{0} & \mathbf{0} & A_2^{*-} \\ \mathbf{0} & \mathbf{0} & A_2^{--} \end{bmatrix},$$

$$A_1 = \begin{bmatrix} \mathbf{0} & A_1^{+*} & \mathbf{0} \\ \mathbf{0} & A_1^{**} & \mathbf{0} \\ \mathbf{0} & A_1^{-*} & \mathbf{0} \end{bmatrix}, \quad E = \begin{bmatrix} A_2^{+-} \\ A_2^{*-} \\ A_2^{--} \end{bmatrix}, \quad C = \begin{bmatrix} \mathbf{1}\boldsymbol{\beta} & \mathbf{0} & \mathbf{0} \end{bmatrix}.$$

All the matrices A_i, $i = 0,1,2$ are of order $(n_s + n_t + 1) \times (n_s + n_t + 1)$, matrix E is of order $(n_s + n_t + 1) \times n_t$, and matrix C is of order $n_t \times (n_s + n_t + 1)$. The block elements are given as follows:

- $A_0^{++} = \sum_{k=0}^{min(n_s,n_t)} (\alpha T^k \mathbf{t}) S^{k+1} = \sum_{k=0}^{min(n_s,n_t)} a_{k+1} S^{k+1}$, of order $n_s \times n_s$
- $A_0^{*+} = \sum_{k=0}^{min(n_s,n_t)} (\alpha T^k \mathbf{t}) \beta S^{k+1} = \sum_{k=0}^{min(n_s,n_t)} a_{k+1} \beta S^{k+1}$, of order $1 \times n_s$
- $A_0^{-+} = \sum_{k=0}^{min(n_s,n_t)} (T^k \mathbf{t})(\beta S^{k+1})$, of order $n_t \times n_s$
- $A_2^{+-} = \sum_{k=0}^{min(n_s,n_t)} (S^k \mathbf{s})(\alpha T^{k+1})$, of order $n_s \times n_t$
- $A_2^{*-} = \sum_{k=0}^{min(n_s,n_t)} (\beta S^k \mathbf{s}) \alpha T^{k+1} = \sum_{k=0}^{min(n_s,n_t)} s_{k+1} \alpha T^{k+1}$, of order $1 \times n_t$
- $A_2^{--} = \sum_{k=0}^{min(n_s,n_t)} (\beta S^k \mathbf{s}) T^{k+1} = \sum_{k=0}^{min(m,n)} s_{k+1} T^{k+1}$, of order $n_t \times n_t$
- $A_1^{+*} = \sum_{k=0}^{min(n_s,n_t)} (\alpha T^k \mathbf{t}) S^k \mathbf{s} = \sum_{k=0}^{min(n_s,n_t)} a_{k+1} S^k \mathbf{s}$, of order $n_s \times 1$
- $A_1^{**} = \sum_{k=0}^{min(n_s,n_t)} (\alpha T^k \mathbf{t})(\beta S^k \mathbf{s}) = \sum_{k=0}^{min(n_s,n_t)} a_{k+1} s_{k+1}$, of order 1×1
- $A_1^{-*} = \sum_{k=0}^{min(n_s,n_t)} (\beta S^k \mathbf{s}) T^k \mathbf{t} = \sum_{k=0}^{min(n_s,n_t)} s_{k+1} (T^k \mathbf{t})$, of order $n_t \times 1$

We later show how to simplify these block matrices for computational purposes. The type of simplification approach adopted depends on whether one uses the remaining time or elapsed time approach of representation for the inter-arrival and service processes.

The stationary vector $\boldsymbol{x}$ is given as

$$\boldsymbol{x} = \boldsymbol{x} P, \quad \boldsymbol{x}\mathbf{1} = 1,$$

with

$$\boldsymbol{x} = [\boldsymbol{x}_0, \boldsymbol{x}_1, \boldsymbol{x}_2, \cdots], \ \boldsymbol{x}_i = [\boldsymbol{x}_i^+, \boldsymbol{x}_i^*, \boldsymbol{x}_i^-], \ i \geq 1, \text{ and } \boldsymbol{x}_0 = \boldsymbol{x}_0^-.$$

The matrix R and the stationary distributions are given as

$$R = A_0 + RA_1 + R^2 A_2, \quad \text{and} \quad \boldsymbol{x}_{i+1} = \boldsymbol{x}_i R, \ i \geq 1.$$

We also know that there is a matrix G which is the minimal non-negative solution to the matrix quadratic equation

$$G = A_2 + A_1 G + A_0 G^2, \tag{5.140}$$

and the relationship between R and G is such that

$$R = A_0 (I - A_1 - A_0 G)^{-1}. \tag{5.141}$$

Based on the structure of A_2 we know that the matrix G is of the form

$$G = \begin{bmatrix} \mathbf{0} & \mathbf{0} & G^{+-} \\ \mathbf{0} & 0 & G^{*-} \\ \mathbf{0} & \mathbf{0} & G^{--} \end{bmatrix},$$

where

$$G^{+-} = A_2^{+-} + A_1^{+*}G^{*-} + A_0^{++}G^{+-}G^{--}, \tag{5.142}$$

$$G^{*-} = A_2^{*-} + A_1^{**}G^{*-} + A_0^{*+}G^{+-}G^{--}, \tag{5.143}$$

and

$$G^{--} = A_2^{--} + A_1^{-*}G^{*-} + A_0^{-+}G^{+-}G^{--}. \tag{5.144}$$

We write $\hat{G} = G^{+-}G^{--}$, and $H = (1 - A_1^{**})^{-1}(A_2^{*-} + A_0^{*+}\hat{G})$. By manipulating Equations (5.142) to (5.144) we have

$$\hat{G} = A_2^{+-}(A_2^{--} + A_1^{-*}H + A_1^{-+}\hat{G}) + (A_1^{+*}H + A_0^{++}\hat{G})(A_2^{--} + A_1^{-*}H + A_0^{-+}\hat{G}). \tag{5.145}$$

This is a matrix equation which can be applied iteratively to compute $\hat{G}$ which is of order $m \times n$. After computing $\hat{G}$ we can then obtain the block matrices of G as follows:

$$G^{*-} = H. \tag{5.146}$$

$$G^{--} = A_2^{--} + A_1^{-*}H + A_0^{-+}\hat{G}. \tag{5.147}$$

$$G^{+-} = A_2^{+-} + A_1^{+*}H + A_0^{++}\hat{G}. \tag{5.148}$$

So it might be easier to compute the matrix G first and then compute R from it because computing R from Equation (5.141) only requires us to know the first block of row of the inverse matrix as a result of the structure of matrix A_0 which has only the block column that is non-zero.

Let us denote $D = (I - A_1 - A_0G)$. It is clear that the structure of D is of the form

$$D = \begin{bmatrix} D_{11} & D_{12} & D_{13} \\ \mathbf{0} & D_{22} & D_{23} \\ \mathbf{0} & D_{32} & D_{33} \end{bmatrix}. \tag{5.149}$$

It's inverse F such that $FD = I$ can be written in block form as

$$F = \begin{bmatrix} F_{11} & F_{12} & F_{13} \\ F_{21} & F_{22} & F_{23} \\ F_{31} & F_{32} & F_{33} \end{bmatrix}. \tag{5.150}$$

After simple algebraic calculations we have

$$F_{11} = D_{11}^{-1}, \tag{5.151}$$

$$F_{12} = [D_{11}^{-1} D_{13} D_{33}^{-1} D_{32} - D_{11}^{-1} D_{12}][D_{22} - D_{23} D_{33}^{-1} D_{32}]^{-1}, \tag{5.152}$$

and

$$F_{13} = -(F_{12} D_{23} + D_{11}^{-1} D_{13}) D_{33}^{-1}. \tag{5.153}$$

From here we know that

$$R = \begin{bmatrix} A_0^{++} F_{11} & A_0^{++} F_{12} & A_0^{++} F_{13} \\ A_0^{*+} F_{11} & A_0^{*+} F_{12} & A_0^{*+} F_{13} \\ A_0^{-+} F_{11} & A_0^{-+} F_{12} & A_0^{-+} F_{13} \end{bmatrix}. \tag{5.154}$$

So the computation of the matrix R involves iterative computations of $n_s \times n_t$ matrix $\hat{G}$ from which the three block matrices G^{+-}, G^{*-} and G^{--} are directly computed, and of partial computation of the inverse matrix F, i.e. we only compute the first block row of this matrix. Computing the R matrix is thus reasonably efficient.

Consider the U matrix associated with this Markov chain, with $U = A_1 + A_0 G = A_1 + RA_2$, we have U given as

$$U = \begin{bmatrix} \mathbf{0} & U^{+*} & U^{+-} \\ \mathbf{0} & U^{**} & U^{*-} \\ \mathbf{0} & U^{-*} & U^{--} \end{bmatrix}. \tag{5.155}$$

Then we have

$$U^{+*} = A^{+*},\ U^{**} = A^{**},\ U^{-*} = A^{-*} \tag{5.156}$$

and

$$U^{+-} = A^{++} G^{+-},\ U^{*+} = A^{*+} G^{*-},\ U^{--} = A^{-+} G^{--}. \tag{5.157}$$

It is clear that because of the structures of T and S we know that $T^j = 0,\ \forall j \geq n$ and $S^i = 0,\ \forall i \geq m$. Let us assume that for both the inter-arrival and service times we adopt the remaining time approach for representation. Let us define the following:

- $I(k,j)$ as a $k \times k$ matrix with zero elements and 1 in locations $(j,1)$, $(j+1,2), \cdots, (k, k-j+1)$. It is immediately clear that the identity matrix I_k of order k can be written as $I_k = I(k,0)$. The expression $S^k = I(n_s, k+1)$ and $T^k = I(n_t, k+1)$. Also $I(k,j)I(k,1) = I(k,j+1)$.
- $b_i = \sum_{j=1}^{min(n_t, n_s - i)} a_j s_{i+j}$.
- $d_i = \sum_{j=1}^{min(n_s, n_t - i)} s_j a_{i+j}$.

Then we have

1.

$$A_0^{++} = \sum_{k=0}^{min(n_s,n_t)} a_{k+1} I(n_s, k+1) = \begin{bmatrix} 0 & 0 & 0 & \cdots & 0 \\ a_1 & 0 & 0 & \cdots & 0 \\ a_2 & a_1 & 0 & \cdots & 0 \\ a_3 & a_2 & a_1 & \cdots & 0 \\ \vdots & \ddots & \ddots & \ddots & \vdots \end{bmatrix}_{n_s \times n_s}$$

2.

$$A_0^{*+} = \sum_{k=0}^{min(n_s,n_t)} a_k \beta S^{k+1} = [b_1,\ b_2,\ \cdots,\ b_{n_s}]$$

3.

$$A_0^{-+} = \sum_{k=0}^{min(n_s,n_t)} (T^k \mathbf{t})(\beta S^{k+1})$$

$$= \begin{bmatrix} s_2 & s_3 & s_4 & \cdots & s_{n_s-2} & s_{n_s-1} & s_{n_s} & 0 \\ s_3 & s_4 & s_5 & \cdots & s_{n_s-1} & s_{n_s} & 0 & 0 \\ s_4 & s_5 & s_6 & \cdots & s_{n_s} & 0 & 0 & \\ \vdots & \vdots & \vdots & \vdots & \vdots & \vdots & \vdots & \vdots \end{bmatrix}_{n_t \times n_s}$$

4.

$$A_2^{+-} = \sum_{k=0}^{min(n_s,n_t)} (S^k \mathbf{s})(\alpha T^{k+1})$$

$$= \begin{bmatrix} a_2 & a_3 & a_4 & \cdots & a_{n-2} & a_{n-1} & a_n & 0 \\ a_3 & a_4 & a_5 & \cdots & a_{n-1} & a_n & 0 & 0 \\ a_4 & a_5 & a_6 & \cdots & a_n & 0 & 0 & \\ \vdots & \vdots & \vdots & \vdots & \vdots & \vdots & \vdots & \vdots \end{bmatrix}_{n_s \times n_t}$$

5.

$$A_2^{*-} = \sum_{k=0}^{min(n_s,n_t)} s_{k+1}(\alpha T^{k+1}) = [d_1,\ d_2,\ \cdots,\ d_{n_t}]$$

6.

$$A_2^{--} = \sum_{k=0}^{min(n_s,n_t)} s_{k+1} I(n_t, k+1) = \begin{bmatrix} 0 & 0 & 0 & \cdots & 0 \\ s_1 & 0 & 0 & \cdots & 0 \\ s_2 & s_1 & 0 & \cdots & 0 \\ s_3 & s_2 & s_1 & \cdots & 0 \\ \vdots & \ddots & \ddots & \ddots & \vdots \end{bmatrix}_{n_t \times n_t}$$

7.

$$A_1^{+*} = \sum_{k=0}^{min(n_s,n_t)} t_{k+1} S^k \mathbf{s} = \begin{bmatrix} a_1 \\ a_2 \\ \vdots \\ a_{min(n_s,n_t)} \\ 0 \\ \vdots \\ 0 \end{bmatrix}_{n_s \times 1}$$

8.

$$A_1^{**} = \sum_{k=0}^{min(n_s,n_t)} (\alpha T^k \mathbf{t})(\beta S^k \mathbf{s}) = \sum_{k=0}^{min(n_s,n_t)} a_k s_k$$

9.

$$A_1^{-*} = \sum_{k=0}^{min(n_s,n_t)} s_{k+1}(T^k \mathbf{t}) = \begin{bmatrix} s_1 \\ s_2 \\ \vdots \\ s_{min(n_s,n_t)} \\ 0 \\ \vdots \\ 0 \end{bmatrix}_{n_t \times 1}$$

It is clear from Latouche and Ramaswami (1997) that it is straightforward to obtain the stationary distributions at arrivals and also at departures from the information about the system at epoch of events. They also show how to obtain the information at arbitrary times. By applying similar arguments used by Latouche and Ramaswami (1997) we can also derive the stationary distributions at arbitrary times from the results of the case of time of events. It is however not necessary to present those results here since the techniques used will practically be the same. But we point out that if the interest is to obtain the system performance at arbitrary times, then it is more efficient to go ahead and apply the results of Alfa and Xue (2007) directly, rather than try to study the system at epochs of events and then recover the arbitrary times results from it.

5.11 PH/PH/1 System with Start-up Times

Consider a regular PH/PH/1 system with arrival process given as a Phase type with representation $(\boldsymbol{\alpha}, T)$ of dimension n and phase service with representation $(\boldsymbol{\beta}, S)$ of dimension m, with mean $E[\mathbf{S}]$. Suppose when the system starts up from empty the service time of the first item is phase type with representation $(\boldsymbol{\beta}^*, S^*)$ also of dimension m and $E[\mathbf{S}^*]$. We assume that $E[\mathbf{S}^*] \geq E[\mathbf{S}]$. Let $\mathbf{s}^* = \mathbf{1} - S^*\mathbf{1}$. The state space of this DTMC is $\{0, K_t \cup (X_t, J_t, K_t, L_t),\ X_t = 1,2,3,\cdots; J_t = 0,1; K_t = 1,2,\cdots,n; L_t = 1,2,\cdots,m\}$ where $J_t = 0$ if the item in service is the first one in the busy period and 1 otherwise, and K_t and L_t are the phases of arrival and service. Note that we allowed the two phase services to have the same dimension; this is not necessary. The associated transition matrix P can be written as

$$P = \begin{bmatrix} B & C & & & \\ E & A_1 & A_0 & & \\ & A_2 & A_1 & A_0 & \\ & & \ddots & \ddots & \ddots \end{bmatrix},$$

where

$$B = T,\ C = [(\mathbf{t}\boldsymbol{\alpha}) \otimes \boldsymbol{\beta}^*,\ 0],\ E = \begin{bmatrix} T \otimes \mathbf{s}^* \\ T \otimes \mathbf{s} \end{bmatrix},$$

$$A_0 = \begin{bmatrix} (\mathbf{t}\boldsymbol{\alpha}) \otimes S^* & 0 \\ 0 & (\mathbf{t}\boldsymbol{\alpha}) \otimes S \end{bmatrix},$$

$$A_1 = \begin{bmatrix} T \otimes S^* & (\mathbf{t}\boldsymbol{\alpha}) \otimes (\mathbf{s}^*\boldsymbol{\beta}) \\ 0 & T \otimes S + (\mathbf{t}\boldsymbol{\alpha}) \otimes (\mathbf{s}\boldsymbol{\beta}) \end{bmatrix},$$

$$A_2 = \begin{bmatrix} 0 & T \otimes (\mathbf{s}^*\boldsymbol{\beta}) \\ 0 & T \otimes (\mathbf{s}\boldsymbol{\beta}) \end{bmatrix}.$$

Once again we can use MAM to carry out the analysis of this system. All other cases of single server systems can be inferred from this model.

5.12 GI/G/1 Queues

The GI/G/1 is the most basic and the most general of single server queues. However, obtaining an exact solution for its stationary distribution of number in the system has been elusive. We have to turn to numerical approaches. Here we present a very efficient algorithmic approach for studying this system of queues.

We let the arrival process be described as follows. Let the inter-arrival times between two consecutive packets be $\mathscr{A} > 0$, and define $a_k = Pr\{\mathscr{A} = k\}$. Similarly we let the service time of each packet be $\mathscr{S} > 0$ and define $b_k = Pr\{\mathscr{S} = k\}$. We assume the inter-arrival and service times are independent of each other. It was pointed out in Alfa (2004) and earlier in Neuts (1975) that any discrete probability distribution with finite support can be represented as a PH distribution. Alfa (2004) also pointed out that even if the support is infinite we can still represent it by the IPH distribution, which is just a PH distribution with infinite states. Hence GI/G/1 system with finite support for inter-arrival and service times can be modelled as a PH/PH/1 system. It was shown in Section 2.5.5 how to represent the general distribution as a PH distribution. In this section we focus mainly on the case where both the inter-arrival and service times have finite supports.

Consider the case where $a_k = 0,\ \forall k > K_a = n_t < \infty$, and $s_j = 0,\ \forall j > K_s = n_s < \infty$. We further write

$$\mathbf{a} = [a_1,\ a_2,\ \cdots,\ a_{n_t}],\ \ \mathbf{b} = [b_1,\ b_2,\ \cdots,\ b_{n_s}].$$

There are two ways of representing each of these two distributions as PH distributions, one is using the elapsed time (ET) approach and the other is using the remaining time (RT) approach. This implies that we can actually model the GI/G/1 system in four different ways as PH/PH/1 system as follows. We write (X,Y) and let X represent inter-arrival times type and Y service times type, then we have four possible selections from: (ET,ET), (RT,RT), (ET,RT) and (RT,ET). We will present only two of these and let the reader develop the other two. We discuss (RT,RT) and (ET,ET).

5.12.1 The (RT,RT) representation for the GI/G/1 queue

Using the results of Section 2.5.5 we can write the inter-arrival times and service times as RT type of PH distributions as follows. For the inter-arrival time with PH written as (α, T) we have

$$\alpha = [a_1,\ a_2,\ a_3,\ \cdots,\ a_{n_t}],\ T = \begin{bmatrix} \mathbf{0}_{n_t-1} & 0 \\ I_{n_t-1} & \mathbf{0}^T_{n_t-1} \end{bmatrix},$$

where $\mathbf{0}_j$ is a j row of zeros and I_k is an identity matrix of order k. Similarly we can write the PH distribution for the service times as (β, S) with

$$\alpha = [b_1,\ b_2,\ b_3,\ \cdots,\ b_{n_s}],\ S = \begin{bmatrix} \mathbf{0}_{n_s-1} & 0 \\ I_{n_s-1} & \mathbf{0}^T_{n_s-1} \end{bmatrix}.$$

We consider a trivariate Markov chain $\{X_n, K_n, J_n\}$, where at time n, X_n is the number of packets in the system, K_n the remaining time to next arrival and J_n the remaining service time of the packet in service. The DTMC representing the number in the system for this GI/G/1 queue is $\Delta = \{(0,k) \cup (i,k,j)\}; k = 1,2,\cdots,n_t; j = 1,2,\cdots,n_s; i \geq 1$, i.e. we have number in the system, phase of arrival and phase of service – in that order. The transition matrix for this system is a QBD and can be written as

$$P = \begin{bmatrix} B & C & & \\ E & A_1 & A_0 & \\ & A_2 & A_1 & A_0 \\ & & \ddots & \ddots & \ddots \end{bmatrix},$$

with

$$B = T,\ C = (\mathbf{t}\alpha) \otimes \beta,\ E = T \otimes \mathbf{s},\ A_0 = (\mathbf{t}\alpha) \otimes S,$$

$$A_1 = (\mathbf{t}\alpha) \otimes (\mathbf{s}\beta) + T \otimes S,\ A_2 = T \otimes (\mathbf{s} \otimes \beta).$$

It is immediately clear that we can apply the matrix-geometric results to this system, in fact we can use the results from the PH/PH/1 case here directly. The special feature of this model is that we have additional structures which we can exploit regarding the block matrices. The first one is that because of our arrangement of the state space we find that our block matrix A_0 has only one block of rows (the first ones) that is non-zero, as such the structure of our R matrix is

$$R = \begin{bmatrix} R_1 & R_2 & R_3 & \cdots & R_{n_t} \\ \mathbf{0} & \mathbf{0} & \mathbf{0} & \cdots & \mathbf{0} \\ \mathbf{0} & \mathbf{0} & \mathbf{0} & \cdots & \mathbf{0} \\ \vdots & \vdots & \vdots & \cdots & \mathbf{0} \\ \mathbf{0} & \mathbf{0} & \mathbf{0} & \cdots & \mathbf{0} \end{bmatrix}. \quad (5.158)$$

Hence we can apply the standard linear algorithm for computing matrix R so as to work in smaller blocks. It is probably easier to work with than any of the quadratically convergent algorithms in this special case. Based on this we have the following equations

$$R_k = a_k S + R_1 a_k(\mathbf{s}\boldsymbol{\beta}) + R_{k+1} S + R_1 R_{k+1}(\mathbf{s}\boldsymbol{\beta}), \quad 1 \le k \le n_t - 1, \tag{5.159}$$

$$R_{n_t} = a_{n_t} S + R_1 a_{n_t}(\mathbf{s}\boldsymbol{\beta}). \tag{5.160}$$

Let $C_0 = S$ and $C_1 = (\mathbf{s}\boldsymbol{\beta})$, and let us write

$$\mathscr{J} = C_0 \otimes I + C_1 \otimes R_1, \quad \mathscr{H} = vecC_0 + (C_1^T \otimes I) vecR_1, \tag{5.161}$$

then we have

$$vecR_{n_t} = a_{n_t} \mathscr{H}, \tag{5.162}$$

and by back substitution we have

$$vecR_{n_t - k} = \sum_{v=0}^{k} a_{n_t - v} \mathscr{J}^{k-v} \mathscr{H}, \; k = 0, 1, 2, \cdots, n_t - 1. \tag{5.163}$$

All we really need to do is compute R_1 iteratively from

$$vecR_1 = \sum_{v=0}^{n_t - 1} a_{n_t - v} \mathscr{J}^{n_t - v - 1} \mathscr{H}, \tag{5.164}$$

starting with $R_1 := 0$ and the remaining R_k, $k = 2, 3, \cdots, n_t$ are computed directly.

After the matrix R has been computed we can then obtain the boundary equations and all the information about the queue.

As a reminder from Section 3.5.2, for a matrix $A = [A_1, A_2, \cdots, A_N]$, where A_v is its vth column, we define $vecA = [A_1^T, A_2^T, \cdots, A_v, \cdots, A_N^T]^T$.

5.12.2 New algorithm for the GI/G/1 system

Alternatively we may arrange the state space as $\Delta = \{(0,k) \cup (i,j,k)\}$; $k = 1, 2, \cdots, n_t; j = 1, 2, \cdots, n_s; i \ge 1$, in which case our block matrices will be of the form

$$B = T, \; C = \boldsymbol{\beta} \otimes (\mathbf{t}\boldsymbol{\alpha}), \; E = \mathbf{s} \otimes T, \; A_0 = S \otimes (\mathbf{t}\boldsymbol{\alpha}),$$

$$A_1 = (\mathbf{s}\boldsymbol{\beta}) \otimes (\mathbf{t}\boldsymbol{\alpha}) + S \otimes T, \; A_2 = (\mathbf{s} \otimes \boldsymbol{\beta}) \otimes T.$$

With this arrangement we can exploit a different aspect of the matrices as follows. An efficient algorithm was developed by Alfa and Xue (2007) for analyzing this type of system. As usual we still need to compute the R matrix given as

$$R = A_0 + RA_1 + R^2 A_2.$$

It is clear that we can write

$$\boldsymbol{x}_0(B + VE) = \boldsymbol{x}_0, \tag{5.165}$$

$$\boldsymbol{x}_0\mathbf{1} + \boldsymbol{x}_0 V(I - R)^{-1}\mathbf{1} = 1, \tag{5.166}$$

where

$$V = B(I - A_1 - RA_2)^{-1} \text{ and} \tag{5.167}$$

$$\boldsymbol{x}_1 = \boldsymbol{x}_0 V. \tag{5.168}$$

A simple iterative method for computing R is

$$R(k+1) := A_0(I - A_1 - R(k)A_2)^{-1}, \tag{5.169}$$

with

$$R(0) := 0. \tag{5.170}$$

Even though this method of iteration may not be efficient under normal circumstances, however for this particular problem it is very efficient once we capitalize on the structure of the problem. The Logarithmic Reduction method by Latouche and Ramaswami (1993) is very efficient but only more efficient than this new linear algorithm at very high traffic cases, otherwise this algorithm is more efficient for medium to low traffic intensities.

Consider a matrix sequence $\mathscr{U} = \{U_i, 0 \leq i \leq n_t - 1\}$, where $U_i \in R^{n_s \times n_s}$, and define

$$f(\mathscr{U}) = \sum_{i=0}^{n_t-1} U_i \otimes ((\mathbf{t}\boldsymbol{\alpha})T^i). \tag{5.171}$$

It was shown in Alfa and Xue (2007) that the matrix sequence $\{R(k),\ k \geq 0\}$ generated from the above iteration is of the form

$$R(k) = f(\mathscr{U}(k)), \tag{5.172}$$

where $\mathscr{U}(k) = \{U_i(k), 0 \le i \le n_t - 1\}$. If we define $\mathscr{V}(k) = \{V_i(k), 0 \le i \le n_t - 1\}$, where

$$V_0(k) = \mathbf{s}\boldsymbol{\beta}, \tag{5.173}$$

$$V_i(k) = V_{i-1}(k)S + U_{i-1}(k)\mathbf{s}\boldsymbol{\beta},\ 1 \le i \le n_t - 1, \tag{5.174}$$

then $\mathscr{U}(k)$ is determined from $\mathscr{V}(k)$ by

$$U_i(k+1) = S^{i+1} + S^*(I - V^*(k))^{-1}V_i(k),\ 0 \le i \le n_t - 1, \tag{5.175}$$

where

$$V^*(k) = \sum_{i=0}^{n_t-1} a_{i+1}V_i(k) \text{ and } S^* = \sum_{i=1}^{K} a_i S^i,$$

and

$$\tilde{K} = min(n_s - 1, n_t - 1).$$

In what follows we present the algorithm for computing the R matrix.

<u>The Algorithm:</u>

1. Set stopping tolerance ε
2. $S^* := a_1 S + a_2 S^2 + \cdot + a_{\tilde{K}} S^{\tilde{K}}$
3. $U_i^{new} := 0,\ i = 0, 1, 2, \cdots, n_t - 1$
4. Do
5. $U_i^{old} := U_i^{new},\ i = 0, 1, 2, \cdots, n_t - 1$
6. $V_0 := \mathbf{s}\boldsymbol{\beta}$
7. For $i = 1 : n_t - 1$
8. $V_i := V_{i-1}S + U_{i-1}^{old}\mathbf{s}\boldsymbol{\beta}$
9. End
10. $V^* := a_1 V_0 + a_2 V_1 + \cdots + a_{n_t} V_{n_t-1}$
11. $V^* := S^*(I - V^*)^{-1}$
12. $V_0 := V^* V_0$
13. For $i = 1 : n_t - 1$
14. $V_i := V_{i-1}S + V^*(U_{i-1}^{old}\mathbf{s})\boldsymbol{\beta}$
15. End
16. $X_i^{new} := S^{i+1} + V_i,\ i = 0, 1, \cdots, n_t - 1$
17. Until $max_{i,j,k}|(U_i^{new})_{kl} - U_i^{old})_{kl}| < \varepsilon$.

5.12.3 The (ET,ET) representation for the GI/G/1 queue

Now we study the GI/G/1 system by considering the inter-arrival and service times as PH distributions using the elapsed time approach. The arrival process is of PH distribution (α, T) and service is PH distribution (β, S) written as

$$\alpha = [1,\ 0,\ 0,\ \cdots,\ 0];\ \beta = [1,\ 0,\ 0,\ \cdots,\ 0],$$

$$T_{i,j} = \begin{cases} \tilde{a}_i, & j = i+1 \\ 0, & otherwise, \end{cases},\quad S_{i,j} = \begin{cases} \tilde{b}_i, & j = i+1 \\ 0, & otherwise, \end{cases},$$

where

$$\tilde{a}_i = \frac{u_i}{u_{i-1}},\ u_i = 1 - \sum_{v=0}^{i} a_v,\ u_0 = 1,\ \tilde{a}_{n_t} = 0,$$

$$\tilde{b}_i = \frac{v_i}{v_{i-1}},\ v_i = 1 - \sum_{u=0}^{i} b_u,\ v_0 = 1,\ \tilde{b}_{n_s} = 0.$$

The DTMC representing the number in the system for this GI/G/1 queue is $\Delta = \{(0,k) \cup (i,k,j)\}; k = 1,2,\cdots,n_t; j = 1,2,\cdots,n_s; i \geq 1$, i.e. we have number in the system, phase of service and phase of arrival – in that order. The transition matrix for this system is also a QBD and can be written as

$$P = \begin{bmatrix} B & C & & \\ E & A_1 & A_0 & \\ & A_2 & A_1 & A_0 \\ & & \ddots & \ddots & \ddots \end{bmatrix},$$

with

$$B = T,\ C = \beta \otimes (\mathbf{t}\alpha),\ E = \mathbf{s} \otimes T,\ A_0 = S \otimes (\mathbf{t}\alpha),$$

$$A_1 = (\mathbf{s}\beta) \otimes (\mathbf{t}\alpha) + S \otimes T,\ A_2 = (\mathbf{s}\beta) \otimes T.$$

It is immediately clear that we can analyze this system using the matrix-geometric results as we did with the [RT,RT] case. However, the special feature of this model which gives us an additional structure which we can exploit is slightly different. We find that our block matrix A_2 has only one block of columns that is non-zero, as such the structure of our G matrix is

$$G = \begin{bmatrix} G_1 & 0 & 0 & \cdots & 0 \\ G_2 & 0 & 0 & \cdots & 0 \\ G_3 & 0 & 0 & \cdots & 0 \\ \vdots & \vdots & \vdots & \cdots & \vdots \\ G_{n_s} & 0 & 0 & \cdots & 0 \end{bmatrix}, \tag{5.176}$$

where G is the minimal non-negative solution to the matrix quadratic equation

$$G = A_2 + A_1 G + A_0 G^2. \tag{5.177}$$

Hence we try and compute the G matrix first using the block structure and then obtain the R matrix from it directly. We write out the block matrix equations of G as follows

$$G_k = b_k^* T + b_k^* (\mathbf{t}\boldsymbol{\alpha}) G_k + \tilde{b}_k T G_{k+1} + \tilde{b}_k (\mathbf{t}\boldsymbol{\alpha}) G_{k+1} G_1, \; 1 \le k < n_s, \tag{5.178}$$

$$G_{n_s} = T + (\mathbf{t}\boldsymbol{\alpha}) G_1, \tag{5.179}$$

where $b_k^* = 1 - \tilde{b}_k$.

For simplicity we write these equations as

$$G_k = b_k^* V_0 + \tilde{b}_k V_0 G_{k+1} + b_k^* V_1 G_1 + \tilde{b}_k V_1 G_{k+1} G_1, \; 1 \le k < n_s, \tag{5.180}$$

$$G_{n_s} = b_{n_s}^* V_0 + b_{n_s}^* V_1 G_1, \tag{5.181}$$

keeping in mind that $b_{n_s}^* = 1$.

Let

$$\mathscr{H} = vecV_0 + (I \otimes V_1) vecG_1, \quad \mathscr{J} = I \otimes V_0 + (G_1)^T \otimes V_1, \tag{5.182}$$

then using the same types of arguments as in the previous case we have

$$vecG_1 = [I - S(n_s-1,1)\mathscr{J}^{n_s-1}(I \otimes V_1)]^{-1}[S(n_s-1,1)\mathscr{J}^{n_s-1} vecV_0$$

$$+ b_1^* \mathscr{H} + \sum_{j=2}^{n_s-1} S(n_s-1,j) b_{n_s-j+1}^* \mathscr{J}^{n_s-1} \mathscr{H}] \tag{5.183}$$

and

$$vecG_{n_s-k} = S(k,1)\mathscr{J}^k [vecV_0 + (I \otimes V_1) vecG_1] + b_{n_s-k}^* \mathscr{H}$$

$$+ \sum_{j=2}^{k} S(k,j) b_{n_s-j+1}^* \mathscr{J}^{k-j+1} \mathscr{H}, \; k = 1, 2, \cdots, n_s - 1, \tag{5.184}$$

where $S(i,j) = \prod_{k=i}^{j} \tilde{b}_{n_s-i}$.

So all we need is to compute G_1 iteratively and then compute the remaining G_k, $k = 2,3,\cdots,n_s$ explicitly as above. A detail discussion of these results are available in Alfa (2004).

5.13 MAP/PH/1 Queues

The MAP/PH/1 queue is a generalization of the PH/PH/1 queue. If we let the arrivals be described by the two sub-stochastic matrices D_0 and D_1 of order n_t, with $D = D_0 + D_1$ being a stochastic matrix. Letting the service times have a phase type distribution with representation $(\boldsymbol{\beta}, S)$ of dimension n_s, then we can simply analyze this system by replacing the block matrices of the transition matrix of the PH/PH/1 queue as follows. Wherever T appears in the PH/PH/1 results, it should be replaced with D_0 and wherever $\mathbf{t}\alpha$ appears in the PH/PH/1 queue it should be replaced with D_1. Consider the transition probability matrix of the PH/PH/1 system written as

$$P = \begin{bmatrix} B & C & & & \\ E & A_1 & A_0 & & \\ & A_2 & A_1 & A_0 & \\ & & \ddots & \ddots & \ddots \end{bmatrix},$$

then the block matrices for this MAP/PH/1 has the following expressions

$$B = D_0, \ C = D_1 \otimes \boldsymbol{\beta}, \ E = D_0 \otimes \mathbf{s},$$

$$A_0 = D_1 \otimes S, \ A_1 = D_0 \otimes S + D_1 \otimes (\mathbf{s}\boldsymbol{\beta}), \ A_2 = D_0 \otimes (\mathbf{s}\boldsymbol{\beta}).$$

The rest of the analysis is left to the reader as an excercise.

5.14 Batch Queues

Batch queues are systems in which either at each arrival epoch the number of items that arrive could be more than one and/or the number of service items served could be more than at each service completion time. First we consider the case of just batch arrivals. This is very common in telecommunication systems where packets arrive in batches. Another example is in the ATM system of communication, packets of different sizes arrive at the system and then the packets re-grouped into several cells of equal sizes – these cells will be the batch arrivals in this case. Also we may have batch services where more than one item are served together, such as public transportation system where customers are served in groups depending on the number waiting, the server capacity and protocols.

5.14.1 Geo^X/Geo/1 queue

Usually because we are working in discrete time and the size of the time interval, otherwise known as the time slot in telecommunications, affects how we account for the arrival. There is a high chance that if there is an arrival in a time slot that arrival could be more than one. Hence considering a queueing system with batch arrivals is very important in telecommunications.

Here we consider a case where arrival is Bernoulli with $a = \Pr\{\text{an arrival in a time slot}\}$ and $\bar{a}$ its complement. However when arrivals occur they could be in batch. So we define $\theta_i = \Pr\{\text{number of arrivals in a time slot is } i\}$, given there is an arrival, with $i \geq 1$. We let the mean batch size be $\bar{\theta} = \sum_{i=1}^{\infty} i\theta_i$. Let us further write $c_0 = \bar{a}$ and $c_i = a\theta_i,\ i \geq 1$. We therefore have a vector $\mathbf{c} = [c_0,\ c_1,\ \cdots]$ which represents arrivals in a time slot. The mean arrival rate is thus given as $\lambda = a\bar{\theta}$. We also let the service be geometric with parameter b. Consider the DTMC $\{X_n,\ n \geq 0\}$ with state space $\{0, 1, 2, \cdots\}$, where X_n is the number of items in the system at time n. It is straightforward to show that the transition matrix of this DTMC is

$$P = \begin{bmatrix} c_0 & c_1 & c_2 & c_3 & \cdots \\ a_0 & a_1 & a_2 & a_3 & \cdots \\ & a_0 & a_1 & a_2 & \cdots \\ & & a_0 & a_1 & \cdots \\ & & & \ddots & \ddots \end{bmatrix}, \tag{5.185}$$

where the c_is are as defined above and

$$a_0 = bc_0,\ a_i = bc_i + \bar{b}c_{i-1},\ i \geq 1.$$

This is a DTMC of the M/G/1 type and all the matrix-analytic results for the M/G/1 apply. Provided the system is stable, i.e. provided $\lambda < b$, then we have a unique vector $\boldsymbol{x} = [x_0,\ x_1,\ x_2,\ \cdots]$ such that

$$\boldsymbol{x} = \boldsymbol{x}P,\ \ \boldsymbol{x}\mathbf{1} = 1.$$

We have the general relationship

$$x_i = x_0 c_i + \sum_{j=1}^{i+1} x_j a_{i-j+1},\ \ i \geq 0. \tag{5.186}$$

We can apply $z-$transform approach or the matrix-analytic approach to analyze this problem.

5.14.1.1 Transform Approach

If we define

$$X^*(z) = \sum_{i=0}^{\infty} z^i x_i, \quad c^*(z) = \sum_{i=0}^{\infty} z^i c_i, \text{ and } a^*(z) = \sum_{i=0}^{\infty} z^i a_i, \quad |z| \le 1,$$

then we have

$$X^*(z) = \frac{x_0[zc^*(z) - a^*(z)]}{z - a^*(z)}. \tag{5.187}$$

We know that $X^*(z)|_{z\to 1} = 1$. However, we have to apply l'Hôpital's rule here. It is straightforward to show that

$$zc^*(z) = z(\bar{a} + a\theta^*(z)), \tag{5.188}$$

hence

$$\frac{d(zc^*(z))}{dz}|_{z\to 1} = 1 + a\bar{\theta},$$

implying that $\bar{c} = \sum_{i=1}^{\infty} ic_i = a\bar{\theta}$. We can also show that

$$\frac{d(a^*(z))}{dz}|_{z\to 1} = b\bar{c} + \bar{b}\bar{c} + \bar{b} = a\bar{\theta} + \bar{b}. \tag{5.189}$$

Now we have

$$X^*(z)|_{z\to 1} = 1 = x_0[1 + a\bar{\theta} - \bar{b} - a\bar{\theta}][1 - \bar{b} - a\bar{\theta}]^{-1} = x_0 b[1 - a\bar{\theta} - \bar{b}]^{-1}, \tag{5.190}$$

which gives

$$x_0 = \frac{b - a\bar{\theta}}{b} = 1 - \frac{a\bar{\theta}}{b}. \tag{5.191}$$

Hence we have

$$X^*(z) = \frac{(b - a\bar{\theta})[zc^*(z) - a^*(z)]}{b(z - a^*(z))}. \tag{5.192}$$

5.14.1.2 Examples

- **Example 1: Uniform batch size:** We consider the case in which the batch size varies from k_1 to k_2 with θ_i being the probability that the arriving batch size is i with $\theta_i = 0,\ i < k_1; i > k_2$ and both $(k_1, k_2) < \infty$.

For this example, $\theta^*(z) = \sum_{j=k_1}^{k_2} z^i \theta_i$ and $\bar{\theta} = \sum_{j=k_1}^{k_2} j\theta_j$. For example, when $k_1 = k_2 = 3$ we have $\theta^*(z) = z^3$ and $\bar{\theta} = 3$. For this we have

$$X^*(z) = \frac{(b-3a)[z^4 - a^*(z)]}{b(z-a^*(z))}.$$

- **Example 2: Geometric batch size:** We consider the case in which the batch size has the geometric distribution with parameter θ and $\theta_i = \theta(1-\theta)^{i-1}$, $i \geq 1$. In this case $\theta^*(z) = z(1-\theta)(1-z\theta)^{-1}$ and $\bar{\theta} = \theta^{-1}$. For this example we have

$$X^*(z) = \frac{(b-\frac{a}{\theta})[z^2(1-\theta) - a^*(z)(1-z\theta]}{b(z-a^*(z))}.$$

In both examples we simply substitute $\theta^*(z)$ appropriately to the general transform results presented earlier, Equation 5.192, and then obtain $X^*(z)$.

5.14.1.3 Matrix-analytic Approach

By virtue of the fact that if this system is stable, then the G matrix which is a scalar here and we call it g, will be unity. In this case we can easily apply the recursion by Ramaswami (1988) as follows.

For any size $k \geq 1$ we write the transition matrix

$$P(k) = \begin{bmatrix} c_0 & c_1 & c_2 & \cdots & c_{k-1} & \sum_{j=k}^{\infty} c_j \\ a_0 & a_1 & a_2 & \cdots & a_{k-1} & \sum_{j=k}^{\infty} a_j \\ & a_0 & a_1 & \cdots & a_{k-2} & \sum_{j=k-1}^{\infty} a_j \\ & & a_0 & \cdots & a_{k-3} & \sum_{j=k-2}^{\infty} a_j \\ & & & \ddots & \ddots & \vdots \\ & & & & a_0 & \sum_{j=1}^{\infty} a_j \end{bmatrix}, \quad (5.193)$$

which can be written as

$$P(k) = \begin{bmatrix} c_0 & c_1 & c_2 & \cdots & c_{k-1} & \tilde{c}_k \\ a_0 & a_1 & a_2 & \cdots & a_{k-1} & \tilde{a}_k \\ & a_0 & a_1 & \cdots & a_{k-2} & \tilde{a}_{k-1} \\ & & a_0 & \cdots & a_{k-3} & \tilde{a}_{k-2} \\ & & & \ddots & \ddots & \vdots \\ & & & & a_0 & \tilde{a}_1 \end{bmatrix}, \quad (5.194)$$

where $\tilde{c}_k = \sum_{j=k}^{\infty} c_j$, $\tilde{a}_{k-v} = \sum_{j=k-v}^{\infty} a_j$.

Hence we can write

$$x_k = (1-\tilde{a}_1)^{-1}[x_0\tilde{c}_k + \sum_{j=1}^{k-1} x_j\tilde{a}_{k-j+1}], \quad k = 1,2,\cdots. \tag{5.195}$$

Keeping in mind that the x_0 is known then the recursion starts from $k = 2$ and continues to the point of truncation.

5.14.1.4 Waiting Time Distribution

In this section we consider the waiting time in the queue. Let us define y_k, $k = 0,1,2,\cdots$, as the probability that a tagged customer sees y customers ahead of it at steady-state. Then we have

$$y_0 = \frac{1}{\bar{\theta}}[x_0 \sum_{i=1}^{\infty} \theta_i + x_1 \sum_{i=1}^{\infty} \theta_i b],$$

$$y_k = \frac{1}{\bar{\theta}}[\sum_{j=1}^{k} x_j \sum_{i=k-j+1}^{\infty} \theta_i(1-b) + x_0 \sum_{i=k+1}^{\infty} \theta_i + \sum_{j=1}^{k+1} x_j \sum_{i=k-j+2}^{\infty} \theta_i b], \; k = 1,2,\cdots.$$

Let W_q the waiting time in the queue and $w_j^{(q)} = Pr\{W_q = j\}$, then we have

$$w_0^{(q)} = y_0,$$

$$w_r^{(q)} = \sum_{k=1}^{r} y_k \binom{r-1}{k-1} (1-b)^{r-k} b^k, \; r \geq 1.$$

The waiting time in the system can be easily derived from these.

5.14.2 The BMAP/PH/1 System

Consider a system with batch Markovian arrival process (BMAP) represented by the matrices D_k, $k = 0,1,2,\cdots$, of order r, with arrival rate given as λ. The service time is phase type with representation $(\boldsymbol{\beta}, S)$ of dimension m. Once again we let $\mathbf{s} = \mathbf{1} - S\mathbf{1}$. Let the state space of the associated Markov chain be $\{((0 \cup k) \cup (i,j,k)), i \geq 0, 1 \leq j \leq m, 1 \leq k \leq r\}$, where i is the number in the system, j is the phase of service and k is the phase of the BMAP, then we can write the associated transition matrix, P, as

$$P = \begin{bmatrix} B_0 & B_1 & B_2 & B_3 & \cdots \\ C & A_1 & A_2 & A_3 & \cdots \\ & A_0 & A_1 & A_2 & \cdots \\ & & A_0 & A_1 & \cdots \\ & & & \ddots & \ddots & \ddots \end{bmatrix},$$

where

$$B_0 = D_0,\ B_i = D_i \otimes \boldsymbol{\beta},\ i = 1,2,\cdots,\ C = D_0 \otimes \mathbf{s},$$

$$A_0 = D_0 \otimes (\mathbf{s}\boldsymbol{\beta}),\ A_i = D_i \otimes (\mathbf{s}\boldsymbol{\beta}) + D_{i-1} \otimes S,\ i = 1,2,\cdots.$$

First we need to obtain the stationary distribution of this DMTC. Let us have $\boldsymbol{x} = \boldsymbol{x}P,\ \boldsymbol{x}\mathbf{1} = 1$, with $\boldsymbol{x} = [\boldsymbol{x}_0, \boldsymbol{x}_1, \boldsymbol{x}_2, \cdots],\ \boldsymbol{x}_i = [\boldsymbol{x}_{i,1},\ \boldsymbol{x}_{i,2},\ \cdots,\ \boldsymbol{x}_{i,r}], i \geq 0$ and $\boldsymbol{x}_{i,k} = [x_{i,k,1},\ x_{i,k,2},\ ,x_{i,k,m}],\ i \geq 1,\ \boldsymbol{x}_{0,k} = x_{0,k},\ k = 1,2,\cdots,r$.

Because of the C at the boundary we re-block the transition matrix P as

$$\tilde{P} = \begin{bmatrix} \tilde{B}_0 & \tilde{B}_1 & \tilde{B}_2 & \tilde{B}_3 & \cdots \\ \tilde{A}_0 & \tilde{A}_1 & \tilde{A}_2 & \tilde{A}_3 & \cdots \\ & \tilde{A}_0 & \tilde{A}_1 & \tilde{A}_2 & \cdots \\ & & \tilde{A}_0 & \tilde{A}_1 & \cdots \\ & & & \ddots & \ddots & \ddots \end{bmatrix},$$

where

$$\tilde{B}_0 = \begin{bmatrix} B_0 & B_1 \\ C & A_1 \end{bmatrix},\quad \tilde{B}_k = \begin{bmatrix} B_{2k} & B_{2k+1} \\ A_{2k} & A_{2k+1} \end{bmatrix},\ k = 1,2,\cdots,$$

$$\tilde{A}_0 = \begin{bmatrix} 0 & A_0 \\ 0 & 0 \end{bmatrix},\quad \tilde{A}_k = \begin{bmatrix} A_{2k-1} & A_{2k} \\ A_{2k-1} & A_{2k} \end{bmatrix},\ k = 1,2,\cdots.$$

Using the results in Chapter 3 and section 4.8.4, we can obtain the solution to

$$\tilde{\boldsymbol{x}} = \tilde{\boldsymbol{x}}\tilde{P},\ \tilde{\boldsymbol{x}}\mathbf{1} = 1.$$

Keeping in mind that

$$\tilde{\boldsymbol{x}}_k = [\boldsymbol{x}_{2k},\ \boldsymbol{x}_{2k+1}],\ k = 0,1,2,\cdots,$$

then the stationary distribution of P is now known. We can now use this obtain the waiting time distribution in the queue. First we obtain the vector $\mathbf{y}_k,\ k = 0,1,2,\cdots$, which captures the steady state probability vector that at arrival a tagged customer sees k customers ahead of it in the queue. The vector $\mathbf{y}_k, k = 1,2,\cdots$ is of order $1 \times rm$ and $\mathbf{y}_0$ is of order $1 \times r$.

By definition we have

$$\mathbf{y}_0 = \frac{1}{\lambda}[\boldsymbol{x}_0 \sum_{i=1}^{\infty} D_i + \boldsymbol{x}_1 \sum_{i=1}^{\infty} D_i \otimes \mathbf{s}],$$

$$\mathbf{y}_k = \frac{1}{\lambda}[\boldsymbol{x}_0 \sum_{i=k+1}^{\infty} D_i \otimes \boldsymbol{\beta} + \sum_{j=1}^{k+1} \boldsymbol{x}_j \sum_{i=k-j+2}^{\infty} D_i \otimes (\mathbf{s}\boldsymbol{\beta}) + \sum_{j=1}^{k} \boldsymbol{x}_j \sum_{i=k-j+1}^{\infty} D_i \otimes S], \ k \geq 1.$$

Let W_q be the waiting time in the queue and $w_j^{(q)} = Pr\{W_q = j\}$, then we have

$$w_0^{(q)} = \mathbf{y}_0 \mathbf{1},$$

$$w^{(q)} j = \sum_{k=1}^{j} \mathbf{y}_k (\mathbf{1} \otimes I) B_j^{(k)} \mathbf{1}, \ r \geq 1,$$

where $B_j^{(k)}$ is the matrix defined in Section 5.8.1.

5.14.3 *Geo/Geo^Y/1 queue*

In some systems packets are served in batches, rather than individually. Here we consider a case where arrival is Bernoulli with $a = \Pr\{\text{an arrival in a time slot}\}$ and $\bar{a}$ its complement. Service is geometric with parameter b and also in batches of size $\mathscr{B}$, with $\theta_j = Pr\{\mathscr{B} = i, \ i \geq 1\}$, with $\sum_{j=1}^{\infty} \theta_j = 1$. We can write $b_0 = 1 - b$ and $b_i = b\theta_i$. We therefore have a vector $\mathbf{b} = [b_0, \ b_1, \ \cdots]$ which represents the number of packets served in a time slot. The mean service rate is thus given as $\mu = b\bar{\theta}$, where $\bar{\theta} = \sum_{j=1}^{\infty} j\theta_j$. Consider the DTMC $\{X_n, \ n \geq 0\}$ with state space $\{0, 1, 2, \cdots\}$, where X_n is the number of items in the system at time n. It is straightforward to show that the transition matrix of this DTMC is

$$P = \begin{bmatrix} c_0 & d_1 & & & & \\ c_1 & d_1 & d_0 & & & \\ c_2 & d_2 & d_1 & d_0 & & \\ c_3 & d_3 & d_2 & d_1 & d_0 & \\ \vdots & \vdots & \vdots & \vdots & \vdots & \ddots \end{bmatrix},$$

where

$$c_0 = 1 - a, \ d_0 = a, \ d_j = ab_j + (1-a)b_{j-1}, \ j \geq 1, \ c_j = 1 - \sum_{k=0}^{j} d_k, \ j \geq 1.$$

It is immediately clear that we have the GI/M/1 type DTMC and we can apply the related results for analyzing this problem. First we find the variable r which is the minimal non-negative solution to the non-linear equation

$$r = \sum_{j=0}^{\infty} a_j r^j.$$

If the system is stable, i.e. $a < b\bar{\theta}$ then we have a solution

$$\boldsymbol{x} = \boldsymbol{x}P, \ \ \boldsymbol{x}\mathbf{1} = 1,$$

where $\boldsymbol{x} = [x_0, \ x_1, \ x_2, \ \cdots]$, and

$$x_{i+1} = x_i r.$$

The boundary equation is solved as

$$x_0 = \sum_{j=0}^{\infty} x_j c_j = x_0 \sum_{j=0}^{\infty} r^j c_j.$$

This is then normalized by $x_0(1-r)^{-1} = 1$.

5.15 Problems

1. Question 1: Consider a Geo/Geo/1 system with special service for the first customer to be served in a busy period. For this system the arrival has parameter a, and the regular service has parameter b_r. The service time of the customer that initiates a busy period is geometric with parameter b_b.
   a. Set up the transition matrix that captures the behaviour of the number in the system for this model.
   b. What are the stability conditions?
   c. Show how to compute the stationary distribution of this system.
   d. How would you obtain the waiting time (in the system and also in the queue) distribution?
2. Question 2: Answer the same question as in Question 1 for the case where the system is a PH/PH/1 with arrivals represented by (α, T) of dimension n_r, the regular service by (β_r, S_r) of dimension m_r and the other service by (β_b, S_b) of dimension m_b.
3. Question 3: A Geo/Geo/1 system has arrival an service parameters given as a and b, respectively. After a customer finishes service it may return to the queue, with probability θ, for another service, which is also geometric with parameter b.

There is no restriction as to how many times a customer may return for service. Find the stability condition of this system and the stationary distribution of number in the system.

4. Question 4: The probability of arrival in a single server system is geometric with parameter a and given that there is arrival, the number of items in that arrival instance is a random variable X with $c_j = Pr\{X = j\}$, $j = 1, 2, \cdots$. Service is carried out in groups of size $Y \geq 1$ with parameter b_j representing the probability that the service time of a group of size j is b_j. Let the minimum group size served be K and maximum be L, i.e. $b_j = 0, j = 0, 1, \cdots, K-1$ and $b_j = 0, j = L+1, L+2, \cdots$. Set up a DTMC that captures the number of items in the system at any arbitrary time n. Under what conditions is this DTMC positive recurrent? How would you study the waiting time distribution in this system?
5. Question 5: Consider the age process presented in Section 5.10.2.2, use the technique in van Houdt and Blondia (2002) to develop a procedure for obtaining the waiting time distribution.
6. Question 6: Consider a Geo/Geo/1 system with arrival parameter a and three service modes with parameters b_0, b_1 and b_2, with $b_2 = 0 < b_1 < b_0 < 1$. Let S_n be the server's state of service at time n. The server can switch modes at any time and according to a DTMC with transition matrix S with elements $s_{i,j}$, where $s_{i,j} = Pr\{S_{n+1} = j | S_n = i\}, (i,j) = 0, 1, 2$. Set up a DTMC that captures the number of items in the system at arbitrary times. What are the conditions under which the DTMC is positive recurrent? Show how to obtain the waiting distribution for items in this system.

References

Alfa AS (2004) Markov chain representations of discrete distributions applied to queueing models. Computers and Operations Research 31:2365–2385

Alfa AS, Xue J (2007) Efficient computations for the discrete GI/G/1 system. INFORMS Journal on Computing 19:480–484

Chaudhry ML (2000) On numerical computations of some discrete-time queues. In: Grassmann WK (ed) Computational Probability, Kluwer's International Series, pp 365–408

Dafermos SC, Neuts MF (1971) A single server queue in discrete time. Cahiers Centre d'Etudes Rech Oper 13:23–40

Frigui I, Alfa AS (1995) Analysis and computation of the busy period for the discrete MAP/PH/1 queue with vacations. Internal report, University of Manitoba

Gross D, Shortle JF, Thompson JM, Harris CM (2008) Fundamentals of Queueing Theory, 2nd edn. John Wiley and Sons, New York

Heimann D, Neuts MF (1973) The single server queue in discrete time numerical analysis IV. Naval Research Logistics Quarterly 20:753–766

Hunter JJ (1983) Mathematical Techniques of Applied Probability - Volume 2, Discrete Time Model: Techniques and Applications. Academic Press, New York

Kendall DG (1953) Stochastic processes occurring in the theory of queues and their analysis by the method of imbedded Markov chains. Ann Math Statist 24:338–354

Kleinrock L (1975) Queueing Systems, Volume 1: Theory. John Wiley and Sons, New York

Klimko MF, Neuts MF (1973) The single server queue in discrete time - numerical analysis, II. Naval Res Logist Quart 20:304–319

Latouche G, Ramaswami V (1993) A logarithmic reduction algorithm for quasi-birth-and-death processes. J Appli Probab 30:650–674

Latouche G, Ramaswami V (1997) The PH/PH/1 queue at epochs of queue size change. Queueing Systems, Theory and Appl 25:97–114

Latouche G, Ramaswami V (1999) Introduction to matrix analytic methods in stochastic modeling. In: Pennsylvania, Philadelphia, Pennsylvania

Minh DL (1978) The discrete-time single-server queue with time-inhomogeneous compound Poisson input. Journal of Applied Probability 15:590–601

Neuts MF (1975) Probability distributions of phase type. In: Liber Amicorum Prof. Emeritus H. Florin, University of Louvain, Belgium, pp 173–206

Ramaswami V (1988) A stable recursion for the steady state vector in markov chains of M/G/1 type. Stochastic Models 4:183–188

Shi D, Guo J, Liu L (1996) SPH-distributions and the rectangle-iterative algorithm. In: Chakravarthy S, Alfa AS (eds) Matrix-analytic Methods in Stochastic Models, Marcel Dekker, pp 207–224

Tweedie RL (1982) Operator-geometric stationary distributions for Markov chains, with applications to queueing models. Adv Appl Proab 14:368–391

van Houdt B, Blondia C (2002) The delay distribution of a type k customer in a first-come-first-served MMAP[K]/PH[K]/1 queue. Journal of Applied Probability 39:213–223

Chapter 6
Basic Single Node Queueing Models with Finite Buffers

6.1 Introduction

Now we consider finite buffer queues. The methods used for analyzing them are different from those used for infinite buffer systems.

6.2 Geo/Geo/1/K Queues

The Geo/Geo/1/K system is the same as the Geo/Geo/1 system studied earlier, except that now we have a finite buffer $K < \infty$. Hence when a packet arrives to find the buffer full, i.e. K packets in the buffer (implying $K+1$ packets in the system including the one currently receiving service), the arriving packet is dropped. There are cases also where in such a situation an arriving packet is accepted and maybe the packet at the head of the queue is dropped. We will consider that later, even though the two different policies do not affect the number in the system, just the waiting time distribution is affected. We can view this system as a special case of the BD process or of the Geo/Geo/1 system. We consider the case when $a_i = a$, $0 \le i \le K$, $a_i = 0$, $\forall i \ge K+1$ and $b_i = b$, $\forall i \ge 1$. In this case, the transition matrix P becomes

$$P = \begin{bmatrix} \bar{a} & a & & & \\ \bar{a}b & \bar{a}\bar{b}+ab & a\bar{b} & & \\ & \bar{a}b & \bar{a}\bar{b}+ab & a\bar{b} & \\ & & \ddots & \ddots & \ddots \\ & & & \bar{a}b & \bar{a}\bar{b}+a \end{bmatrix}. \tag{6.1}$$

A.S. Alfa, *Applied Discrete-Time Queues*, DOI 10.1007/978-1-4939-3420-1_6

This is a finite space DTMC and methods for obtaining its stationary distribution have been presented in Section 4.3. Even though we can use several methods to analyze this Markov chain, we present only the algebraic approach, which is effective here by nature of its structure – tridiagonal structure with repeating rows– and briefly discuss the z-transform approach.

$$x_0 = x_0\bar{a} + x_1\bar{a}b, \tag{6.2}$$

$$x_1 = x_0 a + x_1\left(\bar{a}\bar{b} + ab\right) + x_2\bar{a}b, \tag{6.3}$$

$$x_i = x_{i-1}a\bar{b} + x_i\left(\bar{a}\bar{b} + ab\right) + x_{i+1}\bar{a}b,\ 2 \le i \le K, \tag{6.4}$$

$$x_{K+1} = x_k a\bar{b} + x_{K+1}\left(\bar{a}\bar{b} + a\right). \tag{6.5}$$

If we let $\theta = (a\bar{b})(\bar{a}b)^{-1}$, then we have

$$x_i = (\bar{b})^{-1}\theta^i x_0,\ 1 \le i \le K \tag{6.6}$$

$$x_{K+1} = \bar{a}b(a\bar{b})^{-1}x_K = \theta^K a(\bar{a}b)^{-1}x_0 \tag{6.7}$$

After normalization, we obtain

$$x_0 = \left[1 + a\theta^K(\bar{a}b)^{-1} + \sum_{i=1}^{K}\theta^i(\bar{b})^{-1}\right]^{-1} \tag{6.8}$$

We can also use the $z-$transform approach to analyze this system.

6.2.1 Busy Period

Using the same idea as in the case of the Geo/Geo/1 case but with a slight modification we show that the duration of the busy period for the Geo/Geo/1/K case is a PH type distribution. We begin our presentation as follows: consider an absorbing Markov chain with $K+1$ transient states $\{1, 2, \cdots, K+1\}$ and one absorbing state $\{0\}$. Let T be the sub-stochastic matrix representing the transient states transitions and given by

$$T = \begin{bmatrix} \bar{a}\bar{b} + ab & a\bar{b} & & \\ \bar{a}b & \bar{a}\bar{b} + ab & a\bar{b} & \\ & \ddots & \ddots & \ddots \\ & & \bar{a}b & \bar{a}\bar{b} + a \end{bmatrix}. \tag{6.9}$$

Let $\mathbf{t} = \mathbf{1} - T\mathbf{1}$ and also let $\mathbf{y} = [1, 0, \ldots]$ then b_i the probability that the busy period duration is i is given as

$$b_i = \mathbf{y}T^{i-1}\mathbf{t},\ i \ge 1. \tag{6.10}$$

The argument here is that the busy period is initiated by the DTMC starting from state $\{1\}$, i.e. due to an arrival and the busy period ends when the system goes into state $\{0\}$, the absorbing state.

The mean duration of the busy period $E[B]$ is simply given as

$$\mu_B = \mathbf{y}(I-T)^{-1}\mathbf{1}. \tag{6.11}$$

We can also study the distribution of the number of packets served during a busy period as a phase type distribution.

6.2.2 *Waiting Time in the Queue*

Distribution of the waiting time in the queue can also be studied as a PH distribution. Since we are dealing with a FCFS system of service only the items that were ahead of the targeted item get served ahead of it. Because this is a finite buffer system a packet may not receive service at all if it arrives to find the buffer full. The probability of not receiving service is x_{K+1}. Now we focus only on those packets that receive service. Let T be the sub-stochastic matrix representing the transient states transitions, i.e. states $\{1,2,\cdots,K\}$, with state $\{0\}$ as the absorbing state. Then we have

$$T = \begin{bmatrix} \bar{b} & & & \\ b & \bar{b} & & \\ & \ddots & \ddots & \\ & & b & \bar{b} \end{bmatrix}. \tag{6.12}$$

The arriving packet that gets served arrives to find i packets ahead of it with probability $\frac{x_i}{1-x_{K+1}}$. Let us write $\boldsymbol{\alpha} = (1-x_{K+1})^{-1}[x_1,\ x_2,\ \ldots, x_K]$. Let W_q be the waiting time in the queue for items that get served and define $w_i^{(q)} = Pr\{W_q = i\}$ we then have

$$w_0^{(q)} = (1-x_{K+1})^{-1}x_0, \tag{6.13}$$

and

$$w_i^{(q)} = \boldsymbol{\alpha} T^{i-1}\mathbf{t},\ i \geq 1. \tag{6.14}$$

The mean waiting time in the queue for those packets that get served is given as

$$E[W_q] = \boldsymbol{\alpha}(I-T)^{-1}\mathbf{1}. \tag{6.15}$$

6.2.3 *Departure Process*

The departure process from a Geo/Geo/1/K system can be easily represented using the Markovian arrival process (MAP). Consider the transition matrix P associated with this DTMC. It is clear that at each transition there is either a departure or no departure. The matrix P can be partitioned into two matrices D_0 and D_1 as follows.

$$D_0 = \begin{bmatrix} \bar{a} & a & & & \\ & \bar{a}\bar{b} & a\bar{b} & & \\ & & \bar{a}\bar{b} & a\bar{b} & \\ & & & \ddots & \ddots \\ & & & & \bar{a}\bar{b}+a\bar{b} \end{bmatrix}, \quad D_1 = \begin{bmatrix} 0 & 0 & & & \\ \bar{a}b & ab & & & \\ & \bar{a}b & ab & & \\ & & \ddots & \ddots & \\ & & & \bar{a}b & ab \end{bmatrix}.$$

It is clear that $D = D_0 + D_1 = P$. The elements of matrices $(D_k)_{i,j}$ represent the probability of $k = 0, 1$ departures with transitions from i to j. Hence the departure process in a Geo/Geo/1/K system is a MAP and the results from MAP can be used to study this process.

For the rest of this chapter we will let the buffer size be $K-1$, $1 \leq K < \infty$. This simply makes the notations easier to work with, especially writing the matrices.

6.3 Geo/G/1/K-1 Queues

This is the Geo/G/1 system with finite buffer. Let the buffer size be $1 \leq K < \infty$. We can use the imbedded Markov chain approach to analyze this system.

Let X_n be the number of customers left behind by the departing nth customer, and A_n be the number of arrivals during the service of the nth customer. Then we have

$$X_{n+1} = ((X_n - 1)^+ + A_{n+1})^- \tag{6.16}$$

where $(U)^-$ =min$\{K,\ U\}$ and $(V)^+$ =max$\{0, V\}$.

This model is classified as in Hunter (1983) under *the late arrival with delayed access system.*

Let a_v =Pr$\{A_n = v,\ v \geq 0\}$, $\forall n$. Then a_v is given as

$$a_v = \sum_{k=v}^{\infty} s_k \binom{k}{v} a^v (1-a)^{k-v},\ v \geq 0. \tag{6.17}$$

Let P be the transition matrix describing this Markov chain, then it is given as

$$P = \begin{bmatrix} a_0 & a_1 & a_2 & \cdots & a_{K-1} & \tilde{a}_K \\ a_0 & a_1 & a_2 & \cdots & a_{K-1} & \tilde{a}_K \\ & a_0 & a_1 & \cdots & a_{K-2} & \tilde{a}_{K-1} \\ & & a_0 & \cdots & a_{K-3} & \tilde{a}_{K-2} \\ & & & \ddots & \ddots & \vdots \\ & & & & a_0 & \tilde{a}_1 \end{bmatrix}, \quad (6.18)$$

where $\tilde{a}_j = \sum_{v=j}^{\infty} a_v$.

Letting the stationary distribution be $\boldsymbol{x} = [x_0, x_1, \cdots, x_K]$, we have

$$x_j = a_j x_0 + \sum_{v=1}^{j+1} a_{j-v+1} x_v, \; 0 \le j \le K-1, \quad (6.19)$$

and

$$x_K = \tilde{a}_K x_0 + \sum_{v=1}^{K} \tilde{a}_{K-v+1} x_v. \quad (6.20)$$

We can apply the standard finite Markov chain analysis methods discussed in Section 4.3 for finding the stationary distribution, i.e. applying

$$\boldsymbol{x} = \boldsymbol{x}P, \;\; \boldsymbol{x}\mathbf{1} = 1.$$

A very interesting thing to note about this system is that if its traffic intensity is less than 1, then we can actually use the Ramaswami's recursion idea for the M/G/1. This is because the g given as the minimal non-negative solution to the equation $g = \sum_{k=0}^{\infty} g^k a_k$ has a solution that $g = 1$ and by applying the recursion we have

$$x_i = (1-\tilde{a}_1)^{-1}[x_0 \tilde{a}_i + (1-\delta_{i,1}) \sum_{v=1}^{i-1} x_v \tilde{a}_{i+1-v}], \; 1 \le i \le K. \quad (6.21)$$

However, we need to determine x_0 first but this is done using the normalization equation.

6.4 GI/Geo/1/K-1 Queues

This is the finite buffer case of the GI/Geo/1 system. So let Y_n be the number of packets in the system at the arrival time of the nth packet, and B_n as the number of packets served during the inter-arrival time of the nth packet. Then we have

$$Y_{n+1} = (Y_n + 1 - B_n)^-, \; Y_n \ge 0, \; B_n \le Y_n + 1, \quad (6.22)$$

where $(V)^- = \min(K,V)$. Further, let us define b_j as the probability that j packets are served during the $(n+1)$th inter-arrival time if the system is found empty by the $(n+1)$th arriving packet and a_j is the corresponding probability if this packet arrives to find the system not empty with j packets served during its inter-arrival time. Then we have

$$a_j = \sum_{k=j}^{\infty} g_k \binom{k}{j} s^j (1-s)^{k-j}, \; j \geq 0 \tag{6.23}$$

$$b_j = 1 - \sum_{i=0}^{j} a_i, \; j \geq 0 \tag{6.24}$$

The transition matrix describing this Markov chain is given by P as follows

$$P = \begin{bmatrix} b_0 & a_0 & & & & \\ b_1 & a_1 & a_0 & & & \\ b_2 & a_2 & a_1 & a_0 & & \\ \vdots & \vdots & \vdots & \vdots & \ddots & \\ b_{K-1} & a_{K-1} & a_{K-2} & \cdots & a_1 & a_0 \\ b_K & a_K & a_{K-1} & \cdots & a_2 & a_1 + a_0 \end{bmatrix} \tag{6.25}$$

Let the stationary distribution be $\boldsymbol{x} = [x_0, \, x_1, \, \cdots, \, x_K]$, then we have

$$\boldsymbol{x} = \boldsymbol{x}P, \; \boldsymbol{x}\mathbf{1} = 1.$$

Once more, this can be solved using standard Markov chain techniques presented in Section 4.3.

6.5 GI/G/1/K-1 Queues

For this class of queues we present two modelling techniques, labelled Model I and Model II.

6.5.1 GI/G/1/K-1 Queues – Model I

In the following model, we represent the inter-arrival times as remaining time and interpret GI/G/1/K as a special case of M/G/1. At time $n (n \geq 0)$, let L_n be the remaining inter-arrival time, X_n be the number of customers in the system and J_n be the phase of the service time. Consider the state space

$$\Delta = \{(L_n, 0) \cup (L_n, X_n, J_n), L_n = 1, 2, \cdots, n_t; X_n = 1, 2, \cdots, K; J_n = 1, 2, \cdots, n_s\}.$$

We also let $(L_n, X_n, J_n)|_{n\to\infty} = (L, X, J)$. In what follows, we refer to level $i (i \geq 1)$ as the set $\{((i, 0) \cup (i, X, J), X = 1, 2, \cdots, K; J = 1, 2, \cdots, n_s\}$, and the second and third variables as the sub-levels.

This is indeed a DTMC. The transition matrix P_a representing the Markov chain for the GI/G/1/K-1 system is

$$P_a = \begin{bmatrix} B_1 & B_2 & B_3 & \cdots & B_{n_t-1} & B_{n_t} \\ V & & & & & \\ & V & & & & \\ & & V & & & \\ & & & \ddots & & \\ & & & & V & 0 \end{bmatrix}, \tag{6.26}$$

where

$$B_j = a_j \begin{bmatrix} 0 & \boldsymbol{\beta} & & & \\ & \mathbf{s}\boldsymbol{\beta} & S & & \\ & & \ddots & \ddots & \\ & & & \mathbf{s}\boldsymbol{\beta} & S \\ & & & & B \end{bmatrix}, \quad V = \begin{bmatrix} 1 & & & & \\ \mathbf{s} & S & & & \\ & \mathbf{s}\boldsymbol{\beta} & S & & \\ & & \ddots & \ddots & \\ & & & \mathbf{s}\boldsymbol{\beta} & S \end{bmatrix}$$

$B = \mathbf{s}\boldsymbol{\beta} + S$, $\mathbf{s} = \mathbf{1} - S\mathbf{1}$, and the matrices B_i, V are of dimensions $(Kn_s+1)\times(Kn_s+1)$.

The Markov chain is level-dependent and is a special case of the queue M/G/1 type. Let the stationary distribution of P_a be $\boldsymbol{x} = [\boldsymbol{x}_1, \boldsymbol{x}_2, \cdots, \boldsymbol{x}_{n_t}]$, where

$$\boldsymbol{x}_k = [x_{k,0}, \boldsymbol{x}_{k,1}, \cdots, \boldsymbol{x}_{k,K}] \text{ and } \boldsymbol{x}_{k,i} = [x_{k,i,1}, x_{k,i,2}, \cdots, x_{k,i,n_s}], i = 1, 2, \cdots, K.$$

The entry $x_{k,0}$ is the probability that the remaining inter-arrival time is k with no customers in the system; $x_{k,i,j}$ is the probability that the remaining inter-arrival time is k, the number of customers in the system is i and the phase of service of the customer in service is j.

Suppose that the Markov chain is positive recurrent, then the stationary vector $\boldsymbol{x}$ exists and satisfies

$$\boldsymbol{x}P_a = \boldsymbol{x}, \qquad \boldsymbol{x}\mathbf{1} = \mathbf{1}.$$

Writing $B_i = a_i U$, the equation $\boldsymbol{x}P_a = \boldsymbol{x}$ yields the recursion

$$\boldsymbol{x}_{n_t} = a_{n_t}\boldsymbol{x}_1 U \text{ and } \boldsymbol{x}_k = \boldsymbol{x}_{k+1}V + a_k\boldsymbol{x}_1 U, \quad k = n_t - 1, \cdots, 2, 1. \tag{6.27}$$

With this, we have the following result.

If the Markov chain is positive recurrent, we have

$$\boldsymbol{x}_k = \boldsymbol{x}_1 U \hat{a}_k(V),\ k = 1,2,3,\cdots,n_t, \tag{6.28}$$

where $\hat{a}_k(V) = \sum_{j=0}^{n_t-k} a_{k+j} V^j$, and $\boldsymbol{x}_1$ is normalized such that

$$\boldsymbol{x}_1 \mathbf{1} = (\sum_{k=1}^{n_t} c_k)^{-1}, \qquad c_k = \sum_{i=k}^{n_t} a_i.$$

This is based on the following induction. For $k = n_t$, it is trivial. Suppose that Equation (6.28) is true for $k = s$, then for $k = s-1$, we derive from (6.27) that

$$\boldsymbol{x}_{s-1} = \boldsymbol{x}_s V + a_{s-1}\boldsymbol{x}_1 U = \boldsymbol{x}_1 U \hat{a}_s(V) V + a_{s-1}\boldsymbol{x}_1 U = \boldsymbol{x}_1 U \hat{a}_{s-1}(V).$$

Thus from the induction process, Equation (6.28) is true for $k = 2,3,\cdots,n_t$.

For evaluating $\boldsymbol{x}_1$, we substitute the formula for $\boldsymbol{x}_2,\cdots,\boldsymbol{x}_{n_t}$ into $\boldsymbol{x}P_a = \boldsymbol{x}$ and obtain

$$\boldsymbol{x}_1 = \boldsymbol{x}_1 B_1 + \boldsymbol{x}_2 V = \boldsymbol{x}_1 U[a_1 I + \hat{a}_2(V)V] = \boldsymbol{x}_1 U \hat{a}_1(V),$$

and the constraint $\boldsymbol{x}\mathbf{1} = 1$ yields

$$\boldsymbol{x}_1(I + c_2 U + c_3 UV + \cdots + c_{n_t} UV^{n_t-2})\mathbf{1} = 1.$$

Noting that $c_1 = 1$ and U, V are stochastic, then

$$\boldsymbol{x}_1 \mathbf{1} = (\sum_{k=1}^{n_t} c_k)^{-1}, \qquad c_k = \sum_{i=k}^{n_t} a_i.$$

Note that $\boldsymbol{x}_1$ here has similar structure as what Alfa presented in Alfa (2003). The reader can refer to Alfa (2003) for the efficient computation of $\boldsymbol{x}_1$. In practical computations, we can use Equation (6.27) instead of Equation (6.28) to calculate $\boldsymbol{x}_k, k = 2,\cdots,n_t$.

Further let $\mathbf{y}_k = \boldsymbol{x}_k \mathbf{1}$, then $\mathbf{y} = [\mathbf{y}_1,\cdots,\mathbf{y}_{n_t}]$ satisfies

$$\mathbf{y} = \mathbf{y}(T + \mathbf{t}\boldsymbol{\alpha}) \quad \text{and} \quad \mathbf{y}\mathbf{1} = 1,$$

where $(\boldsymbol{\alpha}, T) = (\boldsymbol{\alpha}_r, T_r)$. This is a good set of criteria for checking whether $\boldsymbol{x}$ is correctly calculated.

6.5.1.1 Queue length

Recall the variable $x_{k,0}$ is the probability that the remaining inter-arrival time is k with no customers in the system; $x_{k,i,j}$ is the probability that the remaining inter-arrival time is k, the number of customers in the system is i and the phase of service of the customer in service is j. Hence, if we let q_i be the probability of having i customers in the system, then

$$q_0 = \sum_{k=1}^{n_t} x_{k,0} = \sum_{k=1}^{n_t} \boldsymbol{x}_k \mathbf{e}_1, \tag{6.29}$$

$$q_i = \sum_{k=1}^{n_t} \sum_{j=1}^{n_s} x_{k,i,j} = \sum_{k=1}^{n_t} \boldsymbol{x}_k \hat{\mathbf{e}}_i, \quad i = 1, 2, \cdots, K, \tag{6.30}$$

where $\hat{\mathbf{e}}_i = \sum_{j=2+(i-1)n_s}^{1+in_s} \mathbf{e}_j$.

6.5.1.2 Waiting times

In order to study the waiting time distribution, we need to obtain the state of the system just after the arrival of an arbitrary customer. Let $\mathscr{U}$ be the number of customers in the system as seen by an arriving and unblocked customer at the steady state. Let

$$\psi_{k,0} = Pr\{L = k, \mathscr{U} = 0\}, \qquad \psi_{k,i,j} = Pr\{L = k, \mathscr{U} = i, J = j\}$$

Further let

$$\boldsymbol{\psi}_0 = [\psi_{1,0}, \psi_{2,0}, \cdots, \psi_{n_t,0}], \boldsymbol{\psi}_i = [\boldsymbol{\psi}_{1,i}, \boldsymbol{\psi}_{2,i}, \cdots, \boldsymbol{\psi}_{n_t,i}],$$

and

$$\boldsymbol{\psi}_{k,i} = [\psi_{k,i,1}, \psi_{k,i,2}, \cdots, \psi_{k,i,n_s}],$$

then by conditioning on the state of the observation of the system at arrivals, we obtain

$$\boldsymbol{\psi}_0 = h^{-1}[\boldsymbol{x}_0(\mathbf{t}\boldsymbol{\alpha}) + \boldsymbol{x}_1(\mathbf{t}\boldsymbol{\alpha}) \otimes \mathbf{s}], \tag{6.31}$$

$$\boldsymbol{\psi}_i = h^{-1}[\boldsymbol{x}_i((\mathbf{t}\boldsymbol{\alpha}) \otimes S) + \sum_{i=1}^{K-1} \boldsymbol{x}_{i+1}(\mathbf{t}\boldsymbol{\alpha}) \otimes \mathbf{s}\boldsymbol{\beta}], \tag{6.32}$$

where $\boldsymbol{x}_0 = [x_{1,0}, x_{2,0} \cdots, x_{n_t,0}], \boldsymbol{x}_i = [\boldsymbol{x}_{1,i}, \boldsymbol{x}_{2,i}, \cdots, \boldsymbol{x}_{n_t,i}]$, and

$$h = [\boldsymbol{x}_0(\mathbf{t}\boldsymbol{\alpha}) + \boldsymbol{x}_1(\mathbf{t}\boldsymbol{\alpha}) \otimes \mathbf{s}]\mathbf{1} + [\boldsymbol{x}_i((\mathbf{t}\boldsymbol{\alpha}) \otimes S) + \sum_{i=1}^{K-1} \boldsymbol{x}_{i+1}(\mathbf{t}\boldsymbol{\alpha}) \otimes (\mathbf{s}\boldsymbol{\beta})]\mathbf{1}.$$

Let $\mathscr{W}$ be the waiting time of a customer. Define

$$W_r = Pr\{\mathscr{W} \leq r\}, \qquad w_r = Pr\{\mathscr{W} = r\}.$$

The waiting time of an arbitrary customer is the workload in the system as seen by him, at his arrival. Hence, we have

$$w_0 = \boldsymbol{\psi}_0 \mathbf{1}, \tag{6.33}$$

$$w_r = \sum_{i=1}^{r} \boldsymbol{\psi}_i (\mathbf{1}_{n_t} \otimes I_{n_s}) B_r^{(i)} \mathbf{1}_{n_s}, \quad r \geq 1, \tag{6.34}$$

where $B_r^{(i)}$ satisfy

$$B_0^{(0)} = I_{n_s}, \quad B_r^{(r)} = (\mathbf{s}\boldsymbol{\beta})^r, \quad B_r^0 = 0, \quad B_r^{(1)} = S^{r-1}(\mathbf{s}\boldsymbol{\beta}),\ r \geq 1,$$
$$B_r^{(i)} = (\mathbf{s}\boldsymbol{\beta}) B_{r-1}^{(i-1)} + S B_{r-1}^{(i)},\ 2 \leq i \leq r.$$

where the matrix $B_r^{(i)}$ represents the probability distribution of the work in the system at the arrival of a customer who finds i customers ahead of him, and $(B_r^i)_{u,v}$ represents the probability that the phase of the service of the customer in service at his arrival of the arbitrary customer is u and at the completion of the i customer service it is in phase v, and that the workload associated with the i customers in the system is r units of time.

To derive the distribution of W_r, we need to define $\hat{X}_n$, the number of customers ahead of the arbitrary customer in the queue, and $\hat{J}_n$, the service phase of the customer in service. Then the state space of the absorbing Markov chain is given as

$$\Delta_w = \{(0) \cup (\hat{X}_n, \hat{J}_n), \hat{X}_n = 1, 2, \cdots, K-1; \hat{J}_n = 1, 2, \cdots, n_s\},$$

and the associated transition matrix is just the $((K-1)n_s + 1) \times ((K-1)n_s + 1)$ principal sub-matrix of V. We denote it as V_w. Let $\mathbf{z}^{(0)} = [z_0^{(0)}, \mathbf{z}_1^{(0)}, \mathbf{z}_2^{(0)}, \cdots, \mathbf{z}_{K-1}^{(0)}]$, where $z_0^{(0)} = \Psi_0 \mathbf{1}_{n_t}, \mathbf{z}_i^{(0)} = \boldsymbol{\psi}_i^{(0)}(\mathbf{1}_{n_t} \otimes I_{n_s})$, then we have

$$\mathbf{z}^{(i)} = \mathbf{z}^{(i-1)} V_w = \mathbf{z}^{(0)} V_w^i, \quad i \geq 1,$$

and

$$W_r = z_0^{(r)}.$$

6.5.2 GI/G/1/K-1 Queue – Model II

In this section, we consider the GI/G/1/K-1 system with the inter-arrival time with representation (α, T) and service times with representation (β, S). We use the remaining service time as the first variable, i.e. the level, for the state space of GI/G/1/K system. Let L_n be the phase of the arrival, X_n be the number of the customers in the system, J_n be the remaining service time of the customer in service at time n. Consider the state space

$$\Delta = \{(0, L_n) \cup (J_n, X_n, L_n), J_n = 1, 2, \cdots, n_s; X_n = 1, \cdots, K; L_n = 1, \cdots, n_t\}.$$

It is obvious that Δ is indeed a Markov chain with transition matrix P_{sr} as

$$P_{sr} = \begin{bmatrix} T & E & & & & & \\ D_0 & D_1 & D_2 & D_3 & \cdots & D_{n_s-1} & D_{n_s} \\ & H & & & & & \\ & & H & & & & \\ & & & H & & & \\ & & & & \ddots & & \\ & & & & & H & 0 \end{bmatrix},$$

where $E = [\mathbf{t}\alpha, \hat{\mathbf{0}}]$, $D_0 = \begin{bmatrix} T \\ \hat{\mathbf{0}}' \end{bmatrix}$ with $\hat{\mathbf{0}}$ of dimension $n_t \times (K-1)n_t$, and for $j \geq 1$,

$$D_j = b_j \begin{bmatrix} \mathbf{t}\alpha & & & \\ T & \mathbf{t}\alpha & & \\ & \ddots & \ddots & \\ & & T & \mathbf{t}\alpha \end{bmatrix}, \quad H = \begin{bmatrix} T & \mathbf{t}\alpha & & \\ & \ddots & \ddots & \\ & & T & \mathbf{t}\alpha \\ & & & D \end{bmatrix},$$

where $D = \mathbf{t}\alpha + T$ and $\mathbf{t} = \mathbf{1} - T\mathbf{1}$, and the matrices $D_i (i \geq 1)$ and H are of dimensions $(Kn_t) \times (Kn_t)$.

Obviously, P_{sr} is also a special case of M/G/1 type, and it is easy to show that P_{sr} is positive recurrent always. Hence the associated stationary distribution exists. Let the stationary distribution of P_{sr} be $\mathbf{y} = [\mathbf{y}_0, \mathbf{y}_1, \mathbf{y}_2, \cdots, \mathbf{y}_{n_s}]$, where

$$\mathbf{y}_0 = [y_{0,1}, y_{0,2}, \cdots, y_{0,n_t}], \quad \mathbf{y}_k = [\mathbf{y}_{k,1}, \cdots, \mathbf{y}_{k,K}],$$

and $\mathbf{y}_{k,i} = [y_{k,i,1}, y_{k,i,2}, \cdots, y_{k,i,n_t}], i = 1, 2, \cdots, K$.

Writing $D_j = b_j F, j = 1,2,\cdots,m$, we have

$$\mathbf{y}_k = \mathbf{y}_1 F \hat{b}_k(H), \quad k = 2,3,\cdots,n_s, \tag{6.35}$$

where $\hat{b}_k(H) = \sum_{j=0}^{n_s-k} b_{k+j} H^j$, and $[\mathbf{y}_0, \mathbf{y}_1]$ is the left eigenvector which corresponds to the eigenvalue of 1 for the stochastic matrix

$$\begin{bmatrix} T & E \\ D_0 & D_{11} \end{bmatrix},$$

where $D_{11} = \sum_{j=1}^{n_s} b_j F H^{j-1}$ records the probability, starting from level 1, of reaching level 1 before level 0, and $[\mathbf{y}_0, \mathbf{y}_1]$ is normalized with

$$\mathbf{y}_0 \mathbf{1} + \mathbf{y}_1 [I + (\sum_{k=2}^{n_s} \hat{c}_k) F] \mathbf{1} = 1, \tag{6.36}$$

where $\hat{c}_k = \sum_{v=k}^{n_s} b_v$.

The arguments leading to this result are as follows. The equality $\mathbf{y} P_{sr} = \mathbf{y}$ yields

$$\mathbf{y}_{n_s} = \mathbf{y}_1 D_{n_s} \quad \text{and} \quad \mathbf{y}_k = \mathbf{y}_1 D_k + \mathbf{y}_{k+1} H, \quad k = 2,3,\cdots,n_s.$$

From this recursion and induction process, it is easy to show that Equation (6.35) is true for $k = 2,3,\cdots,n_s$.

Also from Equation (6.35) for $\mathbf{y}_2$ and the equation

$$[\mathbf{y}_0 \quad \mathbf{y}_1] = [\mathbf{y}_0 \quad \mathbf{y}_1 \quad \mathbf{y}_2] \begin{bmatrix} T & E \\ D_0 & D_1 \\ 0 & H \end{bmatrix} = [\mathbf{y}_0 \quad \mathbf{y}_1] \begin{bmatrix} T & E \\ D_0 & D_{11} \end{bmatrix},$$

we obtain

$$D_{11} = D_1 + F \hat{b}_2(H) H = D_1 + D_2 H + \cdots + D_{n_s} H^{n_s-1} = \sum_{j=1}^{n_s} b_j F H^{j-1}.$$

Now the equation $\mathbf{y}\mathbf{1} = 1$ leads to

$$\mathbf{y}_0 \mathbf{1} + \mathbf{y}_1 \mathbf{1} + \sum_{k=2}^{n_s} \mathbf{y}_1 F \hat{b}_k(H) \mathbf{1} = 1.$$

We consider the same GI/G/1/K-1 system but with J_n representing the elapsed service time of the customer in service at time n, L_n be the phase of the arrival, X_n be the number of the customers in the system. We present only Model II and the corresponding transition matrix associated with the DTMC. The transition matrix P_{se} should be

$$P_{se} = \begin{bmatrix} T & E & & & \\ C_1 & B_1 & A_1 & & \\ C_2 & B_2 & & A_2 & \\ \vdots & \vdots & & & \ddots \\ C_{n_s-1} & B_{n_s-1} & & & & A_{n_s-1} \\ C_{n_s} & B_{n_s} & & & & 0 \end{bmatrix},$$

where $B_j = (1-\tilde{b}_j)F$, $A_j = \tilde{b}_j H$, $C_j = (1-\tilde{b}_j)D_0$, and F, H, D_0, E are defined earlier in this section. Let the stationary distribution of P_{se} be $\mathbf{y} = [\mathbf{y}_0, \mathbf{y}_1, \mathbf{y}_2, \cdots, \mathbf{y}_{n_s}]$, where $\mathbf{y}_i$ are defined earlier. Similarly, we have the following result.

For this we have

$$\mathbf{y}_k = \mathbf{y}_{k-1} A_{k-1},\ k = 2, 3, \cdots, n_s, \tag{6.37}$$

and $[\mathbf{y}_0, \mathbf{y}_1]$ is the left eigenvector which corresponds to the eigenvalue of 1 for the stochastic matrix

$$\begin{bmatrix} T & E \\ \hat{b}_1(H)D_0 & \hat{b}_1(H)F \end{bmatrix},$$

where $\hat{b}_1(H) = \sum_{j=1}^{n_s-1} b_j H^{j-1}$ and $[\mathbf{y}_0, \mathbf{y}_1]$ is normalized such that

$$\mathbf{y}_0 \mathbf{1} + (\sum_{k=1}^{n_s} \hat{c}_k) \mathbf{y}_1 \mathbf{1} = 1,$$

where $\hat{c}_k = \sum_{v=k}^{n_s} b_v$.

Note that the most computationally involved step in Model I-II is evaluating the matrices $\hat{a}_1(V)$ and $\hat{b}_1(H)$, respectively. Therefore, Model I is more appropriate when $n_t > n_s$, and Model II is good when $n_t < n_s$. When $n_t = n_s$, either method is appropriate.

The idea of re-arranged state space is easily extended to the $GI^X/G/1/K-1$ system. For details see Liu et al (2010).

6.6 Queues with Very Large Buffers

The analysis of finite buffer queues depend heavily on standard linear algebra tools that have been adapted to finite Markov chains. Those tools are quite efficient. However, one notices that as the buffer size gets to be extremely large even those tools do consume a lot of computational efforts. In this section we present techniques for approximating the distribution of number in the system and blocking probabilities in finite queues with very large buffer sizes. We focus mainly on the PH/PH/1/K-1 system.

Consider a PH/PH/1/K-1 system of Section 5.10, but now with finite buffer of size $K-1$ where K is very large. Recollect that the arrival process PH is represented by (α, T) of dimension n_t and PH service process is represented by $(\boldsymbol{\beta}, S)$ of dimension n_s. The transition matrix representing this system is

$$P_F = \begin{bmatrix} B & C & & & \\ E & A_1 & A_0 & & \\ & A_2 & A_1 & A_0 & \\ & & \ddots & \ddots & \ddots \\ & & & A_2 & F \end{bmatrix},$$

where

$$B = T,\ E = T \otimes \mathbf{s},\ C = (\mathbf{t}\boldsymbol{\alpha}) \otimes \boldsymbol{\beta},\ A_0 = (\mathbf{t}\boldsymbol{\alpha}) \otimes S,$$
$$A_1 = (\mathbf{t}\boldsymbol{\alpha}) \otimes (\mathbf{s}\boldsymbol{\beta}) + T \otimes S,\ A_2 = T \otimes (\mathbf{s}\boldsymbol{\beta}) \text{ and } F = A_1 + A_0.$$

When $K = \infty$ we know that the system is stable only if $\lambda < \mu$, where λ and μ are the arrival and service rates, respectively. There are two main scenarios of interest to us, viz: i) when $\lambda < \mu$, i.e. traffic intensity less than 1, and ii) when $\lambda > \mu$, i.e. traffic intensity greater than 1. We do not consider the case of $\lambda = \mu$. Our interest is to find the $\boldsymbol{x}(F) = \boldsymbol{x}(F)P_F,\ \boldsymbol{x}(F)\mathbf{1} = 1$.

6.6.1 *When Traffic Intensity is Less Than* 1

When traffic intensity is less than 1 we can approximate this system by letting $K \to \infty$ and use the results of PH/PH/1 in Section 5.10.

For the stable PH/PH/1 system (see Section 5.10), we have that $\boldsymbol{x} = \boldsymbol{x}P,\ \boldsymbol{x}\mathbf{1} = 1$ and $\boldsymbol{x}_{i+1} = \boldsymbol{x}_i R$.

For the finite case we have now that $\boldsymbol{x}_j = 0,\ j > K$, hence we approximate

$$\boldsymbol{x}(F) = \boldsymbol{x}U,$$

where

$$U = [\sum_{j=K+1} \boldsymbol{x}_j \mathbf{1}]^{-1}.$$

Hence

$$\boldsymbol{x}(F)_{i+1} = \boldsymbol{x}(F)_i R,\quad 1 \le i \le K-1,$$

where $\boldsymbol{x}_1(F) = \boldsymbol{x}_1 U$.

6.6.2 *When Traffic Intensity is More Than* 1

When the traffic intensity exceeds 1, we have to adopt a different type of approximation based on duality by rotation (Alfa and Zhao (2000)). First we define two new PH distributions $(\boldsymbol{\alpha}^*, T^*)$ of dimension n_t and $(\boldsymbol{\beta}^*, S^*)$ of dimension n_s. These are related to the first sets of PH distributions as follows:

$$\alpha_i^* = \alpha_{n_t-i+1} \text{ and } T_{i,j}^* = T_{n_t-i+1,n_t-j+1},\ 1 \leq (i,j) \leq n_t$$

and

$$\beta_i^* = \beta_{n_s-i+1} \text{ and } S_{i,j}^* = S_{n_s-i+1,n_t-j+1},\ 1 \leq (i,j) \leq n_s.$$

We also have

$$\mathbf{t}^* = \mathbf{1} - T^*\mathbf{1} \text{ and } \mathbf{s}^* = \mathbf{1} - S^*\mathbf{1}.$$

Now let us consider a PH/PH/1/K-1 system with arrivals $(\boldsymbol{\beta}^*, S^*)$ and service $(\boldsymbol{\alpha}^*, T^*)$. We notice that the resulting transition matrix of this new system given by P_F^* is a 180 degrees rotation of matrix P_F, and the associated queueing system has traffic intensity less than 1. The matrix P_F^* has the following structure

$$P_F^* = \begin{bmatrix} F^* & A_2^* & & & \\ A_0^* & A_1^* & A_2^* & & \\ & \ddots & \ddots & \ddots & \\ & & A_0^* & A_1^* & E^* \\ & & & C^* & B^* \end{bmatrix},$$

where

$$B^* = T^*,\ E^* = T \otimes \mathbf{s}^*,\ C^* = (\mathbf{t}^*\boldsymbol{\alpha}^*) \otimes \boldsymbol{\beta}^*,\ A_0^* = (\mathbf{t}^*\boldsymbol{\alpha}^*) \otimes S^*,$$
$$A_1^* = (\mathbf{t}^*\boldsymbol{\alpha}^*) \otimes (\mathbf{s}^*\boldsymbol{\beta}^*) + T^* \otimes S^*,\ A_2^* = T^* \otimes (\mathbf{s}^*\boldsymbol{\beta}^*) \text{ and } F^* = A_1^* + A_0^*.$$

Let the solution to stationary distribution for this system be

$$\boldsymbol{x}(F)^* = \boldsymbol{x}(F)^* P_F^*,\ \boldsymbol{x}(F)^*\mathbf{1} = 1.$$

This new system is actually studying the number of empty spaces in the buffer including the server position, as such it is a stable system. Our traffic intensity, so to speak, is μ/λ which is less than 1. We now apply the same idea as in the case of traffic intensity less than 1 to obtain the stationary distribution of this system as $K \rightarrow \infty$. We assume that the associated R^* matrix for this system given by the solution to

$$R^* = A_2^* + R^* A_1^* + (R^*)^2 A_0^*$$

has been determined, and we have obtained

$$\boldsymbol{x}_{i+1}^* = \boldsymbol{x}_i^* R^*$$

is also determined and the boundary vectors $\boldsymbol{x}_0^*$ and $\boldsymbol{x}_1^*$ are also determined. Then the solution to $\boldsymbol{x}(F)^*$ is given as

$$\boldsymbol{x}^*(F) = \boldsymbol{x}^* U^*,$$

where

$$U^* = [\sum_{j=K+1} \boldsymbol{x}_j^* \mathbf{1}]^{-1}.$$

The probability that the buffer is full, i.e. $\boldsymbol{x}_K(F)\mathbf{1}$, which we call σ is approximated by

$$\sigma = \boldsymbol{x}_0^*(F)\mathbf{1}.$$

Let y_j be the probability that the number in the system is j, then the approximation is

$$y_j = \boldsymbol{x}_{K-j}^*(F)\mathbf{1},\ j = 0, 1, 2, \cdots, K.$$

If we are interested in the joint distribution of the number in the system and the phases of arrival and services, they can be obtained as

$$x_{i,j,r} = x_{K-i,n_t-j,n_s-r}^*,\quad 1 \le i \le K;\ 1 \le j \le n_t;\ 1 \le r \le n_s,$$

where $x_{i,j,r}$ is the probability that the number in the system is i, the arrival is in phase j and service is in phase r. Keep in mind that when there is no customer in the system then we have only $x_{0,j}$ and the phases have to be modified accordingly.

6.7 Problems

1. Question 1:Consider three storages (1, 2 and 3). Each storage is of size $K = 1$. Items arrive to each storage according to the Bernoulli process with parameter $a_i,\ i = 1, 2, 3$. At each time slot the server inspects all the three storages and if each of them has one packet in it the server picks each of the items from each

storage to form a package and sends out this package instantaneously. A package consists of one item from each of the three storages. If the server can not make up a complete package, because not all storages have one packet waiting, it waits until it is possible. Consider the number of items in the storages at time n and set up this system of the three storages and a server as a discrete time Markov chain and show its state space. Develop the transition matrix that represents this system. State your assumptions clearly. Given that the system starts at time n with all storages empty, how would you obtain the probability that the system returns to the state of all storages empty for the first time at time $n+m$?

2. Question 2: Consider a data communication system with a infinite-capacity buffer for temporarily storing messages. The messages arrive according to a Bernoulli process with parameter a. When a message arrives, it could be a type 1 message with probability θ_1 or type 2 with probability θ_2, where $\theta_1 + \theta_2 = 1$. The transmission times for both types of messages are the same and have a geometric distribution with parameter b. The system can transmit only one message at a time. The following protocol for the admission of messages is used. As soon as the number of messages in the buffer is K_1 messages of type 2 are no longer admitted in the system however those of type 1 will be admitted. As soon as the number in the system drops to K_2 or below it, the admission of both types 1 and 2 is resumed, where $0 < K_2 \leq K_1 < \infty$. We are interested in studying the distribution of the number of packets in the system at arbitrary times. Set this up as a DTMC for the case of $K_2 = 3,\ K_1 = 6$ by displaying the associated transition matrix.
3. Question 3: A telecommunication switch manufacturer *Switcher* makes a class of switches which stores packets in a buffer of size K, $K < \infty$. Packets are removed from this buffer at discrete times one-by-one, i.e. every unit of time one packet is transmitted, if there is a packet in the buffer. If a packet arrives during a time slot to find the buffer full then the packet is lost. Let A_n be the number of packets that arrive during the nth slot, and X_n the number of packets in the buffer at time n. We assume that $A_n = A,\ \forall n$ and $a_i = Pr\{A = i\},\ i = 0, 1, \cdots, J,\ J < \infty$.

   a. Set up a discrete time Markov chain that represents X_n and display the associated transition matrix.
   b. Let $K = 10,\ a_0 = 0.3,\ a_1 = 0.4,\ a_2 = 0.2,\ a_3 = 0.1,\ a_i = 0,\ \forall i \geq 4$.

      i. Find the stationary distribution of the number of packets in the system at arbitrary times.
      ii. What is the probability of the buffer being full?
      iii. Suppose the buffer is empty at time 0, compute the expected number of slots that have empty buffer at the end of the next 50 slots.

References

Alfa AS (2003) An alternative approach for analyzing finite buffer queues in discrete time. Performance Evaluation 53:75–92

Alfa AS, Zhao YQ (2000) Overload analysis of the PH/PH/1/K queue and the queue of the M/G/1 type with very large *K*. Asia-Pacific J Oper Res 17:123–135

Hunter JJ (1983) Mathematical Techniques of Applied Probability - Volume 2, Discrete Time Model: Techniques and Applications. Academic Press, New York

Liu Q, Alfa AS, Xue J (2010) Analysis of the discrete-time GI/G/1/K using remaining time approach. J Appl Math and Informatics 28(1-2):153–162

Chapter 7
Multiserver Single Node Queueing Models

7.1 Introduction

Multiserver queueing systems are very useful in modelling telecommunication systems. Usually in such systems we have several channels that are used for communications. These are considered as parallel servers in a queueing system. Throughout this chapter we will be dealing with cases where the parallel servers are identical. The case of heterogeneous servers will not be covered in this book.

Two classes of multiserver queues will be dealt with in detail; the Geo/Geo/k and the PH/PH/k. We will show later that most of the other multiserver queues to be considered are either special cases of the PH/PH/k or general cases of the Geo/Geo/k systems.

One key difference between discrete time and continuous time analyses of queueing systems comes when studying multiserver systems. As long as a single server queue has single arrivals and only one item is served at a time by each server then for most basic continuous time models the system is BD such as the M/M/s or QBD such as the M/PH/s. This is not the case with the discrete time systems. As will be seen in this section, the simple Geo/Geo/k system is not a BD. It is a special case of the GI/M/1 type but may be converted to a QBD. This key difference comes because it is possible to have more than one job completion at any time, when there is more than one job receiving service. This difference makes the discrete time multiserver queues more involved in terms of analysis.

7.2 Geo/Geo/k Systems

In the Geo/Geo/k queue we have packets arriving according to the Bernoulli process, i.e. the inter-arrival times have geometric distribution with parameter $a = Pr\{an\ arrival\ at\ any\ time\ epoch\}$, with $\bar{a} = 1 - a$. There are $k < \infty$ identical servers in parallel each with a geometric service distribution with parameter b and

A.S. Alfa, *Applied Discrete-Time Queues*, DOI 10.1007/978-1-4939-3420-1_7

$\bar{b} = 1 - b$. Since the severs are all identical and have geometric service processes. We know that if there are j packets in service, then the probability of i $(i \leq j)$ of those packets having their service completed in one time slot, written as b_i^j, is given by a binomial distribution, i.e.

$$b_i^{(j)} = \binom{j}{i} b^i (1-b)^{j-i}, \quad 0 \leq i \leq j. \tag{7.1}$$

Consider the DTMC $\{X_n, \; n \geq 0\}$ which has the state space $\{0, 1, 2, 3, \cdots\}$ where X_n is the number of packets in the system at time slot n. If we define $p_{i,j} = Pr\{X_{n+1} = j | X_n = i\}$, then we have

$$p_{0,0} = \bar{a}, \tag{7.2}$$

$$p_{0,1} = a, \tag{7.3}$$

$$p_{0,j} = 0, \; \forall j \geq 2, \tag{7.4}$$

and

$$p_{i,j} = \begin{cases} ab_0^{(i)}, & j = i+1, \; i < k \\ \bar{a}b_i^{(i)}, & j = 0, \quad i < k \\ ab_{i-j+1}^{(i)} + \bar{a}b_{i-j}^{(i)}, & 1 \leq j \leq i \\ 0, & j > i+1, \; i < k \end{cases}, \tag{7.5}$$

and

$$p_{i,j} = \begin{cases} c_0 = ab_0^{(k)}, & j = i+1, \quad i \geq k \\ c_{k+1} = \bar{a}b_k^{(k)}, & j = i-k, \quad i \geq k \\ c_{i-j+1} = ab_{i-j+1}^{(k)} + \bar{a}b_{i-j}^{(k)}, & 1 \leq j \leq i, \; i \geq k, \\ 0, & j > i+1, \quad i \geq k \\ 0, & i > k, \quad j \leq i-k \end{cases} \tag{7.6}$$

It is immediately clear that the DTMC is of the GI/M/1 type, with $p_{i,j} = 0, \; i > k, \; j \leq i-k$. Hence we can easily use the results from the GI/M/1 to analyze this system.

7.2.1 As GI/M/1 type

We display the case of $k = 3$ first and then show the general case in a block matrix form. For $k = 3$ the transition matrix of this DTMC is given as

$$P = \begin{bmatrix} \bar{a} & a & & & & & \\ \bar{a}b_1^{(1)} & ab_1^{(1)} + \bar{a}b_0^{(1)} & ab_0^{(1)} & & & & \\ \bar{a}b_2^{(2)} & ab_2^{(2)} + \bar{a}b_1^{(2)} & ab_1^{(2)} + \bar{a}b_0^{(2)} & ab_0^{(2)} & & & \\ \bar{a}b_3^{(3)} & ab_3^{(3)} + \bar{a}b_2^{(3)} & ab_2^{(3)} + \bar{a}b_1^{(3)} & ab_1^{(3)} + \bar{a}b_0^{(3)} & ab_0^{(3)} & & \\ & \bar{a}b_3^{(3)} & ab_3^{(3)} + \bar{a}b_2^{(3)} & ab_2^{(3)} + \bar{a}b_1^{(3)} & ab_1^{(3)} + \bar{a}b_0^{(3)} & ab_0^{(3)} & \\ & & \bar{a}b_3^{(3)} & ab_3^{(3)} + \bar{a}b_2^{(3)} & ab_2^{(3)} + \bar{a}b_1^{(3)} & ab_1^{(3)} + \bar{a}b_0^{(3)} & ab_0^{(3)} \\ & & & \ddots & \ddots & \ddots & \ddots \quad \ddots \end{bmatrix},$$

which we write as

$$P = \begin{bmatrix} p_{0,0} & p_{0,1} & & & & & \\ p_{1,0} & p_{1,1} & p_{1,2} & & & & \\ p_{2,0} & p_{2,1} & p_{2,2} & p_{2,3} & & & \\ c_4 & c_3 & c_2 & c_1 & c_0 & & \\ & c_4 & c_3 & c_2 & c_1 & c_0 & \\ & & c_4 & c_3 & c_2 & c_1 & c_0 \\ & & & \ddots & \ddots & \ddots & \ddots \quad \ddots \end{bmatrix}. \tag{7.7}$$

We consider a stable system. Our interest is to solve for $\boldsymbol{x} = \boldsymbol{x}P$, $\boldsymbol{x}\mathbf{1} = 1$, where $\boldsymbol{x} = [x_0,\ x_1,\ x_2,\ \cdots]$. We first present the GI/M/1 approach for this problem.

It is clear that the rows of the matrix P are repeating after the $(k+1)^{st}$ row. Let the repeating part be written as $[c_{k+1},\ c_k,\ \cdots,\ c_0]$. It is known from that this DTMC is stable if

$$\beta > 1,$$

where $\beta = \sum_{v=0}^{k+1} v c_v$ which gives $\beta = kb + \bar{a} > 1$. Hence provided that

$$kb > a \tag{7.8}$$

then the system is stable. For a stable system we have a scalar r which is the minimum non-negative solution to the equation

$$r = \sum_{j=0}^{k+1} r^j c_j. \tag{7.9}$$

It is easy to see that r is the solution to the polynomial equation

$$r = (br + \bar{b})^k (a + \bar{a}r). \tag{7.10}$$

Using the GI/M/1 results we have

$$x_{i+k} = x_k r^i,\ i \geq 0. \tag{7.11}$$

However, we still need to solve for the boundary values, i.e. $[x_0,\ x_1,\ \cdots,\ x_k]$. Solving the boundary equations reduces to a finite DTMC problem. For the case of $k=3$ we display the transition matrix corresponding to this boundary behaviour, using $p_{i,j}$ as elements of this finite DTMC. We call the transition matrix P_b and write it as follows:

$$P_b = \begin{bmatrix} p_{0,0} & p_{0,1} & & \\ p_{1,0} & p_{1,1} & p_{1,2} & \\ p_{2,0} & p_{2,1} & p_{2,2} & p_{2,3} \\ c_4 & \sum_{i=3}^4 r^{i-3} c_i & \sum_{i=2}^4 r^{i-2} c_i & \sum_{i=1}^4 r^{i-1} c_i \end{bmatrix}. \tag{7.12}$$

Now going back to the more general case from here we can write this equation as

$$p_b = \begin{bmatrix} q_{0,0} & q_{0,1} & & & & \\ q_{1,0} & q_{1,1} & q_{1,2} & & & \\ q_{2,0} & q_{2,1} & q_{2,2} & q_{2,3} & & \\ \vdots & \vdots & \vdots & \vdots & \ddots & \\ q_{k,0} & q_{k,1} & q_{k,2} & q_{k,3} & \cdots & q_{k,k+1} \\ q_{k+1,0} & q_{k+1,1} & q_{k+1,2} & q_{k+1,3} & \cdots & q_{k+1,k+1} \end{bmatrix}, \tag{7.13}$$

and solve for the boundary values as

$$[x_0\ x_1\ x_2\ \cdots\ x_k] = [x_0\ x_1\ x_2\ \cdots\ x_k]P_b. \tag{7.14}$$

This can be solved easily starting from

$$x_k = \frac{x_{k+1}(1-q_{k+1,k+1})}{q_{k,k+1}}, \tag{7.15}$$

and carrying out a backward substitution recursively from $j=k-1$ up to $j=1$ and into

$$x_j = \sum_{v=j-1}^{k+1} x_v q_{v,j}, \quad 1 \le j \le k-1, \tag{7.16}$$

$$x_0 = \sum_{v=0}^{k+1} x_v q_{v,0}. \tag{7.17}$$

This is then normalized as

$$\sum_{j=0}^{\infty} x_j = 1, \quad \to x_k(1-r)^{-1} + \sum_{j=0}^{k-1} x_j = 1. \tag{7.18}$$

The distribution of $X_n|_{n\to\infty}$ is now fully determined, using the standard method.

We point out that Artalejo and Hernández-Lerma (2003) did discuss the above-mentioned recursion for obtaining the boundary values.

Next we consider studying this problem as a QBD type DTMC.

7.2.2 As a QBD

There are more results for QBDs than any other DTMC with block structures in the literature. So whenever it is possible to convert a problem to the QBD structure it is usually an advantage. We study the Geo/Geo/k problem as a QBD. Consider a bivariate DTMC $\{(Y_n, J_n), n \geq 0\}$, $Y_n = 0, 1, 2, \cdots$, $J_n = 0, 1, 2, \cdots, k-1$. At anytime n we let $kY_n + J_n$ be the number of packets in the system, i.e. $kY_n + J_n = X_n$. For example, for the case of $k = 3$ the situation with $Y_n = 5, J_n = 2$ implies that we have $3 \times 5 + 2 = 17$ packets in the system, including the ones receiving service. What exactly we have done is let Y_n be a number of groups of size k in the system and J_n is the remaining number in the system that do not form a group of size k. In order words, $J_n = X_n \; mod \; k$. The transition matrix associated with this DTMC can be written in block form as

$$P = \begin{bmatrix} B & C & & & \\ A_2 & A_1 & A_0 & & \\ & A_2 & A_1 & A_0 & \\ & & \ddots & \ddots & \ddots \end{bmatrix}, \tag{7.19}$$

with

$$B = \begin{bmatrix} p_{0,0} & p_{0,1} & & \\ p_{1,0} & p_{1,1} & p_{1,2} & \\ \vdots & \vdots & \vdots & \ddots \\ p_{k-1,0} & p_{k-1,1} & \cdots & p_{k-1,k-1} \end{bmatrix},$$

$$C = \begin{bmatrix} 0 & 0 & & & \\ 0 & 0 & 0 & & \\ \vdots & \vdots & \vdots & \ddots & \\ p_{k-1,k} & 0 & \cdots & 0 & \end{bmatrix},$$

$$A_2 = \begin{bmatrix} c_{k+1} & c_k & \cdots & c_2 \\ & c_{k+1} & \cdots & \vdots \\ & & \ddots & \vdots \\ & & & c_{k+1} \end{bmatrix}, \quad A_0 = \begin{bmatrix} 0 & 0 & & & \\ 0 & 0 & 0 & & \\ \vdots & \vdots & \vdots & \ddots & \\ c_0 & 0 & \cdots & 0 & \end{bmatrix},$$

$$A_1 = \begin{bmatrix} c_1 & c_0 & & \\ c_2 & c_1 & c_0 & \\ \vdots & \vdots & \cdots & \ddots \\ c_k & c_{k-1} & \cdots & c_1 \end{bmatrix},$$

where the entries c_j are as given above.

This system is now set up to use the QBD results. Provided the system is stable we have a matrix R which is the minimal non-negative solution to the matrix quadratic equation

$$R = A_0 + RA_1 + R^2 A_2.$$

Because of the structure of A_0, i.e. only its bottom left corner element is non-zero, then we have an R matrix which has only the bottom row elements that are non-zero. The R matrix has the following structure

$$R = \begin{bmatrix} 0 & 0 & & \\ 0 & 0 & 0 & \\ \vdots & \vdots & \vdots & \ddots \\ r_0 & r_1 & \cdots & r_{k-1} \end{bmatrix}. \tag{7.20}$$

Now we have the stationary distribution of this system given as

$$\mathbf{y} = \mathbf{y}P, \ \mathbf{y}\mathbf{1} = 1,$$

where $\mathbf{y}=[\mathbf{y}_0, \mathbf{y}_1, \ldots,]$, $\mathbf{y}_i=[\mathbf{y}_{i,0}, \mathbf{y}_{i,1}, \ldots, \mathbf{y}_{i,k-1}]=[x_{k(i-1)}, x_{k(i-1)+1}, \ldots, x_{ki-1}]$. It is straightforward to show that

$$r_j = (r_0)^{j+1}, \tag{7.21}$$

where $r_0 = r$ that was calculated from the GI/M/1 version. Hence computing R simply reduces to actually solving for the r in the GI/M/1 system.

From here we apply the boundary conditions as

$$\mathbf{y}_0 = \mathbf{y}_0 B[R], \ \ \mathbf{y}_0 \mathbf{1} = 1,$$

where $B[R] = B + RA_2$. The normalization equations are then applied to obtain

$$\mathbf{y}_0 (I - R)^{-1} \mathbf{1} = 1.$$

Now we can apply the matrix-geometric result, i.e.

$$\mathbf{y}_{i+1} = \mathbf{y}_i R, \ i \geq 0.$$

Given $\mathbf{y}$ we can determine $\boldsymbol{x}$.

7.2.3 An Example

Consider an example with $a = .25,\ b = .65,\ k = 3$, we have $kb = 1.95 > .25 = a$. Therefore the system is stable. We proceed with computing the r which is the solution to the polynomial equation

$$r = (.35 + .65r)^3(.25 + .75r).$$

The minimal non-negative solution to this equation is given as

$$r = 0.0118.$$

Next we compute R matrix. First we compute $A_0,\ A_1$ and A_2, as

$$A_0 = \begin{bmatrix} & & \\ & & \\ 0.0107 & 0 & 0 \end{bmatrix},\ A_1 = \begin{bmatrix} 0.0919 & 0.0107 & \\ 0.2901 & 0.0919 & 0.0107 \\ 0.4014 & 0.2901 & 0.0919 \end{bmatrix},$$

$$A_2 = \begin{bmatrix} 0.206 & 0.4014 & 0.2901 \\ & 0.206 & 0.4014 \\ & & 0.206 \end{bmatrix}.$$

The R matrix is obtained as

$$R = \begin{bmatrix} 0 & 0 & 0 \\ 0 & 0 & 0 \\ 0.0118 & 1.3924 \times 10^{-004} & 1.6430 \times 10^{-006} \end{bmatrix}.$$

We notice that $r_j = r^{j+1}$.

7.2.4 Waiting Times

We can study this using direct probability arguments (we call this the traditional approach) or we can use the absorbing Markov chain idea. Now we work on the basis that the vector $\boldsymbol{x} = \boldsymbol{x}P,\ \boldsymbol{x}\mathbf{1} = 1$ has been obtained. We consider the FCFS system. We will only present the waiting time in the queue; the waiting time in the system can be easily inferred from the waiting time in the queue.

We define W_q as the waiting time in the queue for a packet, and let $W_i^{(q)} = Pr\{W_q \leq i\},\ i \geq 0$.

- **Traditional approach:** This approach uses probability arguments directly. It is straightforward to show that

$$W_0^{(q)} = \sum_{j=0}^{k-1} x_j. \tag{7.22}$$

The remaining $W_j^{(q)}$, $j \geq 1$ are obtained as follows: Let $H_w^{(v-k+1)}$ be the probability that it takes no more than w units of time to complete the services of $v-k+1$ packets in the system when there are v packets in the system. Then

$$W_j^{(q)} = \sum_{v=k}^{j} H_j^{(v-k+1)} x_v, \; j \geq 1. \tag{7.23}$$

The term $H_i^{(j)}$ can be obtained recursively as follows:
We know that

$$H_i^{(1)} = 1 - (1-b)^i, \; i \geq 1, \tag{7.24}$$

then we have

$$H_i^{(j)} = \sum_{v=1}^{i-j+1} b^v H_{i-v}^{(j-1)}, \; j \leq i \leq \infty. \tag{7.25}$$

Next we develop the z-transform of the waiting time, but before we go further first let us develop some useful transform results that will be helpful. Let h_i be defined as the probability that the service time of a packet is no more than i and $h^*(z)$ as the corresponding z-transform, then we have

$$h^*(z) = \sum_{i=1}^{\infty} z^i h_i = \sum_{i=1}^{\infty} z^i (1-(1-b)^i), \; |z| < 1. \tag{7.26}$$

It is straightforward to see that

$$h^*(z) = \frac{zb}{(1-z)(1-(1-b)z)}, \; |z| < 1. \tag{7.27}$$

We know that $f^*(z) = \frac{zb}{1-z(1-b)}$ is the z-transform of the distribution of service time of a packet. If we define $h_i^{(j)}$ as the probability that the total service times of j packets is less than or equal to $i \geq j$ then we can easily show that for the case of 2 packets

$$h_i^{(2)} = f_1 h_{i-1} + f_2 h_{i-2} + \cdots + f_{i-1} h_1, \; i \geq 2. \tag{7.28}$$

Taking the z-transform of this and letting it be defined as $h^{(2)}(z)$ we have

$$h^{(2)^*}(z) = h(z)f(z) = \frac{(bz)^2}{(1-z)(1-z(1-b))^2}. \tag{7.29}$$

After repeating this process incrementally we have

$$h^{(j)}(z) = h^{(j-1)}(z)f(z) = \frac{(bz)^j}{(1-z)(1-z(1-b))^j}, \quad |z| < 1. \tag{7.30}$$

Now if we let $W^{(q)*}(z) = \sum_{v=0}^{\infty} z^v W_v^{(q)}, \; |z| < 1$, then we have

$$W^{(q)*}(z) = \sum_{j=0}^{k-1} x_j + \sum_{v=k}^{\infty} x_v \frac{(bz)^{v-k+1}}{(1-z)(1-z(1-b))^{v-k+1}}, \quad |z| < 1. \tag{7.31}$$

This reduces to

$$W^{(q)*}(z) = (x_k r^{-1}) \sum_{v=k}^{\infty} \frac{(bzr)^{v-k+1}}{(1-z)(1-z(1-b))^{v-k+1}} + \sum_{j=0}^{k-1} x_j, \quad |z| < 1. \tag{7.32}$$

- **Absorbing Markov chain approach:** This is based on the matrix-analytic approach. Since this is a FCFS system, once an arbitrary packet joins the queue, only the packets ahead of it at that time complete service ahead of it. Hence by studying the DTMC that keeps track of packets ahead of this arbitrary packet we end up with a transition matrix P_w, where

$$P_w = \begin{bmatrix} I & & & \\ \tilde{A}_2 & \tilde{A}_1 & & \\ & \tilde{A}_2 & \tilde{A}_1 & \\ & & \ddots & \ddots \end{bmatrix}, \tag{7.33}$$

with

$$\tilde{A}_2 = \begin{bmatrix} b_k^{(k)} & b_{k-1}^{(k)} & \cdots & b_1^{(k)} \\ & b_k^{(k)} & \cdots & b_2^{(k)} \\ & & \ddots & \vdots \\ & & & b_k^{(k)} \end{bmatrix}, \quad \tilde{A}_1 = \begin{bmatrix} b_0^{(k)} & & & \\ b_1^{(k)} & b_0^{(k)} & & \\ \vdots & \vdots & \ddots & \\ b_{k-1}^{(k)} & b_{k-2}^{(k)} & \cdots & b_0^{(k)} \end{bmatrix}.$$

Let us define a vector $\mathbf{u}^{(0)} = \mathbf{y}$, with corresponding smaller vectors $\mathbf{u}_i^{(0)} = \mathbf{y}_i$, and let

$$\mathbf{u}^{(i)} = \mathbf{u}^{(i-1)} P_w = \mathbf{u}^{(0)} P_w^i, \quad i \geq 1.$$

We have

$$W_i^{(q)} = Pr\{W_q \leq i\} = \mathbf{u}_0^{(i)}\mathbf{1},\ i \geq 1, \tag{7.34}$$

with

$$W_0^{(q)} = \mathbf{u}_0^{(0)}\mathbf{1}. \tag{7.35}$$

7.3 GI/Geo/k Systems

This system can be studied using different techniques. However we are going to present the matrix-analytic approach which is easier to explain and the display of its matrix makes it easy to understand. The matrix approach uses the remaining (residual) inter-arrival time as a supplementary variable.

We assume that this system has inter-arrival times that have a general distribution vector $\mathbf{a} = [a_1,\ a_2,\ \cdots]$. The service times follow the geometric distribution with parameter b, and b_i^j is the probability that the services of i out of the j in service are completed in a time epoch. We have $k < \infty$ identical servers in parallel.

7.3.1 *Using MAM - Same as using supplementary variables*

Consider the DTMC $\{(X_n, J_n), n \geq 0\}$, $X_n \geq 0$, $J_n = 1, 2, \cdots$. At time n, we define X_n as the number of packet in the system and J_n as the remaining time before the next packet arrives. The transition matrix representing this DTMC is given as

$$P = \begin{bmatrix} B_{0,0} & B_{0,1} & & & & & & \\ B_{1,0} & B_{1,1} & B_{1,2} & & & & & \\ \vdots & \vdots & \vdots & \ddots & & & & \\ B_{k,0} & B_{k,1} & \cdots & B_{k,k} & B_{k,k+1} & & & \\ & A_{k+1} & A_k & A_{k-1} & \cdots & A_0 & & \\ & & A_{k+1} & A_k & A_{k-1} & \cdots & A_0 & \\ & & & \ddots & \ddots & \ddots & \ddots & \ddots \end{bmatrix}, \tag{7.36}$$

where the block matrices are defined as follows. First we define new matrices to help us make a compact representation of the blocks. Define I_∞ as an identity matrix of infinite order, $\mathbf{0}_\infty^T$ as an infinite order row vector of zeros and 0_∞ as a infinite order square matrix of zeros. Now we can present the block matrices as

$$A_0 = \begin{bmatrix} b_0^{(k)}\mathbf{a} \\ 0_\infty \end{bmatrix}, \; A_{k+1} = \begin{bmatrix} \mathbf{0}_\infty^T \\ b_k^{(k)} I_\infty \end{bmatrix}, \; A_i = \begin{bmatrix} b_i^{(k)}\mathbf{a} \\ b_{i-1}^{(k)} I_\infty \end{bmatrix}, \; 1 \leq i \leq k,$$

$$B_{0,0} = \begin{bmatrix} \mathbf{0}_\infty^T \\ I_\infty \end{bmatrix}, \; B_{0,1} = \begin{bmatrix} \mathbf{a} \\ 0_\infty \end{bmatrix},$$

$$B_{i,0} = \begin{bmatrix} \mathbf{0}_\infty^T \\ b_i^{(i)} I_\infty \end{bmatrix}, \; 1 \leq i \leq k, \; B_{i,i+1} = \begin{bmatrix} b_0^{(i)}\mathbf{a} \\ 0_\infty \end{bmatrix}, \; 1 \leq i \leq k,$$

$$B_{i,j} = \begin{bmatrix} b_j^{(i)}\mathbf{a} \\ b_{j-1}^{(i)} I_\infty \end{bmatrix}, \; 1 \leq i \leq k, \; 1 \leq j \leq i \leq k.$$

As an example, we display one of the blocks in detail. Consider the block matrix A_i, $1 \leq i \leq k$ we write it out as

$$A_i = \begin{bmatrix} b_i^{(k)}a_1 & b_i^{(k)}a_2 & b_i^{(k)}a_3 & b_i^{(k)}a_4 & b_i^{(k)}a_5 & \cdots \\ b_{i-1}^{(k)} & & & & & \\ & b_{i-1}^{(k)} & & & & \\ & & b_{i-1}^{(k)} & & & \\ & & & b_{i-1}^{(k)} & & \\ & & & & b_{i-1}^{(k)} & \\ & & & & & \ddots \end{bmatrix}.$$

We assume that the system is stable and let the $x_{i,j} = Pr\{X_n = i, J_n = j\}|_{n\to\infty}$, and further let $\boldsymbol{x}_i = [x_{i,1}, \; x_{i,2}, \; \cdots]$ and $\boldsymbol{x} = [\boldsymbol{x}_0, \; \boldsymbol{x}_1, \; \cdots]$. Then we know that $\boldsymbol{x}$ is given as the unique solution to

$$\boldsymbol{x} = \boldsymbol{x}P, \;\; \boldsymbol{x}\mathbf{1} = 1.$$

We also know that there is a matrix R which is the non-negative solution to the matrix polynomial equation

$$R = \sum_{j=0}^{k+1} R^j A_j.$$

This R matrix has the following structure, by virtue of the structure of matrix A_0,

$$R = \begin{bmatrix} r_1 & r_2 & r_3 & \cdots \\ 0 & 0 & 0 & \cdots \\ 0 & 0 & 0 & \cdots \\ \vdots & \vdots & \vdots & \vdots \end{bmatrix}, \qquad (7.37)$$

where $r_i,\ i = 1,2,\cdots$, are scalar elements.

Clearly, to obtain the matrix R we will have to truncate it at some point. It is intuitive that if the elements of $\mathbf{a} = [a_1,\ a_2,\ a_3,\ \cdots]$ are decreasing from a point $K < \infty$ onwards then the elements of $\mathbf{r} = [r_1,\ r_2,\ r_3,\ \cdots]$ will also be decreasing from around that point. In such situations, which are the ones encountered in practice, we can assume that an appropriate truncation would be acceptable. After obtaining the R matrix we simply go ahead and apply the standard matrix-geometric results to this problem.

Now we consider a special case where the inter-arrival times have finite support.

Case of finite support for inter-arrival times:

When the inter-arrival time has a finite support $K_a = n_t < \infty$, then this problem becomes easier to handle. There is no truncation needed and capitalizing on the structure of A_0 still helps tremendously. First let us define an identity matrix I_j of order j a row vector of zeros $\mathbf{0}_j^T$ of order j and a corresponding column vector $\mathbf{0}_j$. The structure of the transition matrix remains the same, however the blocks need to be defined more specifically for dimension. We have

$$A_0 = \begin{bmatrix} b_0^{(k)}\mathbf{a} \\ \mathbf{0}_{n_t-1}\mathbf{0}_{n_t}^T \end{bmatrix}, \quad A_{k+1} = \begin{bmatrix} \mathbf{0}_{n_t-1}^T & 0 \\ b_k^{(k)} I_{n_t-1} & \mathbf{0}_{n_t-1} \end{bmatrix},$$

$$A_i = \begin{bmatrix} b_i^{(k)}\mathbf{a} \\ \tilde{A}_i \end{bmatrix}, \quad \tilde{A}_i = [b_{i-1}^{(k)} I_{n_t-1},\ \mathbf{0}_{n_t-1}],\ 1 \le i \le k,$$

$$B_{0,0} = \begin{bmatrix} \mathbf{0}_{n_t-1}^T & 0 \\ I_{n_t-1} & \mathbf{0}_{n_t-1} \end{bmatrix},$$

$$B_{0,1} = \begin{bmatrix} \mathbf{a} \\ \mathbf{0}_{n_t-1}\mathbf{0}_{n_t}^T \end{bmatrix}, \quad B_{i,0} = \begin{bmatrix} \mathbf{0}_{n_t-1}^T & 0 \\ b_i^{(i)} I_{n_t-1} & \mathbf{0}_{n_t-1} \end{bmatrix}, \quad 1 \le i \le k,$$

$$B_{i,i+1} = \begin{bmatrix} b_0^{(i)}\mathbf{a} \\ \mathbf{0}_{n_t-1}\mathbf{0}_{n_t}^T \end{bmatrix}, \quad 1 \le i \le k,$$

$$B_{i,j} = \begin{bmatrix} b_j^{(i)}\mathbf{a} \\ \tilde{B}_{i,j} \end{bmatrix}, \quad \tilde{B}_{i,j} = b_{j-1}^{(i)} I_{n_t-1},\ \mathbf{0}_{n_t-1}],\ 1 \le i \le k,\ 1 \le j \le i \le k.$$

7.3.2 Numerical Example

Consider the case where $k = 3$, $b = 0.75$. We assume that the inter-arrival time has a finite support $n_t = 4$, with $\mathbf{a} = [0.1,\ 0.3,\ 0.5,\ 0.1]$. The mean arrival rate is $\lambda = (\sum_{j=1}^{4} j a_j)^{-1} = 0.3846$ and service rate is $\mu = kb = 2.25$. We only display A_0, A_1, A_2, A_3, and A_4.

$$A_0 = \begin{bmatrix} a_1 b_0^{(3)} & a_2 b_0^{(3)} & a_3 b_0^{(3)} & a_4 b_0^{(3)} \\ 0 & 0 & 0 & 0 \\ 0 & 0 & 0 & 0 \\ 0 & 0 & 0 & 0 \end{bmatrix} = \begin{bmatrix} .0016 & .0047 & .0078 & .0016 \\ 0 & 0 & 0 & 0 \\ 0 & 0 & 0 & 0 \\ 0 & 0 & 0 & 0 \end{bmatrix},$$

$$A_1 = \begin{bmatrix} a_1 b_1^{(3)} & a_2 b_1^{(3)} & a_3 b_1^{(3)} & a_4 b_1^{(3)} \\ b_0^{(3)} & 0 & 0 & 0 \\ 0 & b_0^{(3)} & 0 & 0 \\ 0 & 0 & b_0^{(3)} & 0 \end{bmatrix} = \begin{bmatrix} .0141 & .0422 & .0703 & .0141 \\ .0156 & 0 & 0 & 0 \\ 0 & .0156 & 0 & 0 \\ 0 & 0 & .0156 & 0 \end{bmatrix},$$

$$A_2 = \begin{bmatrix} a_1 b_2^{(3)} & a_2 b_2^{(3)} & a_3 b_2^{(3)} & a_4 b_2^{(3)} \\ b_1^{(3)} & 0 & 0 & 0 \\ 0 & b_1^{(3)} & 0 & 0 \\ 0 & 0 & b_1^{(3)} & 0 \end{bmatrix} = \begin{bmatrix} .0422 & .1266 & .2109 & .0422 \\ .1406 & 0 & 0 & 0 \\ 0 & .1406 & 0 & 0 \\ 0 & 0 & .1406 & 0 \end{bmatrix},$$

$$A_3 = \begin{bmatrix} a_1 b_3^{(3)} & a_2 b_3^{(3)} & a_3 b_3^{(3)} & a_4 b_3^{(3)} \\ b_2^{(3)} & 0 & 0 & 0 \\ 0 & b_2^{(3)} & 0 & 0 \\ 0 & 0 & b_2^{(3)} & 0 \end{bmatrix} = \begin{bmatrix} .0422 & .1266 & .2109 & .0422 \\ .4219 & 0 & 0 & 0 \\ 0 & .4219 & 0 & 0 \\ 0 & 0 & .4219 & 0 \end{bmatrix},$$

$$A_4 = \begin{bmatrix} 0 & 0 & 0 & 0 \\ b_3^{(3)} & 0 & 0 & 0 \\ 0 & b_3^{(3)} & 0 & 0 \\ 0 & 0 & b_3^{(3)} & 0 \end{bmatrix} = \begin{bmatrix} 0 & 0 & 0 & 0 \\ .4219 & 0 & 0 & 0 \\ 0 & .4219 & 0 & 0 \\ 0 & 0 & .4219 & 0 \end{bmatrix}.$$

The resulting R matrix is given as

$$R = \begin{bmatrix} .0017 & .0049 & .0080 & .0016 \\ 0 & 0 & 0 & 0 \\ 0 & 0 & 0 & 0 \\ 0 & 0 & 0 & 0 \end{bmatrix}.$$

7.4 PH/PH/k Systems

The PH/PH/k system is a queueing system in which the arrival process is phase type with parameters (α, T) of dimension n_t and service is phase type with parameters (β, S) of order n_s, with $k < \infty$ identical servers in parallel. We will see later that this system is really a reasonably general discrete time single node queue with multiple servers. Several other multiserver queues such as the Geo/Geo/k, GI/Geo/k (with finite support for arrival process) and Geo/G/k (with finite support for service times) are all special cases of this PH/PH/k system.

A very important set of matrices that are critical to this system relate to the number of service completions among ongoing services. To that effect let us study the following matrix sequence $\{B_i^{(j)},\ 0 \le i \le j \le k < \infty\}$. This matrix records the probability of i service completions in a time slot when there are j items in service $(i \le j)$. We present a recursion for computing this matrix sequence. First we re-display the matrix sequence $\{B_j^{(i)},\ 0 \le i \le j < \infty\}$ which was defined in Chapter 5.

$$B_0^{(1)} = S,\ \ B_1^{(1)} = (\mathbf{s}\beta),$$

$$B_0^{(j)} = B_0^{(j-1)} \otimes S,\ \ B_j^{(j)} = B_{j-1}^{(j-1)} \otimes (\mathbf{s}\beta),$$

$$B_i^{(j)} = B_{i-1}^{(j-1)} \otimes (\mathbf{s}\beta) + B_i^{(j-1)} \otimes S,\ 1 \le i \le j-1.$$

Let us further consider another matrix sequence $\{\tilde{B}_i^{(j)},\ 0 \le i \le j \le k < \infty\}$, with

$$\tilde{B}_1^{(1)} = \mathbf{s}, \tag{7.38}$$

and

$$\tilde{B}_i^{(j)} = \tilde{B}_i^{(j-1)} \otimes S + B_{i-1}^{(j-1)} \otimes \mathbf{s},\ 1 \le i \le j-1. \tag{7.39}$$

Now consider a DTMC $\{(X_n, L_n, J_n(X_n)), n \ge 0\}$ where at time n, X_n is the number of packets in the system, L_n is the phase of arrival and $J_n(X_n)$ is the set of phases of service of the items in service out of the X_n items in the system. The state space is $\{(0,u) \cup (i,u,v(i)),\ i \ge 0, u = 1,2,\cdots,n_t, v(i) = \{1,2,\cdots,n_s\}^{min\{i,k\}}\}$. Note that if there are u items in service then $v(u) = \{1,2,\cdots,n_s\}^{v}$, because we have to keep track of the phases of service of u items. The transition matrix of this Markov chain can be written as

$$P = \begin{bmatrix} C_{0,0} & C_{0,1} & & & & & \\ C_{1,0} & C_{1,1} & C_{1,2} & & & & \\ \vdots & \vdots & \vdots & \ddots & & & \\ C_{k,0} & C_{k,1} & \cdots & C_{k,k} & C_{k,k+1} & & \\ & A_{k+1} & A_k & A_{k-1} & \cdots & A_0 & \\ & & A_{k+1} & A_k & A_{k-1} & \cdots & A_0 \\ & & & \ddots & \ddots & \ddots & \ddots & \ddots \end{bmatrix}, \quad (7.40)$$

where the block matrices are defined as follows:

$$A_0 = (\mathbf{t}\boldsymbol{\alpha}) \otimes B_0^{(k)},\ A_i = (\mathbf{t}\boldsymbol{\alpha}) \otimes B_i^{(k)} + T \otimes B_{i-1}^{(k)},\ 1 \le i \le k,$$

$$A_{k+1} = T \otimes B_k^{(k)},\ C_{0,0} = T,\ C_{0,1} = (\mathbf{t}\boldsymbol{\alpha}) \otimes \boldsymbol{\beta},$$

$$C_{i,0} = T \otimes \tilde{B}_i^{(i)},\ C_{i,i+1} = (\mathbf{t}\boldsymbol{\alpha}) \otimes B_0^{(i)} \otimes \boldsymbol{\beta},\ 1 \le i \le k-1,$$

$$C_{i,v} = T \otimes \tilde{B}_{i-v}^{(i)} + (\mathbf{t}\boldsymbol{\alpha}) \otimes \tilde{B}_{i-v+1}^{(i)} \otimes \boldsymbol{\beta},\ 1 \le v \le i \le k-1,$$

and

$$C_{k,j} = A_{k+1-j},\ 0 \le j \le k+1.$$

We can now apply matrix-analytic method (MAM) to analyze this system. The procedure is straightforward. The only challenge is that of dimensionality. The R matrix is of dimension $n_t {n_s}^k \times n_t {n_s}^k$. This could be huge. As such this problem can only be analyzed using MAM for a moderate size number of servers. For example, if we have an arrival phase type of order $n_t = 2$, service phase of order $n_s = 2$ and number of servers $k = 5$, then our R matrix is of order 64×64, which is easily manageable. However, if the number of servers goes to $k = 10$ we have an R matrix of order 8192×8192 – a big jump indeed.

For a moderate size R we apply the GI/M/1 results of the matrix-geometric approach. We calculate R which is given as the minimal non-negative solution to the equation

$$R = \sum_{v=0}^{k+1} R^v A_v.$$

Given that the system is stable, then we have that

$$\boldsymbol{x} = \boldsymbol{x}P,\ \boldsymbol{x}\mathbf{1} = 1,$$

with

$$\boldsymbol{x} = [\boldsymbol{x}_0, \boldsymbol{x}_1, \boldsymbol{x}_2, \cdots], \quad \boldsymbol{x}_0 = [x_{0,1}, x_{0,2}, \cdots, x_{0,n_t}],$$

$$\boldsymbol{x}_i = [\boldsymbol{x}_{i,1}, \boldsymbol{x}_{i,2}, \cdots, \boldsymbol{x}_{i,n}], \boldsymbol{x}_{i,j} = [x_{i,j,1}, x_{i,j,2}, \cdots, x_{i,j,n_t(i)}],$$

where $n_t(i) = min\{in_s, kn_s\}$.

From the matrix-geometric results we have

$$\boldsymbol{x}_{i+1} = \boldsymbol{x}_i R, \quad i \geq k.$$

The boundary variables $\boldsymbol{x}_b = [\boldsymbol{x}_0, \boldsymbol{x}_1, \cdots, \boldsymbol{x}_k]$ are obtained by solving

$$\boldsymbol{x}_b = \boldsymbol{x}_b B[R], \ \boldsymbol{x}_b \mathbf{1} = 1,$$

where

$$B[R] = \begin{bmatrix} C_{0,0} & C_{0,1} & & & \\ C_{1,0} & C_{1,1} & C_{1,2} & & \\ \vdots & \vdots & \vdots & \ddots & \\ C_{k-1,0} & C_{k-1,1} & C_{k-1,2} & \cdots & C_{k-1,k} \\ C_{k,0} & C_{k,1} + RA_{k+1} & C_{k,2} + \sum_{v=0}^{1} R^{2-v} A_{k-v+1} & \cdots & C_{k,k+1} + \sum_{v=0}^{k} R^{k-v+1} A_{k-v+1} \end{bmatrix}.$$

This is then normalized by

$$\boldsymbol{x}_0 \mathbf{1} + (\sum_{v=1}^{k-1} \boldsymbol{x}_v) \mathbf{1} + \boldsymbol{x}_k (I-R)^{-1} \mathbf{1} = 1.$$

Usually, all we can manage easily when the dimension of R is huge is just to study the tail behaviour of this system.

7.5 Geo/D/k Systems

This type of queue was used extensively in modelling Asynchronous Transfer Mode (ATM) system. The system is as follows. There are $k < \infty$ identical servers in parallel. Packets arrive according to the Bernoulli process with parameter a and each packet requires a constant service time $d < \infty$ which is provided by just one server, any of the k servers, with $k < d$. If we know the total number of packets in the system at time $t \geq 1$, we can easily study the number in the system at times $t, t+d, t+2d, t+3d, \cdots, t+jd, \cdots$. The continuous time analogue of this system is well discussed by Tijms (1994). The first related result on the M/D/k system is by Crommelin (1932).

Let $\mathscr{A}_m \geq 0$ be the number of packets that arrive during the time interval of duration $m \geq 1$, and let

$$a_i^{(m)} = Pr\{\mathscr{A}_m = i\},\ 0 \leq i \leq m. \tag{7.41}$$

Further let X_n be the number of packets in the system at time n and $c_{w,v} = Pr\{X_{n+1} = v|X_n = w\}|_{n\to\infty}$, then we have

$$c_{w,v} = \begin{cases} a_v^{(d)}, & 0 \leq v \leq d;\ 0 \leq w \leq k, \\ a_{v+k-i}^{(d)}, & 1 \leq v \leq d;\ i \geq k+1. \end{cases} \tag{7.42}$$

It is clear that X_n is a DTMC with the transition probability matrix P given as

$$P = \begin{bmatrix} a_0^{(d)} & a_1^{(d)} & a_2^{(d)} & \cdots & a_d^{(d)} & & & \\ a_0^{(d)} & a_1^{(d)} & a_2^{(d)} & \cdots & a_d^{(d)} & & & \\ \vdots & \vdots & \vdots & \cdots & \vdots & & & \\ a_0^{(d)} & a_1^{(d)} & a_2^{(d)} & \cdots & a_d^{(d)} & & & \\ & a_0^{(d)} & a_1^{(d)} & a_2^{(d)} & \cdots & a_d^{(d)} & & \\ & & a_0^{(d)} & a_1^{(d)} & a_2^{(d)} & \cdots & a_d^{(d)} & \\ & & & \ddots & \ddots & \ddots & \ddots & \ddots \end{bmatrix}. \tag{7.43}$$

This P matrix can be re-blocked to have

$$P = \begin{bmatrix} B & C & & & \\ E & A_1 & A_0 & & \\ & A_2 & A_1 & A_0 & \\ & & A_2 & A_1 & A_0 \\ & & & \ddots & \ddots & \ddots \end{bmatrix}, \tag{7.44}$$

where the matrix B is of dimension $k \times k$, C is of dimension $k \times (d-k+1)$, E is of dimension $(d-k+1) \times k$ and A_j, $j = 0,1,2$ are each of dimension $(d-k+1) \times (d-k+1)$. We display only B and C and the rest of the block matrices follow.

$$B = \begin{bmatrix} a_0^{(d)} & a_1^{(d)} & \cdots & a_{k-1}^{(d)} \\ a_0^{(d)} & a_1^{(d)} & \cdots & a_{k-1}^{(d)} \\ \vdots & \vdots & \cdots & \vdots \\ a_0^{(d)} & a_1^{(d)} & \cdots & a_{k-1}^{(d)} \end{bmatrix}, \quad C = \begin{bmatrix} a_k^{(d)} & a_{k+1}^{(d)} & \cdots & a_d^{(d)} \\ a_k^{(d)} & a_{k+1}^{(d)} & \cdots & a_d^{(d)} \\ \vdots & \vdots & \cdots & \vdots \\ a_k^{(d)} & a_{k+1}^{(d)} & \cdots & a_d^{(d)} \end{bmatrix}.$$

We can now apply the matrix-analytic results for QBD to this problem. We can obtain the matrix R which is the minimal nonnegative solution to the matrix

Equation $R = A_0 + RA_1 + R^2A_2$, obtain the solution to $\boldsymbol{x} = \boldsymbol{x}P,\ \boldsymbol{x}\mathbf{1} = 1$ and also solve the boundary equations and normalize. After that we have $\boldsymbol{x}_0$ and $\boldsymbol{x}_1$ determined, where $\boldsymbol{x} = [\boldsymbol{x}_0,\ \boldsymbol{x}_1,\ \boldsymbol{x}_2,\ \cdots]$ and $\boldsymbol{x}_{i+1} = \boldsymbol{x}_i R$.

Finally the distribution of number in the system $x_i = Pr\{X_t = i\}|_{t\to\infty}$ is obtained from the following relationship:

$$\boldsymbol{x}_0 = [x_0,\ x_1,\ \cdots,\ x_{k-1}],\ \boldsymbol{x}_1 = [x_k,\ x_{k+1},\ \cdots,\ x_d],$$

$$\boldsymbol{x}_j = [x_{(j-1)(d-k+1)-k},\ x_{(j-1)(d-k+1)-k+1},\ \cdots,\ x_{j(d-k+1)-k-1}],\ j \geq 2.$$

7.5.1 *Waiting Times*

We can study the waiting time distribution here using a special technique when the service discipline is FCFS. If the service discipline is FCFS, then we can consider this system as one in which we focus only on one server and assign it only the kth arriving packet for service. We thus transform this Geo/D/k system to an NB_k/D/1 system for the sake of studying the FCFS waiting time. Here NB_k stands for a negative binomial arrival process with k phases, i.e. $n_t = k$. This NB_k actually captures the arrival process to a target server. Hence this system now has a PH arrival process $(\boldsymbol{\alpha}, T)$ of dimension k with $\boldsymbol{\alpha} = [1,\ 0,\ 0,\ \cdots,\ 0]$ and $T = \begin{bmatrix} 1-a & a & & \\ & 1-a & a & \\ & & \ddots & \ddots \\ & & & 1-a \end{bmatrix}$. The service time can be represented as a PH distribution $(\boldsymbol{\beta}, S)$ of order $n_s = d$ with $\boldsymbol{\beta} = [1,\ 0,\ 0,\ \cdots,\ 0]$ and $S = \begin{bmatrix} & 1 & & \\ & & 1 & \\ & & & \ddots \\ & & & & 1 \\ & & & & \end{bmatrix}$.

Consider the state space $\Delta = \{(0,\ell) \cup (i,\ell,j),\ i \geq 1, \ell = 1,2,\cdots,k; j = 1,2,\cdots,d\}$, where the first tupple $(0,\ell)$ represents the states with no packet in the system for the target server and arrival is in phase ℓ, i.e. the server is waiting for $(k-\ell)^{th}$ packet to be assigned to it, and the second tupple (i,ℓ,j) represents the states with i packets waiting in the system for the target server, arrival is in phase ℓ and the current packet in service has completed j units of its service. The DTMC representing this system has a transition matrix P of the form

$$P = \begin{bmatrix} B & C & & & \\ E & A_1 & A_0 & & \\ & A_2 & A_1 & A_0 & \\ & & A_2 & A_1 & A_0 \\ & & & \ddots & \ddots & \ddots \end{bmatrix},$$

where

$$B = T,\ C = (\mathbf{t}\boldsymbol{\alpha}) \otimes \boldsymbol{\beta},\ E = T \otimes \mathbf{s},\ A_0 = (\mathbf{t}\boldsymbol{\alpha}) \otimes S,$$

$$A_1 = (\mathbf{t}\boldsymbol{\alpha}) \otimes (\mathbf{s}\boldsymbol{\beta}) + T \otimes S,\ A_2 = T \otimes (\mathbf{s}\boldsymbol{\beta}).$$

By virtue of the structure of matrix A_0 we can see that the associated R matrix for this system has only the last block of rows that is non-zero, i.e. we have

$$R = \begin{bmatrix} 0 & 0 & \cdots & 0 \\ 0 & 0 & \cdots & 0 \\ \vdots & \vdots & \cdots & \vdots \\ 0 & 0 & \cdots & 0 \\ R_1 & R_2 & \cdots & R_{kd} \end{bmatrix}. \tag{7.45}$$

With this structure we can write the equations for the R matrix in block form for each of the kd blocks and save on computations, with

$$R_1 = aS + (1-a)SR_1 + a(\mathbf{s}\boldsymbol{\beta})R_{kd} + (1-a)(\mathbf{s}\boldsymbol{\beta})R_{kd}R_1, \tag{7.46}$$

$$R_j = aSR_{j-1} + (1-a)SR_j + a(\mathbf{s}\boldsymbol{\beta})R_{kd}R_{j-1} + (1-a)(\mathbf{s}\boldsymbol{\beta})R_{kd}R_j,\ 2 \le j \le kd. \tag{7.47}$$

From here on we can apply standard QBD results to obtain $\boldsymbol{x} = \boldsymbol{x}P$, $\boldsymbol{x}\mathbf{1} = 1$, where $\boldsymbol{x} = [\boldsymbol{x}_0,\ \boldsymbol{x}_1,\ \cdots]$, and $\boldsymbol{x}_0 = [x_{0,1},\ x_{0,2},\ \cdots,\ x_{0,k}]$, $\boldsymbol{x}_i = [\boldsymbol{x}_{i,1},\ \boldsymbol{x}_{i,2},\ \cdots,\ \boldsymbol{x}_{i,k}]$ and $\boldsymbol{x}_{i,\ell} = [x_{i,\ell,1},\ x_{i,\ell,2},\ \cdots,\ x_{i,\ell,d}]$. Because this is a special case of the PH/PH/1 we simply apply the results of Section 5.10. However, for the Geo/D/k one notices some additional structures which can be explored. For details see Rahman and Alfa (2009). We can now apply the standard results for the PH/PH/1 discussed in Section 5.10 to obtain the waiting times.

7.6 MAP/D/k Systems

Ii is straightforward to extend the results of the Geo/D/k to the case of MAP/D/k. We simply change the Bernoulli arrival process with parameters a and $1-a$ to a Markovian arrival process (MAP), with matrices D_1 and D_0, respectively, which is the matrix analogue of the Bernoulli process. However, there are some subtle aspects related to the setup of the waiting time distribution because the inter-arrival time is no longer geometric.

Let the associate MAP be represented by the two n square matrices D_0 and D_1, where the elements $(D_k)_{ij}$ represent the probability of $k = 0, 1$ arrivals during the transition from state i to j in a single time interval. We also know that the matrix $D = D_0 + D_1$ is stochastic, with $\boldsymbol{\pi} = \boldsymbol{\pi} D$, $\boldsymbol{\pi}\mathbf{1} = 1$ and arrival rate $\lambda = \boldsymbol{\pi} D_1 \mathbf{1}$. Further let the service time be constant and equal to $d < \infty$. This can be represented by a phase type distribution $(\boldsymbol{\beta}, S)$ of dimension d where $\boldsymbol{\beta} = [1,\ 0,\ 0,\ \cdots,\ 0]$ and $S = \begin{bmatrix} 0 & I_{d-1} \\ \mathbf{0} & \mathbf{0} \end{bmatrix}$, where I_j is an identity matrix of order j.

The assumption here is that customers are served in the FIFO order, in which case there is no overtaking, i.e. an $(i+1)^{st}$ arriving customer will not finish service before the ith arriving customer. In fact, the ith arriving customer will finish service before the $(i+1)^{st}$. With this behaviour of the system, we know that we can just study the arrival of the following customers: $i,\ i+k,\ i+2k,\ \cdots,\ i+jk,\ \cdots$ and consider only one of the servers since all the servers are identical.

Let a group of k customers be grouped together and let the last one in each group be called a super-customer, then the arrival process of the super-customer is represented by another MAP of order kn and represented by two matrices two D_0^* and D_1^*, where

$$D_0^* = \begin{bmatrix} D_0 & D_1 & & \\ & D_0 & D_1 & \\ & & \ddots & \ddots \\ & & & D_0 \end{bmatrix}, \quad \text{and } D_1^* = \begin{bmatrix} & & \\ & & \\ D_1 & & \end{bmatrix}.$$

That arrival rate of the super-customers is $\lambda_k = k^{-1}\lambda$. This can also be obtained as $\lambda_k = \boldsymbol{\pi}_k D_1^* \mathbf{1}$, where $\boldsymbol{\pi}_k = \boldsymbol{\pi}_k (D_0^* + D_1^*)$, $\boldsymbol{\pi}_k \mathbf{1} = 1$. We can now study this system as a special queue $_k$MAP/D/1 system which focusses on just one server. This system can be used to capture the waiting time of an arbitrary customer.

Now consider the following state space $\{L_t, S_t, J_t\}$ where this is on

$$\{(i,s,j); i \geq 0, 1 \leq s \leq d, 1 \leq j \leq kn\},$$

where at time t, L_t is the number of super-customers in the system, S_t the elapsed time of the service of the super-customer who is in service, and J_t the arrival phase of the super-customer. This is a DTMC with its transition matrix P written as

$$P = \begin{bmatrix} B & C & & \\ E & A_1 & A_0 & \\ & A_2 & A_1 & A_0 \\ & & \ddots & \ddots & \ddots \end{bmatrix},$$

where

$$B = D_0^*,\ C = \boldsymbol{\beta} \otimes D_1^*,\ E = \mathbf{s} \otimes D_0^*,\ A_0 = S \otimes D_1^*,$$

$$A_1 = S \otimes D_0^* + (\mathbf{s}\boldsymbol{\beta}) \otimes D_1^*,\ A_2 = (\mathbf{s}\boldsymbol{\beta}) \otimes D_0^*.$$

Due to the special structure of A_2, i.e. it has only one block element that is non-zero, which is at the bottom left corner, the structure of the G matrix, which is the minimal non-negative solution to the matrix quadratic equation $G = A_2 + A_1G + A_0G^2$ is of the form

$$G = \begin{bmatrix} G_1 & \mathbf{0} & \cdots & \mathbf{0} \\ G_2 & \mathbf{0} & \cdots & \mathbf{0} \\ \vdots & \vdots & \cdots & \vdots \\ G_d & \mathbf{0} & \cdots & \mathbf{0} \end{bmatrix},$$

where

$$G_i = D_0^* G_{i+1} + D_1^* G_{i+1} G_i,\ \ 1 \le i \le d1,$$

$$G_d = D_0^* + D_1^* G_1.$$

Let the matrix $F_t(i)$ represent arrivals in t units of time, where $F_t(i)$ is the matrix coefficient corresponding to z^i in $(D_0^* + zD_1^*)^t$, it was shown in Alfa (2003b) that

$$G_1 = \sum_{j=0}^{d-1} F_d(j) G_1^j,\ d \le 2,$$

and

$$G_i = \sum_{v=0}^{d-1} F_{d-i+1}(v) G_1^v,\ 2 \le i \le d-1,\ d \ge 2.$$

The matrices $F_i(j)$ can be computed as

$$F_i(j) = F_{i-1}(j) D_0^* + F_{i-1}(j-1) D_1^*,\ 1 \le i \le d,\ 0 \le j \le i-2,$$

where

$$F_0(0) = I,\ F_i(i-1) = F_{i-1}(i-2) D_1^*.$$

The matrix R which is the minimal non-negative solution to the matrix equation $R = A_0 + RA_1 + R^2A_2$ can now be obtained from the G matrix as

$$R = A_0(I - A_1 - A_0G)^{-1}.$$

Applying the standard matrix geometric technique we can obtain the vector $\boldsymbol{x} = [\boldsymbol{x}_0, \boldsymbol{x}_1, \boldsymbol{x}_2, \cdots]$ which is the solution to $\boldsymbol{x} = \boldsymbol{x}P, \ \boldsymbol{x}\mathbf{1} = 1$.

Now we proceed to study the waiting time of a customer, which is the same as the waiting time of the dth customer of any super-customer. Let $\mathbf{y}_i, \ i \geq 0$ be the stationary distribution vector that a super-customer sees i super-customers ahead of it in the queue at arrival. This vector $\mathbf{y}_i$ can now be written as

$$\mathbf{y}_0 = \frac{k}{\lambda}(\boldsymbol{x}_0 D_1^* + \boldsymbol{x}_1(\mathbf{s} \otimes D_1^*)),$$

and

$$\mathbf{y}_i = \frac{k}{\lambda}(\boldsymbol{x}_i S \otimes D_1^* + \boldsymbol{x}_{i+1}((\mathbf{s}\boldsymbol{\beta}) \otimes D_1^*)), \ i \geq 1.$$

If we further partition $\mathbf{y}_i = [\mathbf{y}_{i,1}, \ \mathbf{y}_{i,2}, \ \cdots, \ \mathbf{y}_{i,d}]$, then we have

$$\mathbf{y}_0 = \frac{k}{\lambda}[\boldsymbol{x}_0 D_1^* + \boldsymbol{x}_{1,d} D_1^*],$$

$$\mathbf{y}_{i,1} = \frac{k}{\lambda}[\boldsymbol{x}_{i+1,d} D_1^*], \ i \geq 1,$$

$$\mathbf{y}_{i,j} = \frac{k}{\lambda}[\boldsymbol{x}_{i,j-1} D_1^*], \ 1 \geq 1, \ 2 \leq j \leq d.$$

The $\mathbf{y}_i$ may be simplified further as

$$\mathbf{y}_0 = \frac{k}{\lambda}[\boldsymbol{x}_0, \ \boldsymbol{x}_{1,d}](\mathbf{1}_2 \otimes D_1^*),$$

$$\mathbf{y}_i = \frac{k}{\lambda}[\boldsymbol{x}_{i+1,d}, \ \boldsymbol{x}_{i,1}, \ \boldsymbol{x}_{i,2}, \ \cdots, \ \boldsymbol{x}_{i,d-1}] \times (I_d \otimes D_1^*), \ i \geq 1.$$

Let w_r be the probability that the waiting time of a customer (i.e. the kth customer of a super-customer) is r, then we have

$$w_0 = \mathbf{y}_0 \mathbf{1}_{kn},$$

$$w_r = \sum_{i=1}^{r} \mathbf{y}_i (I_d \otimes \mathbf{1}_{kn}) \Omega_r^{(i)} \mathbf{1}_{kn}, \ r \geq 1.$$

where $\Omega_r^{(i)}$ is a square matrix of order d and given by

$$\Omega_0^{(0)} = I_d,\ \Omega_r^{(r)} = (\mathbf{s}\boldsymbol{\beta})^r,$$

$$\Omega_r^{(0)} = 0,\ r \neq 0,\ \Omega_r^{(1)} = S^{r-1}(\mathbf{s}\boldsymbol{\beta}), r \geq 1,$$

$$\Omega_r^{(i)} = (\mathbf{s}\boldsymbol{\beta})\Omega_{r-1}^{(i-1)} + S\Omega_{r-1}^{(i)},\ 2 \leq i \leq r,\ \Omega_r^{(i)} = \mathbf{0},\ r < i.$$

Note that here for $i \geq 1$, the element $(\Omega_r^{(i)})_{ij} = 0,\ \forall v \neq 1$, because a customer's service always starts form phase one.

7.7 BMAP/D/k Systems

In this section we extend the results of the MAP/D/k to BMAP/D/k. The same ideas follows, however there is a subtle key difference in the way the super-customer is grouped for the BMAP case.

Let the arrivals be represented by a series of order $n \times n$ matrices $D_0,\ D_1,\ D_2,\ \cdots,$ where as usual, the element $(D_k)_{ij}$ captures the probability of k arrivals when the underlining Markov chain goes through a transition from state i to state j, where $(i,j) = 1, 2, \cdots, n$. Let $D = \sum_{k=0}^{\infty} D_k$; the matrix D is stochastic and we assume that it has an invariant vector $\boldsymbol{\pi}$ given by $\boldsymbol{\pi} = \boldsymbol{\pi} D,\ \boldsymbol{\pi}\mathbf{1} = 1$. The arrival rate in this case is given by $\lambda = \boldsymbol{\pi}\sum_{i=1}^{\infty} iD_i\mathbf{1}$. As in the case of the MAP/D/k we let the service time be constant and equal to $d < \infty$. This can be represented by a phase type distribution $(\boldsymbol{\beta}, S)$ of dimension d where $\boldsymbol{\beta} = [1,\ 0,\ 0,\ \cdots,\ 0]$ and $S = \begin{bmatrix} 0 & I_{d-1} \\ \mathbf{0} & \mathbf{0} \end{bmatrix}$, where I_j is an identity matrix of order j. It is known that this system is stable if $\lambda d < k$, see Alfa Alfa (2003a).

If the matrix sequence D_v is finite, i.e. $D_v = 0,\ \forall v > K$, then our interest in this section in the nontrivial cases where $k < Kd$. Again we make the same assumption of FIFO service as in the MAP/D/k system. Now we consider a super- customer as in the MAP case, except that this super-customer is derived differently. It is part of a batch super-customer system. Let us define what we call a ${}_k$BMAP arrival process describing super-customers by defining new matrix series $C_0,\ C_1,\ C_2,\ \cdots,$ of order $kn \times kn$, with

$$C_0 = \begin{bmatrix} D_0 & D_1 & \cdots & D_{k-1} \\ & D_0 & \cdots & D_{k-2} \\ & & \ddots & \vdots \\ & & & D_0 \end{bmatrix},$$

and

$$C_j = \begin{bmatrix} D_{jk} & D_{jk+1} & \cdots & D_{(j+1)k-1} \\ D_{jk-1} & D_{jk} & \cdots & D_{(j+1)k-2} \\ \vdots & \vdots & \ddots & \vdots \\ D_{(j-1)k+1} & D_{(j-1)k+2} & \cdots & D_{jk} \end{bmatrix}, \; j = 1,2,3\cdots$$

Note that here D_{jk} implies $D_{j\times k}$.

Let $C = \sum_{i=0}^{\infty} C_i$ and $\pi_k = \pi_k C$, $\pi_k \mathbf{1} = 1$. Then the arrival rate of the super-customers is $\lambda_k = \pi_k \sum_{i=1}^{\infty} iC_i\mathbf{1}$, where $\lambda_k = k^{-1}\lambda$.

Since all the servers are identical and there is no overtaking, i.e. strictly speaking an $(i+1)^{st}$ arriving customer will not finish service before the ith arriving customer, we can study the waiting time of an arbitrary customer by studying the ${}_k$BMAP/D/1 system.

Now consider the following state space $\{L_t, S_t, J_t\}$ where this is on

$$\{(i,s,j); i \geq 0, 1 \leq s \leq d, 1 \leq j \leq kn\},$$

where at time t, L_t is the number of super-customers in the system, S_t the elapsed time of the service of the super-customer who is in service, and J_t the arrival phase of the super-customer. This is a DTMC with its transition matrix P written as

$$P = \begin{bmatrix} B_0 & B_1 & B_2 & B_3 & B_4 & \cdots \\ E & A_1 & A_2 & A_3 & A_4 & \cdots \\ & A_0 & A_1 & A_2 & A_3 & \cdots \\ & & A_0 & A_1 & A_2 & \cdots \\ & & & \ddots & \ddots & \ddots \end{bmatrix},$$

where

$$B_0 = C_0,\; B_i = \boldsymbol{\beta} \otimes C_i,\; i = 1,2,3,\cdots,\; E = \mathbf{s} \otimes C_0,$$

$$A_0 = (\mathbf{s}\boldsymbol{\beta}) \otimes C_0,\; A_i = S \otimes C_{i-1} + (\mathbf{s}\boldsymbol{\beta}) \otimes C_i,\;\; i = 1,2,3,\cdots.$$

This is an M/G/1 type Markov chain and it has a matrix G which is the minimal non-negative solution to

$$G = \sum_{j=0}^{\infty} A_j G^j.$$

Because of the special structure of matrix A_0, we find that the matrix G is of the form

$$G = \begin{bmatrix} G_1 & \mathbf{0} & \mathbf{0} & \cdots & \mathbf{0} \\ G_2 & \mathbf{0} & \mathbf{0} & \cdots & \mathbf{0} \\ G_3 & \mathbf{0} & \mathbf{0} & \cdots & \mathbf{0} \\ \vdots & \vdots & \vdots & \cdots & \vdots \\ G_d & \mathbf{0} & \mathbf{0} & \cdots & \mathbf{0} \end{bmatrix}.$$

It is shown in Alfa (2003a) that the block elements of G can be computed as

$$G_1 = \sum_{j=}^{\infty} F_d(j) G_1^j, \; d \geq 2,$$

and

$$G_i = \sum_{v=0}^{\infty} F_{d-i+1}(v) G_1^v, \; 2 \leq i \leq d-1, \; d \geq 2.$$

The matrices $F_i(j)$ are given as

$$F_i(j) = \sum_{v=0}^{j} F_{i-1}(j-v) C_v, \; i \geq 1, j \geq 0,$$

where $F_0(0) = I$, and $F_0(j) = \mathbf{0}, \forall j \neq 0$. We can now use the standard methods for M/G/1 for computing the stationary distribution $\boldsymbol{x}$ of this Markov chain, i.e. $\boldsymbol{x} = \boldsymbol{x}P$, $\boldsymbol{x}\mathbf{1} = 1$, where $\boldsymbol{x} = [\boldsymbol{x}_0, \; \boldsymbol{x}_1, \; \boldsymbol{x}_2, \; \cdots]$,. For details on the approach for solving this see Chapter 5 and also Neuts (1989) and Latouche and Ramaswami (1999).

The waiting time distribution of an arbitrary customer can now be studied as follows. Let $\mathbf{y}_i$, $i \geq 0$, be the stationary probability vector that a super-customer sees i super-customers ahead of it at its time of arrival. We let $y_{i,j,v}$ be the elements of $\mathbf{y}_i$ which represent the probability that a super-customer arrives to fin d i super-customers waiting ahead of it, the on-going service is in phase j and the ${}_k$BMAP is in phase v. We have

$$\mathbf{y}_0 = \frac{k}{\lambda} \left[\boldsymbol{x}_0 \sum_{j=1}^{\infty} C_j + \boldsymbol{x}_1 \sum_{j=1}^{\infty} \mathbf{s} \otimes C_j \right],$$

and for $i \geq 1$, we have

$$\mathbf{y}_i = \frac{k}{\lambda} \left[\boldsymbol{x}_0 \sum_{j=i+1}^{\infty} \boldsymbol{\beta} \otimes C_j + \sum_{j=1}^{i} \boldsymbol{x}_j \sum_{v=i-j+1}^{\infty} S \otimes C_v + \sum j = 1^{i+1} \boldsymbol{x}_j \sum_{v=i-j+2}^{\infty} (\mathbf{s}\boldsymbol{\beta}) \otimes C_v \right].$$

If we partition $\mathbf{y}_i$ such that $\mathbf{y}_i = [\mathbf{y}_{i,1}, \mathbf{y}_{i,2}, \cdots, \mathbf{y}_{i,d}]$, $i \geq 1$, then we have

$$\mathbf{y}_0 = \frac{k}{\lambda}\left[\boldsymbol{x}_0 \sum_{j=1}^{\infty} C_j + \boldsymbol{x}_{1,d} \sum_{j=1}^{\infty} C_j\right],$$

$$\mathbf{y}_{i,1} = \frac{k}{\lambda}\left[\boldsymbol{x}_0 \sum_{j=i+1}^{\infty} C_j + \sum_{j=1}^{i+1} \boldsymbol{x}_{j,d} \sum_{v=i-j+2}^{\infty} C_v\right], \; i \geq 1,$$

$$\mathbf{y}_{i,j} = \frac{k}{\lambda}\left[\sum_{w=1}^{i} \boldsymbol{x}_{w,j-1} \sum_{v=i-w+1}^{\infty} C_v\right], \; i \geq 1, \; 2 \leq j \leq d.$$

Let w_r be the probability that the waiting time of a customer (i.e. the kth customer of a super-customer) is r, then we have

$$w_0 = \mathbf{y}_0 \mathbf{1}_{kn},$$

$$w_r = \sum_{i=1}^{r} \mathbf{y}_i (I_d \otimes \mathbf{1}_{kn}) \Omega_r^{(i)} \mathbf{1}_{kn}, \; r \geq 1.$$

where $\Omega_r^{(i)}$ is a square matrix of order d and given by

$$\Omega_0^{(0)} = I_d, \; \Omega_r^{(r)} = (\mathbf{s}\boldsymbol{\beta})^r,$$

$$\Omega_r^{(0)} = 0, \; r \neq 0, \; \Omega_r^{(1)} = S^{r-1}(\mathbf{s}\boldsymbol{\beta}), r \geq 1,$$

$$\Omega_r^{(i)} = (\mathbf{s}\boldsymbol{\beta})\Omega_{r-1}^{(i-1)} + S\Omega_{r-1}^{(i)}, \; 2 \leq i \leq r, \; \Omega_r^{(i)} = \mathbf{0}, \; r < i.$$

Note that here for $i \geq 1$, the element $(\Omega_r^{(i)})_{ij} = 0$, $\forall v \neq 1$, because a customer's service always starts form phase one.

We can easily obtain the distribution on the number in the system for the original queueing system BMAP/D/k by using the results in Section 8 of Alfa (2003a).

7.8 Geo/G/∞ Systems

We consider a queueing system in which customers arrive singly according to the geometric inter-arrival times with parameter a, i.e. the arrival process is according to the Bernoulli process with a representing the probability of an arrival during a time interval. The service times S follow a general distribution, with $b_i = Pr\{S = i\}$, $i \geq 1$ and the mean service time is given as s, i.e. $s = E[S]$. Further define $B_i = Pr\{S \leq i\}$. There are infinite number of identical servers in parallel. Our interest is to determine the distribution of the number in the system at arbitrary times. We define $\bar{a} = 1 - a$ and $\bar{B}_i = 1 - B_i = Pr\{S > i\}$.

7.8.1 Preliminaries

We define some terms in this section. It is well known that a Bernoulli process consists of a sequence of independent experiments in which there are two possible outcomes $(Success, Failure)$ each time. We let the probability of *Success* be defined as a and *Failure* with probability $\bar{a} = 1 - a$. In n independent consecutive trials the probability of getting $k, (k \leq n)$, *Successes* $r(k,n)$ follows a Binomial distribution and is given as

$$r(k,n) = \binom{n}{k} a^k \bar{a}^{n-k}, \quad 0 \leq k \leq n.$$

Now we define a time-dependent Bernoulli process as that process consisting of independent trials but with each vth trial having a *Success* probability a_v which depends on the time only (i.e. the location of the trial in the sequence) and not the outcome of previous trials. Hence if we carry out n trials the probability of obtaining k *Successes* in the n trials has what is called a "Poissonian-Binomial" distribution, whose $z-$transform is actually a product form of the $z-$transform of several different independent Bernoulli processes. For details on this see (Johnson et al (1992),Page 138). Let $C^*(z,n)$ be the $z-$transform of the number of *Successes* in n trials then we have

$$C^*(z,n) = \prod_{v=1}^{n} [za_v + \bar{a}_v] = \sum_{k=0}^{n} \gamma_k^{(n)} z^k, \ |z| \leq 1, \ n \geq 1. \tag{7.48}$$

where $\gamma_k^{(n)}$ is the probability of getting k successes in n consecutive independent trials, and can be computed recursively as

$$\gamma_k^{(n+1)} = \gamma_{k-1}^{(n)} a_{n+1} + \gamma_k^{(n)} \bar{a}_{n+1}, \ 0 \leq k \leq n, \tag{7.49}$$

where

$$\gamma_0^{(0)} = 1, \text{ and } \gamma_{-1}^{(n)} = \gamma_{n+1}^{(n)} = 0.$$

For any $n > 0$, let $n\tilde{a} = \sum_{v=1}^{n} a_i$. It is known that a Poissonian-Binomial with parameters a_v, $v = 1, 2, \cdots, n$ has the same mean as a Binomial with mean $\tilde{a}$. However, the variance of the Binomial is greater than or equal to the variance of the Poissonian-Binomial. Also, if $\max_n\{a_n\} \to 0$ as $n \to \infty$ and $\sum_{v=1}^{n} a_v = \theta$

remains constant, then the Poissonian-Binomial tends to a Poisson distribution with parameter θ (Feller (1968), Page 282).

We can now proceed to use these definitions and results.

7.8.2 The Number in the System at Arbitrary Times

We study the system at discrete time epochs labelled $n = 0, 1, 2, 3, \cdots$, and adapt the approach used in Latouche and Ramaswami (1999) to study the M/G/∞ system. Let N_n be the number of items in the system at time n and $P^*(z,n)$ be the $z-$transform of N_n given the system starts of empty at time $n = 0$, i.e. $N_0 = 0$ with probability 1. We assume that items are assigned to the server according to the order $1, 2, 3, \cdots$, and this does not affect our results.

Consider times $\cdots, \ n-k, \ n-k+1, \cdots, \ n, n+1, \cdots$. Then there are three possible events by time n regarding the number in the system

- There were no arrivals up to time n, and the probability of this is $\bar{a}^n$ and the associated transform is $z^0 \bar{a}^n = \bar{a}^n$.
- There was an arrival at time $n-k$, but its service was completed by time n. The probability of this event is $\bar{a}^{n-k-1} a B_k$. In this case, the system of the remaining servers $2, 3, \cdots$ behaves like another Geo/G/∞ system, with transform given as $P^*(z,k)$.
- There was an arrival at time $n-k$ whose service was not completed by time k. The probability of this event is $\bar{a}^{n-k-1} a \bar{B}_k$.

It is immediately clear that we have

$$P^*(z,n) = \bar{a}^n + \sum_{k=0}^{n-1} [\bar{a}^{n-k-1} a B_k + z \bar{a}^{n-k-1} a \bar{B}_k] P^*(z,k), \ n \geq 1, \tag{7.50}$$

with

$$P^*(z,0) = 1.$$

Solving this equation recursively leads to

$$P^*(z,n) = \prod_{k=0}^{n-1} [(\bar{a} + a B_k) + z a \bar{B}_k] = \prod_{k=0}^{n-1} [(1 - a \bar{B}_k) + z a \bar{B}_k], \tag{7.51}$$

which is the z-transform of the number in the system at an arbitrary time n. It is quasi-Binomial distribution, for which the parameters keep changing for each trial of the time-dependent Bernoulli process that leads to it. This result was also obtained by Ushakumari and Krishnamoorthy (1995), and they went ahead and presented an equation at the bottom of Page 187 of their paper for computing the distribution of the number of busy servers at any time. That equation is cumbersome to implement. In the next few steps we present an efficient and effective recursive scheme for computing the same distribution.

Let us define $x_i^{(n)} = Pr\{N_n = i|N_0 = 0\}$, such that $P^*(z,n) = \sum_{v=0}^{n} x_v^{(n)} z^v$. Our interest is to develop a recursion for computing x_i^n. Let $\alpha_1^{(k)} = a\bar{B}_k$ and $\alpha_0^{(k)} = 1 - a\bar{B}_k$. The recursive algorithm for computing $x_i^{(n)}$ can be written as

$$x_i^{n+1} = x_{i-1}^n \alpha_1^n + x_i^n \alpha_0^n, \quad 0 \le i \le n+1, \tag{7.52}$$

where

$$x_{-1}^{(n)} = x_{n+1}^{(n)} = 0, \text{ and } x_0^{(0)} = 1.$$

Let us define a vector $\boldsymbol{x}^{(n)} = [x_0^{(n)}, \, x_1^{(n)}, \, \cdots, \, x_n^{(n)}]$ and an $n \times (n+1)$ matrix P_n, which is given as

$$P_n = \begin{bmatrix} \alpha_0^{(n)} & \alpha_1^{(n)} & & & & \\ & \alpha_0^{(n)} & \alpha_1^{(n)} & & & \\ & & \alpha_0^{(n)} & \alpha_1^{(n)} & & \\ & & & \ddots & \ddots & \\ & & & & \alpha_0^{(n)} & \alpha_1^{(n)} \end{bmatrix},$$

then the above result can be written in matrix form as

$$\boldsymbol{x}^{(n+1)} = \boldsymbol{x}^{(n)} P_n, \quad n \ge 0,$$

where $P_0 = [\alpha_0^{(n)}, \, \alpha_1^{(n)}]$ and $\boldsymbol{x}^{(0)} = \mathbf{e}_1^T$ and $\mathbf{e}_i$ is the ith column vector of an identity matrix with $\mathbf{e}_i^T$ as its transpose.

Next we use an example of the Geo/D/∞ case to show that the distribution of the number of busy servers in the Geo/G/∞ system is not necessarily Poisson.

7.8.3 Special Case of Geo/D/∞

Let the service process be constant and equal to $d < \infty$. In this case we have

$$B_v = \begin{cases} 0, & v < d \\ 1 & v \ge d, \end{cases}$$

or

$$\bar{B}_v = \begin{cases} 1, & v < d \\ 0 & v \geq d. \end{cases}$$

Using our recursion we can show that

$$x_i^{(n)} = \binom{n}{i} a^i (\bar{a})^{n-i}, \ n \leq d, \ 0 \leq i \leq d, \tag{7.53}$$

and

$$x_i^{(n+1)} = \begin{cases} x_i^{(n)}, & v \leq d, \ 0 \leq i \leq d \\ 0 & n \leq d, \quad i > d. \end{cases} \tag{7.54}$$

This we know is not Poisson, and confirms our earlier statement.

7.8.4 The Limiting Distribution of the Number in the System

Now we consider the limiting distribution of N_n as $n \to \infty$, i.e. its behaviour at steady state. First we re-write the function (Equation (7.51)) as

$$P^*(z,n) = \prod_{k=0}^{n-1}[(1-a\bar{B}_k) + za\bar{B}_k] = \prod_{k=0}^{n-1}[(1-a(1-z)\bar{B}_k)]. \tag{7.55}$$

It can be shown that provided that $s = \sum_{k=0}^{\infty} \bar{B}_k < \infty$, then this product series converges as $n \to \infty$. We assume that this is the case and use results from Struble (2009) to obtain the upper and lower bounds for this limiting distribution.

Let us write $\alpha_k(z) = a(1-z)\bar{B}_k$, then

$$P^*(z,n) = (1-\alpha_0(z))(1-\alpha_1(z))\cdots(1-\alpha_{n-1}(z)).$$

It is well known that for any y, $1 - y \leq e^{-y}$. This bound is actually tight when we have $0 \leq y \leq 1$. Hence we have

$$P^*(z,n) \leq e^{-(\alpha_0(z)+\alpha_1(z)+\cdots+\alpha_{n-1}(z))},$$

which leads to

$$P^*(z,\infty) \leq e^{-\sum_{k=0}^{\infty}\alpha_k(z)} = e^{-as(1-z)}, \tag{7.56}$$

where $s = \sum_{k=0}^{\infty} \bar{B}_k$ is the mean service time. This is a Poisson limit - an upper bound.

Now we consider the lower bound. Let $F(z) = max_{0 \le k \le \infty} \alpha_k(z)$. We have $F(z) = a(1-z)$. We show that, because $(1-F(z))^{\frac{y}{F(z)}} \le 1-y,\ 0 \le y \le F(z)$, then

$$(1-a(1-z))^{\frac{\alpha_k(z)}{a(1-z)}} \le 1-\alpha_k(z),\ \forall k. \tag{7.57}$$

Hence

$$(1-a(1-z))^{\frac{\sum_{k=0}^{n-1}\alpha_k(z)}{a(1-z)}} \le P^*(z,n) \to (1-a(1-z))^{\frac{sa(1-z)}{a(1-z)}} = (1-a(1-z))^s \le P^*(z,\infty).$$

We therefore have

$$(1-a(1-z))^s \le P^*(z,\infty) \le e^{-as(1-z)}. \tag{7.58}$$

Note that both bounds do not depend on the distribution of the service time S, but only on the mean service time s. Let us write this as

$$L(z) \le P^*(z,\infty) \le U(z).$$

7.8.4.1 A useful property of the bounds

In this section we investigate how tight the first moments of the bounds are, i.e the means of the bounds. Let $E[L] = \frac{dL(z)}{dz}|_{z\to 1}$ and $E[U] = \frac{dU(z)}{dz}|_{z\to 1}$. We immediately see that

$$\frac{dL(z)}{dz}|_{z\to 1} = \frac{dU(z)}{dz}|_{z\to 1} = as,$$

hence

$$E[U] - E[L] = 0.$$

So the bounds coincide in the case of means.

Based on this we conclude that the mean of the number of busy servers is as, which is the same result for the $M/G/\infty$. This is quite intuitive.

Beyond the first moment we were not able to infer additional results related to higher moments.

Next we return to the Geo/D/∞ case and study its limiting behaviour.

7.8.4.2 The Limiting behaviour of the Geo/D/∞ system

Returning to the Geo/D/∞ example of Section 7.8.3, we study its limiting behaviour.

Consider Equations (7.53) and (7.54) and let $n \to \infty$, then we have $x_i^n|_{n\to\infty} = x_i$, leading to

$$x_i = \begin{cases} \binom{d}{i} a^i \bar{a}^{d-i}, \; 0 \le i \le d, \\ 0, \; i > d \end{cases} \tag{7.59}$$

and the $z-$transform of this is

$$x(z) = (az + \bar{a})^d,$$

which is the lower bound found. It is a Binomial distribution.

If we now let $d \to \infty$ while also letting $a \to 0$, at the same time keeping ad constant, then we end up with the well known Poisson approximation to the Binomial distribution with mean ad, i.e. we have

$$x_i = \frac{(ad)^i}{i!} exp(-ad),$$

with its $z-$transform given as

$$x(z) = exp(-ad(1-z)),$$

which is the upper bound given earlier.

So, in the case of the Geo/D/∞ system,

- if the service time is short then the distribution of the number of busy servers at steady state follows the Binomial distribution
- if the service time is very long then the distribution of the number of busy servers at steady state, under certain conditions, follows the Poisson distribution in the limit.

We point out that in Ushakumari and Krishnamoorthy (1995) the authors analyzed the limiting behaviour of the Geo/D/∞ system by assuming that $\bar{B}_n^i \to \bar{B}^i$ as $n \to \infty$. That assumption is only valid in special cases, such as Geo/D/∞ and not in most cases.

7.8.5 Extension to the PH/G/∞ Case

The main result follows very easily for the case when the arrival system is governed by a phase type distribution. Consider a phase arrival system with parameters (α, T) of order m, and general service as defined earlier. Let $\mathbf{t} = \mathbf{1} - T\mathbf{1}$. Now define $G^*(z,n)$ as an $m \times m$ matrix whose elements $(G^*(z,n))_{i,j}$ refer to the $z-$transform of the number in the system at time n with arrival in phase j given that the arival phase was i at time 0. We can easily show that

$$G^*(z,n) = T^n + \sum_{k=0}^{n-1} [T^{n-k-1}\mathbf{t}\alpha B_k + zT^{n-k-1}\mathbf{t}\alpha \bar{B}_k] G^*(z,k),\ n \geq 1.$$

with

$$G^*(z,0) = U = \mathbf{1}\beta,$$

and β is the Perron vector of T.

From this we have

$$G^*(z,n) = \prod_{k=0}^{n-1} [T + \mathbf{t}\alpha B_k + z\mathbf{t}\alpha \bar{B}_k] = \prod_{k=0}^{n-1} [T + \mathbf{t}\alpha - \mathbf{t}\alpha \bar{B}_k + z\mathbf{t}\alpha \bar{B}_k]$$

$$= \prod_{k=0}^{n-1} [T + \mathbf{t}\alpha - (1-z)\mathbf{t}\alpha \bar{B}_k]$$

7.9 Problems

1. Question 1: Consider a single node queue in discrete time. Customers arrive accordion to the geometric distribution with parameter $a,\ 0 < a < 1$. Suppose the system manager have access to using up to $k,\ (1 \leq k < \infty)$ servers. We assume the manager uses only $n < k$ servers as long as the number of customers in the system is not more than $N > k$. Each of the servers has a geometric service distribution with parameter $b,\ 0 < b < a < 1$. Once the number in the system exceeds N the manager adds the remaining $k - n$ servers into the system so they system can have full service.

   a. Set this up as a DTMC with the number of customers in the system as the state space.
   b. Under what conditions is the system stable?
   c. Find the steady state distribution of the number in the system for the case of $k = 3,\ n = 2$ and $N = 5$. Select the values of any values a and b that make the system stable and then compute the distribution of the number in the system.

2. Question 2: Consider a parallel server queue in discrete time with 3 servers. Data collection was carried out and 1000 inter-arrival times were observed over many days. The observed inter-arrival times are $\{1,2,5,7,8,10\}$ *minutes*. The frequencies of the range of the inter-arrival times are, respectively $\{100,250,400,150,50,50\}$. Over the same period of time, 600 service times were observed and all of them were exactly 3 *minutes*. What kind of queueing model will you use for studying this problem? Explain why? Develop a procedure that you will use to compute the waiting time distribution in this system.

References

Alfa AS (2003a) Algorithmic analysis of the BMAP/D/k system in discrete time. Adv Appl Probab 35:1131–1152

Alfa AS (2003b) waiting time distribution of the MAP/D/k system in discrete time – a more efficient algorithm. Operations Research Letters 31:263–267

Artalejo JR, Hernández-Lerma O (2003) Performance analysis and optimal control of the Geo/Geo/c queue. Performance Evaluation 52:15–39

Crommelin CD (1932) Delay probability formulae when the holding times are constant. Post Off Electr Eng J 25:41–50

Feller W (1968) An Introduction to Probability Theory and its Applications, Vol. 1. Wiley, New York

Johnson NL, Kotz S, Kemp AW (1992) Univariate Discrete Distributions. John Wiley & Sons, Inc., New York

Latouche G, Ramaswami V (1999) Introduction to matrix analytic methods in stochastic modeling. In: Pennsylvania, Philadelphia, Pennsylvania

Neuts MF (1989) Structured Stochastic Matrices of M/G/1 Type and Their Applications. Marcel Decker Inc., New York

Rahman MM, Alfa AS (2009) Computational procedures for a class of GI/D/k system in discrete time. J Probab & Statis 2009(Article ID: 716364):18 pages, DOI 10.1155/2009/716364,2009

Struble RA (2009) Infinite products rescued, draft onl. Tech. rep., URL http:www.infiniteproduct.info/struifpr.htm.2009

Tijms HC (1994) Stochastic Models, an Algorithmic Approach. New York, Wiley

Ushakumari PV, Krishnamoorthy A (1995) The queueing system $B_D/G_D/\infty$. Optimization 34:185–193

Chapter 8
Single Node Queueing Models with Server Vacations

8.1 Vacation Queues

Vacation queues are a very important class of queues in real life, especially in telecommunication systems where medium access control (MAC) is a critical component of managing a successful network. By definition, a vacation queue is a queueing system in which the server is available only a portion of the time. At other times it is busy serving other stations or just not available may be due to maintenance activities (either routine or due to a breakdown). Polling systems are a class of queueing systems in which a server goes to attend to different queues based on a schedule, which implies that the server is not always available to a particular queue. Hence when the server is attending to other queues then it is away on vacation as far as the queue not receiving service is concerned. Vacation queues are used extensively to approximate polling systems.

There are two main categories of vacation queues: gated and ungated systems. In a gated system only the packets in the system waiting at the arrival of a server can get served during that visit. All those that arrive after its return from vacation will not get served during that visit. An ungated system does not have that condition. Under each category there are sub-categories which include exhaustive service, number limited service, time limited service and random vacations. In an exhaustive system all the packets waiting in the system (and only those in the gate for a gated system) will all be served before the server goes on another vacation, i.e. until the queue becomes empty for ungated, or the number in the gate is empty for the gated system. For the number limited case the server only processes a pre-determined number of packets during its visit, similarly for time limited the queue is attended to for a fixed length of time. In both cases if the number in the system becomes empty before the limits are reached then the server would go on a vacation. In a random vacation system, the server may stop service at any time and the decision of when to stop serving is random. Generally when a service is interrupted in the time limited and

A.S. Alfa, *Applied Discrete-Time Queues*, DOI 10.1007/978-1-4939-3420-1_8

random cases the point of service resumption after a vacation depends on the rule employed; it could be preemptive resume, preemptive repeat, etc. Finally, vacations could be single or multiple. In a single vacation system when the server returns from a vacation it starts to serve immediately or waits for an arrival of a packet and then starts to serve. In a multiple vacation system, on the other hand, a server will return to a vacation mode if at the end of a vacation there are no waiting packets. In what follows we discuss mainly ungated systems. We focus on exhaustive systems for the Geo/G/1, GI/Geo/1 and MAP/PH/1 for both single and multiple vacations. We then consider the MAP/PH/1 system for the number limited, time limited and random vacations. We study only the multiple vacations for these last categories.

8.2 Geo/G/1 vacation systems

The Geo/G/1 vacation queue is the discrete analogue of the M/G/1 vacation queue that has been used extensively in the literature. In this section we study the exhaustive version of this model for both the single and multiple vacation systems. Consider a single server queue with geometric arrivals with parameter a and general service times S with $b_i = Pr\{S = i\},\ i = 1, 2, \cdots$, with mean service times $b^{'}$, and $b^{''}$ is its second moment about zero. The vacation duration $\mathscr{V}$ has a general distribution given as $v_i = Pr\{\mathscr{V} = i\},\ i = 1, 2, \cdots$, and mean vacation times $v^{'}$, and $v^{''}$ is its second moment about zero. We will study this system as an imbedded Markov chain.

8.2.1 *Single vacation system*

In a single vacation system, the server goes on a vacation as soon as the system becomes empty, i.e. no packets in the system. It returns after a random time $\mathscr{V}$ and starts to serve the waiting packets or waits to serve the first arriving packet if there is no waiting packets at its arrival. This system can be studied using the imbedded Markov chain idea or using the method of supplementary variables. The imbedded Markov chain approach leads to an M/G/1 type Markov chain and modified methods for the M/G/1 models can be applied. The supplementary variable approach is cumbersome to work with in scalar form but when written in matrix form it leads to a QBD all be it with infinite blocks which will have to be truncated. We focus on the imbedded Markov chain approach.

Imbedded Markov chain approach: At time n, let X_n be the number of packets in the system, J_n the state of the vacation, with $J_n = 0$ representing vacation termination and $J_n = 1$ representing service completion. Let A_s and A_v be the number of arrivals during a service time and during a vacation period, respectively. We can write

$$(X_{n+1}, J_{n+1}) = \begin{cases} X_n + A_s - 1, 1; \; X_n \geq 1, \\ A_s, 1; \qquad\qquad (X_n, J_n) = (0,0), \\ A_v, 0; \qquad\qquad (X_n, J_n) = (0,1) \end{cases} . \tag{8.1}$$

Let $a_k^{(y)} = Pr\{A_y = k\}, y = s, v$, then

$$a_k^{(s)} = \sum_{m=k}^{\infty} b_m \binom{m}{k} a^k (1-a)^{m-k}, \; k = 0, 1, 2, \cdots, \text{ and} \tag{8.2}$$

$$a_k^{(v)} = \sum_{m=k}^{\infty} v_m \binom{m}{k} a^k (1-a)^{m-k}, \; k = 0, 1, 2, \cdots. \tag{8.3}$$

Define $G_k^*(az + 1 - a) = (az + 1 - a)^k$, $k \geq 0$ and $A_y^*(z) = \sum_{k=0}^{\infty} a_k^{(y)} z^k$, $y = s, v$, $|z| \leq 1$. Further define $B^*(z) = \sum_{k=1}^{\infty} b_k z^k$, $V^*(z) = \sum_{k=1}^{\infty} v_k z^k$, $|z| \leq 1$. We can write

$$A_s^*(z) = \sum_{k=0}^{\infty} z^k \sum_{m=k}^{\infty} b_m \binom{m}{k} a^k (1-a)^{m-k}. \tag{8.4}$$

After routine algebraic operations we have

$$A_s^*(z) = \sum_{k=0}^{\infty} b_k G_k^*(az + 1 - a), \tag{8.5}$$

$$A_v^*(z) = \sum_{k=0}^{\infty} v_k G_k^*(az + 1 - a). \tag{8.6}$$

These can be written as

$$A_s^*(z) = B^*(G(az + 1 - a)), \tag{8.7}$$

$$A_v^*(z) = V^*(G^*(az + 1 - a)), \; |z| \leq 1. \tag{8.8}$$

Let us assume the system is stable and let

$$x_{i,j} = lim_{n \to \infty} Pr\{X_n = i, \; J_n = j\}, \; i = 0, 1, 2, \cdots, ; j = 0, 1,$$

then we have the following steady state equations

$$x_{i,0} = x_{0,1} a_i^{(v)}, \; i = 0, 1, 2, \cdots \tag{8.9}$$

$$x_{i,1} = x_{0,0} a_i^{(s)} + \sum_{k=1}^{i+1} (x_{k,0} + x_{k,1}) a_{i-k+1}^{(s)}, \; i = 0, 1, 2, \cdots. \tag{8.10}$$

Letting

$$x_1^*(z) = \sum_{i=0}^{\infty} x_{i,1} z^j, \text{ and } x^*(z) = \sum_{i=0}^{\infty} (x_{i,0} + x_{i,1}) z^i, \ |z| \le 1,$$

we have

$$x_1^*(z) = \frac{a_0^{(v)} A_s^*(z)(z-1) + A_s^*(z)[A_v^*(z) - 1]}{z - A_s^*(z)} x_{0,1} \tag{8.11}$$

and

$$x^*(z) = \frac{zA_v^*(z) - A_s^*(z) + a_0^{(v)} A_s^*(z)(z-1)}{z - A_s^*(z)} x_{0,1}. \tag{8.12}$$

By setting $z = 1$ and using l'Hôpital's rule we obtain

$$1 = x(1) = \frac{1 + a\bar{v} - ab' + a_0^{(v)}}{1 - ab'} x_{0,1}, \tag{8.13}$$

$$x_1(1) = \frac{a_0^{(v)} + a\bar{v}}{1 - ab'} x_{0,1} = \frac{a_0^{(v)} + av'}{1 + av' - ab' + a_0^{(v)}}. \tag{8.14}$$

Noting that the mean number in the system (at departure) $E[X]$ is $(x_1(1))^{-1} \frac{dx_1(z)}{dz}|_{z \to 1}$ we obtain

$$E[X] = ab' + \frac{a^2 b''}{2(1 - ab')} + \frac{a^2 v''}{2(a_0^{(v)} + av')}. \tag{8.15}$$

This is the well known decomposition result by several authors (Fuhrman and Cooper (1985), Doshi (1985), and Doshi (1986)). By decomposition we mean the result consists of three component with the first two being due to the Geo/G/1 system without a vacation and the third being due to a vacation.

Supplementary variable approach: For the supplementary variable approach we keep track of the number of packets in the system, the remaining time to service completion when the server is busy and the remaining time to the end of a vacation if the server is on a vacation. We will present the supplementary variable approach in detail later for the case of multiple vacations.

8.2.2 Multiple vacations system

In this system the server goes on a vacation as soon as the system becomes empty. It returns after a random time $\mathscr{V}$ and starts to serve the waiting packets or goes back for another vacation of duration $\mathscr{V}$ if there is no waiting packets at its arrival.

Imbedded Markov chain approach: Considering the imbedded approach and using the same notation as in the case of the single vacation we have

$$(X_{n+1}, J_{n+1}) = \begin{cases} X_n + A_s - 1, 1; \; X_n \geq 1 \\ A_v, 0; \qquad\qquad (X_n, J_n) = (0,1) \end{cases} \tag{8.16}$$

Here the term A_v is the number of arrivals during all the consecutive multiple vacations before the server finds at least one packet in the system. Hence

$$a_j^{(v)} = a_j^{(v)}(1 - a_0^{(v)})^{-1}, \; j = 1, 2, \cdots .$$

Following the same procedure as in the case of the single vacation we have

$$E[X] = ab' + \frac{a^2 b''}{2(1 - ab')} + \frac{av''}{2v'}. \tag{8.17}$$

The decomposition result is also apparent here.

Supplementary variable approach: For the supplementary variable approach we keep track of the number of packets in the system, the remaining time to service completion when the server is busy and the remaining time to the end of a vacation if the server is on a vacation. Consider the state space

$$\Delta = \{(0,l) \cup (i,j) \cup (i,l), i \geq 1, l = 1, 2, \cdots ; j = 1, 2, \cdots .\}$$

The first pair $(0,l)$ refers to the states when the system is empty and the server is on a vacation and the remaining time to the end of the vacation is l. The second pair (i,j) refers to the states when the server is busy serving with i packets in the system and the remaining service time of the packet in service is j. Finally the third pair (i,l) refers to the states when there are i packets in the system and the server is still on vacation which has l more units of time before completion. The transition matrix P of this DTMC can be represented as

$$P = \begin{bmatrix} B & C & & & \\ E & A_1 & A_0 & & \\ & A_2 & A_1 & A_0 & \\ & & \ddots & \ddots & \ddots \end{bmatrix},$$

where

$$A_k = \begin{bmatrix} A_k^{(11)} & A_k^{(12)} \\ A_k^{(21)} & A_k^{(22)} \end{bmatrix},\ k = 0,1,2; C = \begin{bmatrix} C^{(11)} & C^{(12)} \end{bmatrix},\ E = \begin{bmatrix} E^{(11)} \\ E^{(21)} \end{bmatrix}, \text{ with}$$

$$A_0^{(11)} = A_0^{(22)} = \begin{bmatrix} 0 & 0 & 0 & \cdots & \\ a & 0 & \cdots & & \\ 0 & a & 0 & \cdots & \\ & & \ddots & & \end{bmatrix},\ A_0^{(21)} = \begin{bmatrix} ab_1 & ab_2 & ab_3 & ab_4 & \cdots \\ 0 & 0 & 0 & 0 & \cdots \\ 0 & 0 & 0 & 0 & \cdots \\ \vdots & \vdots & \vdots & \vdots & \vdots \end{bmatrix},\ A_0^{(12)} = 0_\infty,$$

$$A_1^{(11)} = \begin{bmatrix} ab_1 & ab_2 & ab_3 & \cdots & \\ 1-a & 0 & \cdots & & \\ 0 & 1-a & 0 & \cdots & \\ & & \ddots & & \end{bmatrix},$$

$$A_1^{(21)} = \begin{bmatrix} (1-a)b_1 & (1-a)b_2 & (1-a)b_3 & (1-a)b_4 & \cdots \\ 0 & 0 & 0 & 0 & \cdots \\ 0 & 0 & 0 & 0 & \cdots \\ \vdots & \vdots & \vdots & \vdots & \vdots \end{bmatrix},$$

$$A_1^{(22)} = \begin{bmatrix} 0 & 0 & 0 & \cdots & \\ 1-a & 0 & \cdots & & \\ 0 & 1-a & 0 & \cdots & \\ & & \ddots & & \end{bmatrix},\ A_1^{(12)} = 0_\infty,$$

$$A_2^{(11)} = \begin{bmatrix} (1-a)b_1 & (1-a)b_2 & (1-a)b_3 & (1-a)b_4 & \cdots \\ 0 & 0 & 0 & 0 & \cdots \\ 0 & 0 & 0 & 0 & \cdots \\ \vdots & \vdots & \vdots & \vdots & \vdots \end{bmatrix},\ A_2^{(12)} = A_2^{(21)} = A_2^{(22)} = 0_\infty,$$

$$B = \begin{bmatrix} (1-a)v_1 & (1-a)v_2 & (1-a)v_3 & \cdots & \\ 1-a & 0 & \cdots & & \\ 0 & 1-a & 0 & \cdots & \\ & & \ddots & & \end{bmatrix},$$

$$C^{(11)} = \begin{bmatrix} av_1 & av_2 & av_3 & \cdots \\ 0 & 0 \cdots & & \\ 0 & 0 & 0 & \cdots \\ & & \ddots & \end{bmatrix}, \quad C^{(12)} = \begin{bmatrix} 0 & 0 & 0 & \cdots \\ a & 0 \cdots & & \\ 0 & a & 0 & \cdots \\ & & \ddots & \end{bmatrix},$$

$$E^{(11)} = \begin{bmatrix} (1-a)b_1 & (1-a)b_2 & (1-a)b_3 & (1-a)b_4 & \cdots \\ 0 & 0 & 0 & 0 & \cdots \\ 0 & 0 & 0 & 0 & \cdots \\ \vdots & \vdots & \vdots & \vdots & \vdots \end{bmatrix}, \quad E^{(21)} = 0_\infty.$$

We explain three of the block matrices in detail and the rest become straightforward to understand. The matrix B refers to a transition that captures the situation of the system that was empty and the server was on vacation and remained on vacation after one time transition with the system still empty. The block matrix C refers to a transition from an empty system to a system with an arrival and it has two components; $C^{(11)}$ refers to the case when a server was on vacation and vacation ends after one transition and service begins; and $C^{(12)}$ is the case of the server on vacation and vacation continued after one transition. The block matrix A_0 refers to a transition from a non-empty system with one arrival, leaving the number in the system having increased by one, and it has four components; $A_0^{(11)}$ refers to a transition from a system in which the server was busy serving and continues to serve after the transition; $A_0^{(12)}$ refers to a transition from a system in which the server was busy serving and then goes on a vacation after the transition (not possible since the system is not empty); $A_0^{(21)}$ refers to a transition from a system in which the server was on a vacation and returns to serve after the transition; and $A_0^{(22)}$ refers to a transition from a system in which the server was on vacation and continues to remain on vacation after the transition.

It is immediately clear that we can apply the matrix-geometric results here. The only challenge we have to deal with is the fact that the block matrices are infinite. If we assume that the system is stable then we can compute the R and G matrices after we truncate the service times and the vacation times distributions. For most practical systems we expect these two quantities to be bounded. So we can go ahead and assume they are bounded. Once we make this assumption applying the results of the QBD is straightforward. First we need to compute the matrix R given by the minimal non-negative solution to the matrix quadratic equation

$$R = A_0 + RA_1 + R^2 A_2.$$

However because matrix A_2 has a nice structure we rather compute the matrix G and then use the result to compute R. The matrix G has a block column of zeros – the left half, because the matrix A_2 has that same structure. So we compute matrix G which is the minimal non-negative solution to the matrix quadratic equation

$$G = A_2 + A_1 G + A_0 G^2,$$

where

$$G = \begin{bmatrix} G_1 & 0 \\ G_2 & 0 \end{bmatrix}. \tag{8.18}$$

All we need to solve for iteratively is G_1 from

$$G_1 = A_2^{(11)} + A_1^{(11)} G_1 + A_0^{(11)} (G_1)^2. \tag{8.19}$$

The matrix G_2 is given as

$$G_2 = A_1^{(21)} G_1 + A_1^{(22)} G_2 + A_0^{(21)} (G_1)^2 + A_0^{(22)} G_2 G_1, \tag{8.20}$$

which we can write as

$$G_2^T = F^T + G_2^T (A_1^{(22)})^T + G_1^T G_2^T (A_0^{(22)})^T, \tag{8.21}$$

where $F = A_1^{(21)} G_1 + A_0^{(21)} (G_1)^2$ and A^T refers to the transpose of the matrix A. The matrix G_2 is then solved for from the linear matrix equation

$$vecG_2^T = vecF^T + (A_1^{(22)} \otimes I) vecG_2^T + (A_0^{(22)} \otimes G_1) vecG_2^T, \tag{8.22}$$

simply as

$$vecG_2^T = vecF^T (I - (A_1^{(22)} \otimes I) - (A_0^{(22)} \otimes G_1))^{-1}. \tag{8.23}$$

8.3 GI/Geo/1 vacation systems

The GI/Geo/1 vacation queue is easily studied using either the imbedded Markov chain approach or the supplementary variable technique. Tian and Zhang (2002) used a combination of the imbedded Markov chain and matrix-geometric approach to analyze this system, similar in principle to the case of the Geo/G/1 vacation queue, except for a minor difference. They studied the system at points of arrivals, as expected, however in addition they distinguished between whether an arrival occurs when the server was busy or when it was on a vacation. Whereas in the case of the GI/Geo/1 vacation the system was studied at departure epochs and at return times from vacations. Details of this GI/Geo/1 approach can be found in that paper by Tian and Zhang (2002). In this book we consider the supplementary variables technique only and show how the matrix-geometric methods can be used to analyze this system. This is presented only for the multiple vacation system. The case of single vacation can be easily inferred from the results provided.

Supplementary variable technique and MGM: We consider the multiple vacation system. Packets arrive with general inter-arrival times $\mathscr{A}$ and $a_i = Pr\{\mathscr{A} = i\}, i = 1,2,\cdots$. The service times follow the geometric distribution with parameter b and the vacation duration $\mathscr{V}$ has a general distribution with $v_j = Pr\{\mathscr{V} = j\}$, $j = 1,2,\cdots$.

The state space for this system can be written as

$$\Delta = \{(0,l,k) \cup (i,k) \cup (i,l,k)\}, l = 1,2,\cdots; i \geq 1; k = 1,2,\cdots.$$

The first tuple $(0,l,k)$ refers to an empty system, with server on a vacation which has l more units of time to be completed and the next packet arrival is k units of time away. The second tuple (i,k) refers to the case with i packets in the system with the server in service and attending to a packet and the next packet arrival is k units of time away. The final tuple (i,l,k) refers to having i packets in the system, with the server on a vacation which has l more units of time before it is completed and k is the remaining time before the next packet arrives into the system. It is immediately clear that the transition matrix P of this DTMC is a QBD type with

$$P = \begin{bmatrix} B & C & & \\ E & A_1 & A_0 & \\ & A_2 & A_1 & A_0 \\ & & \ddots & \ddots & \ddots \end{bmatrix}.$$

If we assume the inter-arrival time and the vacation duration to have finite supports or can be truncated (an essential assumption required for algorithmic tractability) then we can represent the inter-arrival time and also vacation duration by PH distributions. Keeping in mind that a PH is a special case of MAP and geometric distribution is a special case of PH, we can see that any result for a MAP/PH/1 system with PH vacation can be applied to this GI/Geo/1 vacation system. Hence we skip further discussion of the GI/Geo/1 vacation queue and go straight to develop the MAP/PH/1 vacation model in the next section. The only additional feature is that the PH distributions encountered in the GI/Geo/1 vacation have special structures that can be capitalized on. This is left to the reader to explore.

8.4 MAP/PH/1 vacation systems

The model for the MAP/PH/1 with PH vacations, (ϕ, V), is sufficient to cover most of the commonly encountered vacation queues. We will develop this model and point out how it can be used to obtain results for other related vacation models. We let the arrival rate be $\lambda = \boldsymbol{\delta} D\mathbf{1}$, where $\boldsymbol{\delta} = \boldsymbol{\delta} D$, $\boldsymbol{\delta}\mathbf{1} = 1$, the mean service time be $\mu^{-1} = \boldsymbol{\beta}(I-S)^{-1}\mathbf{1}$ and the mean vacation duration be $v^{-1} = \phi(I-V)^{-1}\mathbf{1}$.

Consider a system with MAP arrivals represented by two $n_t \times n_t$ matrices D_0 and D_1. As before we have the elements $(D_k)_{i,j}$ representing transitions from state i to state j with $k = 0, 1$, arrivals. The matrix $D = D_0 + D_1$ is stochastic and irreducible. There is a single server that provides a service with PH distribution represented by $(\boldsymbol{\beta}, S)$ of dimension n_s. We assume that the server takes vacations that have PH durations with representation (ϕ, V) of order n_v. We will be specific about the type of vacation as the models are developed.

8.4.1 *Exhaustive single vacation:*

In this system the server goes on a vacation as soon as the system is empty. After it returns from a vacation it starts to serve the waiting packets or waits for the first arriving packet if there are no waiting packets. Consider the DTMC $\{(X_n, U_n, Z_n, Y_n), n \geq 0\}$, where at time n, X_n is the number of packets waiting in the system, U_n is the arrival phase of the packets, Z_n is the phase of service when the server is in service, and Y_n is the phase of vacation when the server is on a vacation. Essentially, the state space can be written as

$$\Delta = \{(0,k,0) \cup (0,k,l) \cup (i,k,l) \cup (i,k,j);$$

$$k = 1,2,\cdots,n_t; l = 0,1,\cdots,n_v; i = 1,2,\cdots; j = 1,2,\cdots,n_s\}.$$

Here

- The first tuple $(0,k,0)$ refers to an empty system, with arrival in phase k and the server is waiting to serve any arriving packet (i.e. the server is back from a vacation and no waiting packets).
- The second tuple $(0,k,l)$ refers to an empty system, arrival in phase k and server is on vacation which is in phase l.
- The third tuple (i,k,l) refers to having i packets in the system, arrival is in phase k and the server is on vacation which is in phase l.
- The last tuple (i,k,j) refers to having i packets in the system, arrival is in phase k and the server is providing service to packets and the service is in phase j.

The transition matrix P for this DTMC can be written as

$$P = \begin{bmatrix} B & C & & & \\ E & A_1 & A_0 & & \\ & A_2 & A_1 & A_0 & \\ & & A_2 & A_1 & A_0 \\ & & & \ddots & \ddots & \ddots \end{bmatrix},$$

where

$$B = \begin{bmatrix} D_0 & 0 \\ D_0 \otimes \mathbf{v} & D_0 \otimes V \end{bmatrix}, \; C = \begin{bmatrix} D_1 \otimes \boldsymbol{\beta} & 0 \\ D_1 \otimes (\mathbf{v}\boldsymbol{\beta}) & D_1 \otimes V \end{bmatrix},$$

$$E = \begin{bmatrix} 0 & D_0 \otimes (\mathbf{s}\boldsymbol{\phi}) \\ 0 & 0 \end{bmatrix}, \; A_2 = \begin{bmatrix} D_0 \otimes (\mathbf{s}\boldsymbol{\beta}) & 0 \\ 0 & 0 \end{bmatrix},$$

$$A_0 = \begin{bmatrix} D_1 \otimes S & 0 \\ D_1 \otimes (\mathbf{v}\boldsymbol{\beta}) & D_1 \otimes V \end{bmatrix}, \; A_1 = \begin{bmatrix} D_0 \otimes S + D_1 \otimes (\mathbf{s}\boldsymbol{\beta}) & 0 \\ D_0 \otimes (\mathbf{v}\boldsymbol{\beta}) & D_0 \otimes V \end{bmatrix},$$

and $\mathbf{v} = \mathbf{1} - V\mathbf{1}$.

Now we can apply the matrix-geometric method to analyze this system. We know that there is a matrix R which is the minimal non-negative solution to the matrix quadratic equation

$$R = A_0 + RA_1 + R^2 A_2,$$

and the spectral radius of this matrix R is less than one if the system is stable. The system is stable if

$$\boldsymbol{\pi} A_0 \mathbf{1} < \boldsymbol{\pi} A_2 \mathbf{1},$$

where $\boldsymbol{\pi} A = \boldsymbol{\pi}$, $\boldsymbol{\pi}\mathbf{1} = 1$ and $A = A_0 + A_1 + A_2$.

In this particular case we can exploit the structure of the block matrices and compute the matrix G first which is the minimal non-negative solution to the matrix quadratic equation

$$G = A_2 + A_1 G + A_0 G^2.$$

We chose to compute the matrix G first because the matrix A_2 has a block column of zeros which implies that the matrix G is of the form

$$G = \begin{bmatrix} G_1 & 0 \\ G_2 & 0 \end{bmatrix}. \tag{8.24}$$

By writing the G matrix equations in block form we only need to compute $n_t n_s \times n_t n_s + n_t n_r \times n_t n_s$ elements instead of $(n_t n_s + n_t n_r) \times (n_t n_s + n_t n_r)$ elements – a huge saving indeed. The block elements of G can be computed from

$$G_1 = D_0 \otimes (\mathbf{s}\boldsymbol{\beta}) + (D_0 \otimes S + D_1 \otimes (\mathbf{s}\boldsymbol{\beta}))G_1 + (D_1 \otimes S)(G_1)^2, \tag{8.25}$$

$$G_2 = (D_0 \otimes (\mathbf{v}\boldsymbol{\beta}))G_1 + (D_0 \otimes V)G_2 + (D_1 \otimes (\mathbf{v}\boldsymbol{\beta}))(G_1)^2 + (D_1 \otimes V)G_2G_1. \tag{8.26}$$

We note that G_1 is actually the G matrix of a MAP/PH/1 queue with no vacation. So that is easy to obtain. After solving for G_1 we can simply obtain G_2 in a matrix linear equation by using the vectorization transformation. Let us write the second block of equation as

$$G_2 = F_1G_1 + F_2G_2 + F_3(G_1)^2 + F_4G_2G_1. \tag{8.27}$$

Since G_1 is already known we can actually write this equation as

$$G_2 = H_1 + F_2G_2 + F_4G_2F_5, \tag{8.28}$$

where $H_1 = F_1G_1 + F_3(G_1)^2$ and $F_5 = G_1$. Consider an $M \times N$ matrix

$$Y = [Y_{.1},\ Y_{.2},\ \cdots,\ Y_{.j},\ \cdots,\ Y_{.N}],$$

where $Y_{.j}$ is the jth column of the matrix Y, we define a $1 \times NM$ column vector

$$vecY = \begin{bmatrix} Y_{.1} \\ Y_{.2} \\ \vdots \\ Y_{.j} \\ \vdots \\ Y_{.N} \end{bmatrix},$$

then we can write the equation for matrix G_2 as

$$vecG_2 = vecH_1 + ((F_5)^T \otimes F_4)vecG_2 \tag{8.29}$$

which leads to

$$vecG_2 = vecH_1(I - ((F_5)^T \otimes F_4))^{-1}. \tag{8.30}$$

We assume the inverse of $(I - ((F_5)^T \otimes F_4))$ exists. After obtaining the G matrix we can then use it to compute the matrix R using Equation (4.83), i.e.

$$R = A_0(I - A_1 - A_0G)^{-1}.$$

Both the R and G matrices have special structures that are decomposable for efficient computations (see Alfa (2014)).

8.4.2 Stationary Distribution:

Let the steady state vector of this DTMC be $\boldsymbol{x} = [\boldsymbol{x}_0, \boldsymbol{x}_1, \boldsymbol{x}_2, \cdots,]$, with each of the vectors $\boldsymbol{x}_i$ appropriately defined to match the state space described earlier, then we have

$$\boldsymbol{x} = \boldsymbol{x}P, \ \boldsymbol{x}\mathbf{1} = 1,$$

and

$$\boldsymbol{x}_{i+1} = \boldsymbol{x}_i R, \ i \geq 1.$$

We apply the standard matrix-geometric method to obtain the boundary vectors $[\boldsymbol{x}_0, \boldsymbol{x}_1]$ from

$$[\boldsymbol{x}_0, \boldsymbol{x}_1] = [\boldsymbol{x}_0, \boldsymbol{x}_1] \begin{bmatrix} B & C \\ E & A_1 + RA_2 \end{bmatrix}, \ [\boldsymbol{x}_0, \boldsymbol{x}_1]\mathbf{1} = 1,$$

and then normalize with

$$\boldsymbol{x}_0\mathbf{1} + \boldsymbol{x}_1[I - R]^{-1}\mathbf{1} = 1.$$

From here on the steps required to obtain the expected number in the system and waiting times are straightforward.

8.4.3 Example of the Geo/Geo/1 vacation:

Consider the special case of Geo/Geo/1 with geometric vacation. We let arrival have parameter a, service parameter b and vacation parameter c, then our block matrices will be

$$B = \begin{bmatrix} 1-a & 0 \\ (1-a)c & (1-a)(1-c) \end{bmatrix}, \ C = \begin{bmatrix} a & 0 \\ ac & a(1-c) \end{bmatrix},$$

$$E = \begin{bmatrix} 0 & (1-a)b \\ 0 & 0 \end{bmatrix}, \ A_2 = \begin{bmatrix} (1-a)b & 0 \\ 0 & 0 \end{bmatrix},$$

$$A_0 = \begin{bmatrix} a(1-b) & 0 \\ ac & a(1-c) \end{bmatrix}, A_1 = \begin{bmatrix} (1-a)(1-b)+ab & 0 \\ (1-a)c & (1-a)(1-c) \end{bmatrix}.$$

For this example, the G matrix is

$$G = \begin{bmatrix} 1 & 0 \\ 1 & 0 \end{bmatrix}.$$

Hence it is simple to show that the R matrix is explicit, and can be written as

$$R = \frac{1}{\bar{a}b(1-\bar{a}\bar{c})} \begin{bmatrix} a\bar{b}(1-\bar{a}\bar{c}) & 0 \\ ac(1-\bar{a}\bar{c})+a\bar{c}(\bar{a}c+a) & a\bar{a}\bar{c}\bar{b} \end{bmatrix},$$

where $\bar{a} = 1-a$, $\bar{b} = 1-b$, $\bar{c} = 1-c$. This is based on the formula $R = A_0(I - A_1 - A_0G)^{-1}$.

8.4.4 Exhaustive multiple vacations

In this system the server goes on a vacation as soon as the system is empty. After it returns from a vacation it starts to serve the waiting packets but if there are no waiting packets then it proceeds to another vacation. Consider the DTMC $\{(X_n, U_n, Z_n, Y_n), n \geq 0\}$, where at time n, X_n is the number of packets waiting in the system, U_n is the arrival phase of the packets, Z_n is the phase of service when the server is in service, and Y_n is the phase of vacation when the server is on a vacation. Essentially, the state space can be written as

$$\Delta = \{(0,k,l) \cup (i,k,l) \cup (i,k,j);$$

$$k = 1,2,\cdots,n_t; l = 0,1,\cdots,n_v; i = 1,2,\cdots; j = 1,2,\cdots,n_s\}.$$

Here

- The first tuple $(0,k,l)$ refers to an empty system, arrival in phase k and server is on vacation which is in phase l.
- The second tuple (i,k,l) refers to having i packets in the system, arrival is in phase k and the server is on vacation which is in phase l.
- The third tuple (i,k,j) refers to having i packets in the system, arrival is in phase k and the server is providing service to packets and the service is in phase j.

It is immediately clear that the only difference between this multiple vacations and single vacation is at the boundary. We no longer have a server waiting for packets if when it returns from a vacation it finds no waiting packets. The transition matrix P for this DTMC can be written as

$$P = \begin{bmatrix} B & C & & & \\ E & A_1 & A_0 & & \\ & A_2 & A_1 & A_0 & \\ & & A_2 & A_1 & A_0 \\ & & & \ddots & \ddots & \ddots \end{bmatrix},$$

where

$$B = D_0 \otimes (\mathbf{v}\phi + V),\ C = \left[D_1 \otimes (\mathbf{v}\boldsymbol{\beta})\ D_1 \otimes V \right],$$

$$E = \begin{bmatrix} 0 \\ D_0 \otimes (\mathbf{s}\phi) \end{bmatrix},\ A_2 = \begin{bmatrix} D_0 \otimes (\mathbf{s}\boldsymbol{\beta}) & 0 \\ 0 & 0 \end{bmatrix},$$

$$A_0 = \begin{bmatrix} D_1 \otimes S & 0 \\ D_1 \otimes (\mathbf{v}\boldsymbol{\beta}) & D_1 \otimes V \end{bmatrix},\ A_1 = \begin{bmatrix} D_0 \otimes S + D_1 \otimes (\mathbf{s}\boldsymbol{\beta}) & 0 \\ D_0 \otimes (\mathbf{v}\boldsymbol{\beta}) & D_0 \otimes V \end{bmatrix}.$$

Now we can apply the matrix-geometric method to analyze this system. We know that there is a matrix R which is the minimal non-negative solution to the matrix quadratic equation

$$R = A_0 + RA_1 + R^2 A_2,$$

and the spectral radius of this matrix R is less than one if the system is stable. The system is stable if

$$\boldsymbol{\pi} A_0 \mathbf{1} < \boldsymbol{\pi} A_2 \mathbf{1},$$

where $\boldsymbol{\pi} A = \boldsymbol{\pi}$, $\boldsymbol{\pi}\mathbf{1} = 1$ and $A = A_0 + A_1 + A_2$.

Other than the boundary conditions that are different every other aspects of the matrices for the multiple vacations is the same as those for the single vacation. As such we will not repeat the analysis here.

8.5 MAP/PH/1 vacation queues with number limited service

For some polling systems in telecommunications the server attends to each traffic flow (queue) by serving a predetermined number of packets and then goes to another queue and so on. If such a system is approximated by a vacation model then we have a number limited system. This is a system in which a server attends to a queue and serves up to a predetermined number of packets $\mathcal{K} < \infty$ during each visit and then goes on vacation after this number has been served or the system becomes empty (whichever occurs first). If when the server returns from a vacation the system is empty then it proceeds on another vacation, i.e. it is a multiple vacation rule. This kind of queue is easier to study using the matrix-analytic method than standard methods and as such we present that approach.

Consider a MAP/PH/1 vacation queue, where the vacation duration follows a PH distribution with parameters (ϕ, V) of dimension n_v. The arrival follows a MAP and is represented by two matrices D_0 and D_1 of dimension n_t and service is PH with parameters $(\boldsymbol{\beta}, S)$ of dimension n_s. Now consider the following state space

$$\Delta = \{(0,k,l) \cup (i,k,l) \cup (i,u,k,j)\}, k = 1,2,\cdots,n_t; l = 1,2,\cdots,n_v;$$

$$i = 1,2,\cdots; u = 0,1,2,\cdots,\mathscr{K}; j = 1,2,\cdots,n_s.$$

Throughout this state space description, the variable k stands for the phase of arrival of the MAP, l the phase of a vacation, j the phase of service, i the number of packets in the system and u the number of packets that have been served during a particular visit to the queue by the server. The first tuple $(0,k,l)$ represents an empty system with the server on a vacation. The second tuple (i,k,l) represents the case when the server is on a vacation and there are i packets waiting in the system and the third tuple (i,u,k,j) represents the case when the server is attending to the queue and there are i packets in the system, with u packets already served. It is immediately clear that the transition matrix P for this DTMC is a QBD type written as

$$P = \begin{bmatrix} B & C & & \\ E & A_1 & A_0 & \\ & A_2 & A_1 & A_0 \\ & & \ddots & \ddots & \ddots \end{bmatrix},$$

where

$$A_0 = \begin{bmatrix} (\mathbf{t}\boldsymbol{\alpha}) \otimes V & \mathbf{e}_1^T(\mathscr{K}) \otimes (\mathbf{t}\boldsymbol{\alpha}) \otimes (\mathbf{v}\boldsymbol{\beta}) \\ 0 & I_{\mathscr{K}} \otimes (\mathbf{t}\boldsymbol{\alpha}) \otimes S \end{bmatrix},$$

$$A_2 = \begin{bmatrix} 0 & 0 \\ \mathbf{e}_{\mathscr{K}}(\mathscr{K}) \otimes T \otimes \boldsymbol{\phi} \otimes \mathbf{s} & \tilde{I}(\mathscr{K}-1) \otimes T \otimes (\mathbf{s}\boldsymbol{\beta}) \end{bmatrix},$$

$$A_1 = \begin{bmatrix} T \otimes V & \mathbf{e}_1^T(\mathscr{K}) \otimes T \otimes (\mathbf{v}\boldsymbol{\beta}) \\ \mathbf{e}_{\mathscr{K}}(\mathscr{K}) \otimes (\mathbf{t}\boldsymbol{\alpha}) \otimes \boldsymbol{\phi} \otimes \mathbf{s} & I_{\mathscr{K}} \otimes T \otimes S + \tilde{I}(\mathscr{K}-1) \otimes (\mathbf{t}\boldsymbol{\alpha}) \otimes (\mathbf{s}\boldsymbol{\beta}) \end{bmatrix},$$

$$B = T \otimes V, \; C = \begin{bmatrix} (\mathbf{t}\boldsymbol{\alpha}) \otimes V & \mathbf{e}_1^T(\mathscr{K}) \otimes (\mathbf{t}\boldsymbol{\alpha}) \otimes (\mathbf{v}\boldsymbol{\beta}) \end{bmatrix},$$

$$E = \begin{bmatrix} 0 \\ \mathbf{1}(\mathscr{K}) \otimes T \otimes \boldsymbol{\phi} \otimes \mathbf{s} \end{bmatrix},$$

where I_j is an identity matrix of dimension j and $\bar{I}(j) = \begin{bmatrix} 0 & I_j \\ 0 & 0 \end{bmatrix}$, $\mathbf{1}(j)$ is a column vector of ones of dimension j, $\mathbf{e}_j$ is the jth column of an identity matrix and $\mathbf{e}_j^T$ is its transpose; the dimensions are left out to avoid complex notations, but the dimensions are obvious.

We see that this problem is nicely set up as QBD and the results from QBD can be used to analyze this system. We first obtain the associated R matrix and then proceed as discussed in Chapter (3). However, in obtaining the R matrix here we notice that there is no obvious simple structure that can be exploited, hence it is more efficient to use the quadratically converging algorithms such as the Logarithmic Reduction (Latouche and Ramaswami (1993)), or the Cyclic Reduction (Meini (1998)).

8.6 MAP/PH/1 vacation queues with time limited service

In some medium access control protocols the server attends to each traffic flow (queue) for a limited length of time and then goes to another queue and so on. If such a system is approximated by a vacation model then we have a time limited system. This is a system in which a server attends to a queue and serves up to a predetermined length of time $\mathcal{T} < \infty$ during each visit and then goes on vacation after this time has expired or the system becomes empty (whichever occurs first). The server may interrupt the service of a packet that is in service once the time limit is reached. We define a stochastic matrix Q whose elements Q_{j_1',j_1} refer to the probability that a packet's service resumes in phase j_1 after the server returns from a vacation, given that the service was interrupted at phase j_1' at the start of a vacation. If when the server returns from a vacation the system is empty then it proceeds on another vacation, i.e. it is a multiple vacation rule. This kind of queue is also easier to study using the matrix-analytic method.

Consider a MAP/PH/1 vacation queue, where the vacation duration follows a PH distribution with parameters (ϕ, V) of dimension n_v. The arrival follows a MAP and is represented by two matrices D_0 and D_1 of dimension n_t and service is PH with parameters (β, S) of dimension n_s. Now consider the following state space

$$\Delta = \{(0,k,l) \cup (i,k,l,j') \cup (i,u,k,j)\}, k = 1,2,\cdots,n_t; l = 1,2,\cdots,n_v;$$

$$j' = 0,1,2,\cdots,n_s; i = 1,2,\cdots; u = 0,1,2,\cdots,\mathcal{T}; j = 1,2,\cdots,n_s.$$

Throughout this state space description, the variable k again stands for the phase of arrival of the MAP, l the phase of a vacation, j the phase of service, j' the phase at which a packet's service was interrupted (this phase could be zero if no interruption took place), i the number of packets in the system and u the time clock recording the time that has expired since during a particular visit to the queue by the server. The first tuple $(0,k,l)$ represents an empty system with the server on a vacation.

The second tuple (i,k,l) represents the case when the server is on a vacation and there are i packets waiting in the system and the third tuple (i,u,k,j) represents the case when the server is attending to the queue and there are i packets in the system, with u packets already served. It is immediately clear that the transition matrix P for this DTMC is a QBD type written as

$$P = \begin{bmatrix} B & C & & & \\ E & A_1 & A_0 & & \\ & A_2 & A_1 & A_0 & \\ & & \ddots & \ddots & \ddots \end{bmatrix},$$

where

$$A_0 = \begin{bmatrix} (\mathbf{t}\boldsymbol{\alpha}) \otimes V \otimes I_{n_s+1} & \mathbf{e}_1^T(\mathscr{T}) \otimes (\mathbf{t}\boldsymbol{\alpha}) \otimes \mathbf{v} \otimes Q^* \\ \mathbf{e}_{\mathscr{T}}(\mathscr{T}) \otimes (\mathbf{t}\boldsymbol{\alpha}) \otimes \phi \otimes S^* & \bar{I}(\mathscr{T}-1) \otimes (\mathbf{t}\boldsymbol{\alpha}) \otimes S \end{bmatrix},$$

$$A_2 = \begin{bmatrix} 0 & 0 \\ \mathbf{e}_{\mathscr{T}}(\mathscr{T}) \otimes T \otimes \phi \otimes (\mathbf{s}\mathbf{e}_1^T(n_s+1)) & \bar{I}(\mathscr{T}-1) \otimes T \otimes (\mathbf{s}\boldsymbol{\beta}) \end{bmatrix},$$

$$A_1 = \begin{bmatrix} T \otimes V \otimes I_{n_s+1} & \mathbf{e}_1^T(\mathscr{T}) \otimes T \otimes \mathbf{v} \otimes Q^* \\ A_1^{(1)} & A_1^{(2)} \end{bmatrix},$$

$$A_1^{(1)} = \mathbf{e}_{\mathscr{T}}(\mathscr{T}) \otimes [(\mathbf{t}\boldsymbol{\alpha}) \otimes \phi \otimes (\mathbf{s}\mathbf{e}_1^T(n_s+1)) + T \otimes \phi \otimes S^*],$$

$$A_1^{(2)} = \bar{I}(\mathscr{T}-1) \otimes (T \otimes S + (\mathbf{t}\boldsymbol{\alpha}) \otimes (\mathbf{s}\boldsymbol{\beta}),$$

$$B = T \otimes V,\ C = \begin{bmatrix} (\mathbf{t}\boldsymbol{\alpha}) \otimes V & \mathbf{e}_1^T(\mathscr{T}) \otimes (\mathbf{t}\boldsymbol{\alpha}) \otimes (\mathbf{v}\boldsymbol{\beta}) \end{bmatrix},$$

$$E = \begin{bmatrix} 0 \\ \mathbf{1}(\mathscr{T}) \otimes T \otimes \phi \otimes \mathbf{s} \end{bmatrix},$$

where I_j is an identity matrix of dimension j and $\bar{I}(j) = \begin{bmatrix} 0 & I_j \\ 0 & 0 \end{bmatrix}$, $\mathbf{1}(j)$ is a column vector of ones of dimension j, $\mathbf{e}_j$ is the jth column of an identity matrix and $\mathbf{e}_j^T$ is its transpose, $S^* = [\mathbf{0}\ \ S]$ an $(n_s+1) \times n_s$ matrix, $Q^* = \begin{bmatrix} \boldsymbol{\beta} \\ Q \end{bmatrix}$, an $n_s \times (n_s+1)$ matrix. All the zeros in the matrices are dimensioned accordingly; the dimensions are left out to avoid complex notations, but the dimensions are obvious.

We see that this problem is nicely set up as QBD and the results from QBD can be used to analyze this system. We first obtain the associated R matrix and then proceed as discussed in Chapter 2. However, in obtaining the R matrix here we notice that there is no obvious simple structure that can be exploited, hence it is more efficient to use the quadratically converging algorithms such as the Logarithmic Reduction (Latouche and Ramaswami (1993)) or the Cyclic Reduction (Meini (1998)).

8.7 Random Time Limited Vacation Queues/Queues with Server Breakdowns and Repairs

In this section we study a MAP/PH/1 vacation queue with random time limited visits. In this system the server attends to the queue for a random amount of time and then proceeds on a vacation which is of a random time duration also. The visit to a queue has a PH distribution (γ, U) of order κ. With this random time visits the server may interrupt the service of a packet that is in service once a vacation is set to start. We define a stochastic matrix Q whose elements $Q_{j_1' j_1}$ refers to the probability that a packet's service resumes in phase j_1 after server returns from a vacation, given that the service was interrupted at phase j_1' at the start of a vacation. If when the server returns from a vacation the system is empty then it proceeds on another vacation, i.e. it is a multiple vacation rule. This kind of queue is also easier to study using the matrix-analytic method.

Consider a MAP/PH/1 vacation queue, where the vacation duration follows a PH distribution with parameters (ϕ, V) of dimension n_v. The arrival follows a MAP and is represented by two matrices D_0 and D_1 of dimension n_t and service is PH with parameters (β, S) of dimension n_s. Now consider the following state space

$$\Delta = \{(0,k,l) \cup (i,k,l,j') \cup (i,u,k,j)\}, k = 1,2,\cdots,n_t; l = 1,2,\cdots,n_v;$$

$$j' = 0,1,2,\cdots,n_s; i = 1,2,\cdots; u = 0,1,2,\cdots,\kappa; j = 1,2,\cdots,n_s.$$

Throughout this state space description, the variable k again stands for the phase of arrival of the MAP, l the phase of a vacation, j the phase of service, j' the phase at which a packet's service was interrupted (this phase could be zero if no interruption took place), i the number of packets in the system and u is the phase of a particular visit to the queue by the server (this is like the time clock that keeps track of how long a system is being attended to). The first tuple $(0,k,l)$ represents an empty system with the server on a vacation. The second tuple (i,k,l) represents the case when the server is on a vacation and there are i packets waiting in the system and the third tuple (i,u,k,j) represents the case when the server is attending to the queue and there are i packets in the system, with u as the phase of a particular visit - time clock. It is immediately clear that the transition matrix P for this DTMC is a QBD type written as

$$P = \begin{bmatrix} B & C & & \\ E & A_1 & A_0 & \\ & A_2 & A_1 & A_0 \\ & & \ddots & \ddots & \ddots \end{bmatrix},$$

where

$$A_0 = \begin{bmatrix} (\mathbf{t}\alpha) \otimes V \otimes I_{n_s+1} & \gamma \otimes (\mathbf{t}\alpha) \otimes \mathbf{v} \otimes Q^* \\ \mathbf{u} \otimes (\mathbf{t}\alpha) \otimes \phi \otimes S^* & U \otimes (\mathbf{t}\alpha) \otimes S \end{bmatrix},$$

$$A_2 = \begin{bmatrix} 0 & 0 \\ \mathbf{u} \otimes T \otimes \phi \otimes (\mathbf{s}\mathbf{e}_1^T(n_s+1)) & U \otimes T \otimes (\mathbf{s}\beta) \end{bmatrix},$$

$$A_1 = \begin{bmatrix} T \otimes V \otimes I_{n_s+1} & T \otimes (\mathbf{v}\gamma) \otimes Q^* \\ \mathbf{u} \otimes [(\mathbf{t}\alpha) \otimes \phi \otimes (\mathbf{s}\mathbf{e}_1^T(n_s+1)) + T \otimes \phi \otimes S^*] & U \otimes [T \otimes S + (\mathbf{t}\alpha) \otimes (\mathbf{s}\beta)] \end{bmatrix},$$

$$B = T \otimes (V + \mathbf{v}\gamma),\ C = \begin{bmatrix} (\mathbf{t}\alpha) \otimes V & (\mathbf{t}\alpha) \otimes (\mathbf{v}\gamma) \otimes \beta \end{bmatrix},$$

$$E = \begin{bmatrix} 0 \\ (\mathbf{u} + U\mathbf{e}^T) \otimes T \otimes \phi \otimes \mathbf{s} \end{bmatrix},$$

where I_j is an identity matrix of dimension j, $\mathbf{e}_j$ is the jth column of an identity matrix and $\mathbf{e}_j^T$ is its transpose, $S^* = [\mathbf{0}\ \ S]$ an $(n_s+1) \times n_s$ matrix, $Q^* = \begin{bmatrix} \beta \\ Q \end{bmatrix}$, an $n_s \times (n_s+1)$ matrix. The dimensions of the zeros are left out to avoid complex notations, nevertheless they are straightforward to figure out.

If we allow the system to be a single vacation the only thing that is different are the boundary block matrices B, C and E and they are given as

$$B = \begin{bmatrix} T \otimes V & T \otimes (\mathbf{v}\gamma) \\ \mathbf{u} \otimes T \otimes \phi & U \otimes T \end{bmatrix},$$

$$C = \begin{bmatrix} (\mathbf{t}\alpha) \otimes V \otimes \mathbf{e}_1^T(n_s+1) & \gamma \otimes (\mathbf{t}\alpha) \times \mathbf{v} \otimes \beta \\ \mathbf{u} \otimes (\mathbf{t}\alpha) \otimes \phi \otimes \mathbf{e}_1^T(n_s+1) & U \otimes (\mathbf{t}\alpha) \otimes \beta \end{bmatrix},$$

$$E = \begin{bmatrix} 0 & 0 \\ \mathbf{u} \otimes T \otimes \phi \otimes \mathbf{s} & U \otimes T \otimes \mathbf{s} \end{bmatrix}.$$

We see that this problem is nicely set up as QBD and the results from QBD can be used to analyze this system. We first obtain the associated R matrix and then proceed as discussed in Chapter 2. However, in obtaining the R matrix here we notice that there is no obvious simple structure that can be exploited, hence it is more efficient to use the quadratically converging algorithms such as the Logarithmic Reduction (Latouche and Ramaswami (1993)) or the Cyclic Reduction (Meini (1998)).

It is immediately clear that if we let $\gamma = [1,\ 0,\ 0,\ \cdots,\ 0]$ and $U = \begin{bmatrix} \mathbf{0} & 0 \\ I & \mathbf{0}^T \end{bmatrix}$, then we have the time limited MAP/PH/1 vacation system presented in the last section. In addition if we consider a MAP/PH/1 queue in which the server may breakdown during service, go for repairs and resume service later, then we see that the MAP/PH/1 vacation queue with random vacation is actually the same model. The duration of a server's visit to a queue is simply the up time of the server and it's vacation period is the repair period. Hence the MAP/PH/1 vacation queue with random server visits is a very versatile queueing model.

8.8 Problems

1. Question 1: Consider a Geo/Geo/1 vacation queue. The arrival follows a Bernoulli process with parameter a and service is geometric with parameter b. As soon as the number in the system the server goes into a vacation mode. The vacation duration follows the geometric distribution with parameter v. However, during the vacation period if a customer arrives it receives service according to the geometric distribution with parameter $d < b$. Find the stability conditions of the system and develop the equations fore the distribution of the number in the system at arbitrary times.
2. Question 2: Consider Question 1) above. Suppose the service times during the busy period and during vacation both follow the PH distributions with representations (δ_1, V_1) and (δ_2, V_2), respectively. Develop the transition matrix that captures the behaviour of the number in the system.
3. Question 3: A Geo/Geo/1 system with geometric vacation is studied. The arrival, service and vacation parameters are a, b and c, respectively. For this vacation system, the server goes on vacation as soon as the number in the system is zero. After the server returns if the number of customers waiting is less than N, $1 \leq N < \infty$, it would go another vacation with the same duration. It will start to serve only when it finds at least N customers in the system. Set this problem up as a DTMC with the number in the system and other supplementary variables as the state variables. What are the stability conditions for this system? How would you go about obtaining the stationary of the number in the system?Write out the equations for obtaining the distribution of the waiting times in the system, given that you have obtained the stationary distribution of the number in the system.
4. Question 4: Repeat Question 3 for the case of PH/PH/1 with PH vacation for which the PH distributions are represented by (α, T), (β, S), and (δ, V), for arrival, service and vacations. These PH distributions have dimensions n_a, n_s and n_v, respectively.

References

Alfa AS (2014) Decomposition results for a class of vacation queues. Operations Research Letters 42:140–144

Doshi BT (1985) A note on stochastic decomposition in a GI/G/1 queue with vacations or set-up times. J Appl Prob 22:419–428

Doshi BT (1986) Queueing systems with vacations - a survey. Queueing Systems 1:29–66

Fuhrman SW, Cooper RB (1985) Stochastic decomposition in the M/G/1 queue with generalized vacations. Oper Res 33:1117–1129

Latouche G, Ramaswami V (1993) A logarithmic reduction algorithm for quasi-birth-and-death processes. J Appli Probab 30:650–674

Meini B (1998) Solving M/G/1 type markov chains: Recent advances and applications. Stochastic Models 14:479–496

Tian N, Zhang ZG (2002) The discrete-time GI/Geo/1 queue with multiple vacations. Queueing Systems 40:283–294

Chapter 9
Single Node Queueing Models with Priorities

9.1 Introduction

Most of the queue disciplines presented so far are the first-come-first-served systems. As pointed out earlier in the introductory part of single node systems other types of queue disciplines include the priority types, last-come-first-served, service in a random order, and several variants of these. In this chapter we present two types of priority service disciplines – the preemptive and non-preemptive types. We focus mainly on cases with just two classes of packets. The results for such a system can be used to approximate multiple classes systems, if needed. For the two classes presented in this chapter we assume there are two sources sending packets; one source sending one class type and the other sending another class type. The result is that we may have, in one time slot, no packet arrivals, just one packet arrival of either class, or two packet arrivals with one of each class. The other type of situation with no more than one packet arrival during a time slot is a special case of the model presented here. We study the priority queues with geometric arrivals and geometric services in detail.

9.2 Geo/Geo/1 Preemptive

Consider a system with two classes of packets that arrive according to the Bernoulli process with parameters a_k, $k = 1, 2$, where $k = 1$ refers to a high priority (HP) packet and $k = 2$ refers to a lower priority (LP) packet. We consider the preemptive-resume discipline. The service discipline is as follows. No LP packet can start receiving service if there is an HP packet in the system, and if an LP packet is receiving service (when there is no HP packet in the system), this LP packet service will be interrupted if an HP packet arrives before the LP's service is completed. An interrupted service is resumed where it was stopped, even though this information is

A.S. Alfa, *Applied Discrete-Time Queues*, DOI 10.1007/978-1-4939-3420-1_9

irrelevant since service is geometric. Each packet k requires geometric service with parameter b_k. Since we are dealing in discrete time then during a time slot there is the possibility of having up to two packets of each type arriving. Hence we define $a_{i,j}$ as the probability of arrivals of i type HP packet and j arrivals of type LP packet arriving, $i = 0,1; j = 0,1$. Hence we have

$$a_{0,0} = (1-a_1)(1-a_2), \tag{9.1}$$

$$a_{0,1} = (1-a_1)a_2, \tag{9.2}$$

$$a_{1,0} = a_1(1-a_2), \tag{9.3}$$

$$a_{1,1} = a_1 a_2. \tag{9.4}$$

We consider the preemptive priority system, and assume the system is stable. It is simple to show that provided that $\frac{a_1}{b_1} + \frac{a_2}{b_2} < 1$ the system will be stable. For a stable system let $x_{i,j}$ be the probability that we find i of type HP and j of type LP packets in the system, including the one receiving service (if there is one receiving service). We can write the steady state equations as follows:

$$x_{0,0} = x_{0,0}a_{0,0} + x_{1,0}a_{0,0}b_1 + x_{0,1}a_{0,0}b_2, \tag{9.5}$$

$$x_{1,0} = x_{0,0}a_{1,0} + x_{1,0}a_{1,0}b_1 + x_{0,1}a_{1,0}b_2 + x_{1,0}a_{0,0}(1-b_1), \tag{9.6}$$

$$x_{i,0} = x_{i+1,0}a_{0,0}b_1 + x_{i-1,0}a_{1,0}(1-b_1) + x_{i,0}a_{0,0}(1-b_1) + x_{i,0}a_{1,0}b_1, \; i \geq 2, \tag{9.7}$$

$$x_{0,1} = x_{0,2}a_{0,0}b_2 + x_{0,0}a_{0,1} + x_{0,1}a_{0,0}(1-b_2) + x_{0,1}a_{0,1}b_2$$
$$+ x_{1,1}a_{0,0}b_1 + x_{1,0}a_{0,1}b_1, \tag{9.8}$$

$$x_{0,j} = x_{0,j+1}a_{0,0}b_2 + x_{0,j-1}a_{0,1}(1-b_2) + x_{0,j}a_{0,0}(1-b_2) + x_{0,j}a_{0,1}b_2$$
$$+ x_{1,j}a_{0,0}b_1 + x_{1,j-1}a_{0,1}b_1, \; j \geq 2, \tag{9.9}$$

$$x_{1,1} = x_{0,0}a_{1,1} + x_{2,0}a_{0,1}b_1 + x_{2,1}a_{0,0}b_1 + x_{0,1}a_{1,0}(1-b_2)$$
$$+ x_{1,0}a_{0,1}(1-b_1) + x_{1,1}a_{0,0}(1-b_1) + x_{1,1}a_{1,0}b_1, \tag{9.10}$$

$$x_{i,j} = x_{i-1,j-1}a_{1,1}(1-b_1) + x_{i+1,j-1}a_{0,1}b_1 + x_{i+1,j}a_{0,0}b_1 + x_{i-1,j}a_{1,0}(1-b_1)$$
$$+ x_{i,j-1}a_{0,1}(1-b_1) + x_{i,j}a_{0,0}(1-b_1) + x_{i,j}a_{1,0}b_1, \; i > 1, j \geq 1, \; i \geq 1, j > 1. \tag{9.11}$$

Even though we can proceed to analyze this system of equations using the traditional approaches, i.e. the algebraic approach and/or the $z-$transform approach, the effort involved is disproportionately high compared to the results to be obtained.

However, we can use the matrix-geometric approach with less effort and obtain very useful results. We proceed with the matrix-geometric method as follows. First let us write

$$\boldsymbol{x}_i = [x_{i,0},\ x_{i,1},\ x_{i,2},\ \cdots],\ \boldsymbol{x} = [\boldsymbol{x}_0,\ \boldsymbol{x}_1,\ \boldsymbol{x}_2,\ \cdots].$$

Consider the following state space $\Delta = \{(i_1, i_2), i_1 \geq 0, i_2 \geq 0\}$. This state space describes the DTMC that represents this preemptive priority system and the associated transition matrix is of the form

$$P = \begin{bmatrix} B_{0,0} & B_{0,1} & & & & \\ B_{1,0} & B_{1,1} & A_0 & & & \\ & A_2 & A_1 & A_0 & & \\ & & A_2 & A_1 & A_0 & \\ & & & A_2 & A_1 & A_0 \\ & & & & \ddots & \ddots & \ddots \end{bmatrix},$$

where

$$B_{0,0} = \begin{bmatrix} B_{00}^{(00)} & B_{00}^{(01)} & & & & \\ B_{00}^{(10)} & B_{00}^{(1)} & B_{00}^{(0)} & & & \\ & B_{00}^{(2)} & B_{00}^{(1)} & B_{00}^{(0)} & & \\ & & B_{00}^{(2)} & B_{00}^{(1)} & B_{00}^{(0)} & \\ & & & B_{00}^{(2)} & B_{00}^{(1)} & B_{00}^{(0)} \\ & & & & \ddots & \ddots & \ddots \end{bmatrix},$$

with

$$B_{00}^{(00)} = a_{0,0},\ B_{00}^{(01)} = a_{0,1},\ B_{00}^{(10)} = a_{0,0}b_2,\ B_{00}^{(0)} = a_{0,1}(1-b_2),$$

$$B_{00}^{(1)} = a_{0,0}(1-b_2) + a_{0,1}b_2,\ B_{00}^{(2)} = a_{0,0}b_2,$$

$$B_{0,1} = \begin{bmatrix} B_{01}^{(00)} & B_{01}^{(01)} & & & & \\ B_{01}^{(2)} & B_{01}^{(1)} & B_{01}^{(0)} & & & \\ & B_{01}^{(2)} & B_{01}^{(1)} & B_{01}^{(0)} & & \\ & & B_{01}^{(2)} & B_{01}^{(1)} & B_{01}^{(0)} & \\ & & & B_{01}^{(2)} & B_{01}^{(1)} & B_{01}^{(0)} \\ & & & & \ddots & \ddots & \ddots \end{bmatrix},$$

with

$$B_{01}^{(00)} = a_{1,0},\ B_{01}^{(01)} = a_{1,1},\ B_{01}^{(2)} = a_{1,0}b_2,\ B_{01}^{(0)} = a_{1,1}(1-b_2),$$

$$B_{01}^{(1)} = a_{1,0}(1-b_2) + a_{1,1}b_2,$$

$$B_{1,0} = \begin{bmatrix} B_{10}^{(00)} & B_{10}^{(01)} & & & & & \\ & B_{10}^{(1)} & B_{10}^{(0)} & & & & \\ & & B_{10}^{(1)} & B_{10}^{(0)} & & & \\ & & & B_{10}^{(1)} & B_{10}^{(0)} & & \\ & & & & B_{10}^{(1)} & B_{10}^{(0)} & \\ & & & & & \ddots & \ddots \end{bmatrix},$$

with

$$B_{10}^{(00)} = a_{0,0}b_1,\ B_{10}^{(01)} = a_{0,1}b_1,\ B_{10}^{(0)} = a_{0,1}b_1,\ B_{10}^{(1)} = a_{0,0}b_1,$$

$$B_{1,1} = \begin{bmatrix} B_{11}^{(1)} & B_{11}^{(0)} & & & & & \\ & B_{11}^{(1)} & B_{11}^{(0)} & & & & \\ & & B_{11}^{(1)} & B_{11}^{(0)} & & & \\ & & & B_{11}^{(1)} & B_{11}^{(0)} & & \\ & & & & B_{11}^{(1)} & B_{11}^{(0)} & \\ & & & & & \ddots & \ddots \end{bmatrix},$$

with

$$B_{11}^{(1)} = a_{0,0}(1-b_1) + a_{1,0}b_1,\ B_{11}^{(0)} = a_{1,1}b_1 + a_{0,1}(1-b_1),$$

$$A_0 = \begin{bmatrix} A_0^{(1)} & A_0^{(0)} & & \\ & A_0^{(1)} & A_0^{(0)} & \\ & & \ddots & \ddots \end{bmatrix},\ A_0^{(1)} = a_{1,0}(1-b_1),\ A_0^{(0)} = a_{1,1}(1-b_1),$$

$$A_2 = \begin{bmatrix} A_2^{(1)} & A_2^{(0)} & & \\ & A_2^{(1)} & A_2^{(0)} & \\ & & \ddots & \ddots \end{bmatrix},\ A_2^{(1)} = a_{0,0}b_1,\ A_2^{(0)} = a_{0,1}b_1,$$

$$A_1 = \begin{bmatrix} A_1^{(1)} & A_1^{(0)} & & \\ & A_1^{(1)} & A_1^{(0)} & \\ & & \ddots & \ddots \end{bmatrix},\ A_1^{(1)} = a_{0,0}(1-b_1) + a_{1,0}b_1,\ A_1^{(0)} = a_{0,1}(1-b_1) + a_{1,1}b_1.$$

9.2.1 Stationary Distribution

If the system is stable we know that there is a relationship between the vector $\boldsymbol{x}$ and the matrix P of the form

$$\boldsymbol{x} = \boldsymbol{x}P, \ \ \boldsymbol{x}\mathbf{1} = 1,$$

and there exists a matrix R for which

$$\boldsymbol{x}_{i+1} = \boldsymbol{x}_i R,$$

where R is the minimal nonnegative solution to the matrix quadratic equation

$$R = A_0 + RA_1 + R^2 A_2.$$

It was shown by Alfa (1998) and earlier by Miller (1981) for the continuous time case that the matrix R has the following structure

$$R = \begin{bmatrix} r_0 & r_1 & r_2 & r_3 & \cdots \\ & r_0 & r_1 & r_2 & \cdots \\ & & r_0 & r_1 & \cdots \\ & & & r_0 & \ddots \\ & & & & \ddots \end{bmatrix}. \tag{9.12}$$

The arguments leading to this result are as follow. The matrix R is the expected number of visits to the HP state $i+1$ before first returning to the HP state i, given that the process started from state i. By this definition of the matrix R it is straightforward to see how the upper triangular pattern arises, i.e. because the LP number in the system can not reduce when there is an HP in the system. Since the number of LP arrivals is independent of the state of the number of the HP in the system, the rows of R will be repeating. This structure makes it easier to study the matrix R.

Solving for R involves a recursive scheme as follows:

Let

$$f = 1 - a_{0,0}(1-b_1) - a_{1,0}b_1,$$

then we have

$$r_0 = \frac{f - \sqrt{(f^2 - 4a_{1,0}b_1 a_{0,0}(1-b_1))}}{2a_{0,0}b_1}. \tag{9.13}$$

The term r_1 is obtained as

$$r_1 = \frac{a_{1,1}(1-b_1)+r_0(a_{0,1}(1-b_1)+a_{1,1}b_1)+(r_0)^2a_{0,1}b_1}{f-2r_0a_{0,0}b_1}, \tag{9.14}$$

and the remaining r_j, $j \geq 2$ are obtained as

$$r_j = \frac{ur_{j-1}+v\sum_{k=0}^{j-1} r_{j-k-1}r_k + w\sum_{k=1}^{j-1} r_{j-1}r_1}{f-2r_0a_{0,0}b_1}, \tag{9.15}$$

where $u = a_{0,1}(1-b_1)+a_{1,1}b_1$, $v = a_{0,1}b_1$, $w = a_{0,0}b_1$. It was pointed out in Alfa (1998) that $r_{j+1} \leq r_j$, $j \geq 1$, hence we can truncate R at an appropriate point of accuracy desired.

We need to obtain the boundary results $[\boldsymbol{x}_0, \boldsymbol{x}_1]$ and this is carried out using the boundary equation that

$$[\boldsymbol{x}_0, \boldsymbol{x}_1] = [\boldsymbol{x}_0, \boldsymbol{x}_1]B[R], \ [\boldsymbol{x}_0, \boldsymbol{x}_1]\mathbf{1} = 1,$$

where

$$B[R] = \begin{bmatrix} B_{00} & B_{01} \\ B_{10} & B_{11}+RA_2 \end{bmatrix}.$$

This can be solved using censored Markov chain idea to obtain $\boldsymbol{x}_0$, and after routine algebraic operations we have

$$\boldsymbol{x}_1 = \boldsymbol{x}_0 B_{01}(I-B_{11}-RA_1)^{-1}.$$

Note that for most practical situations the matrix $(I-B_{11}-RA_1)$ is non-singular. The vector $\boldsymbol{x}_0$ is normalized as

$$\boldsymbol{x}_0\mathbf{1}+\boldsymbol{x}_0[B_{01}(I-B_{11}-RA_1)^{-1}(I-R)^{-1}]\mathbf{1} = 1.$$

Both the matrices $(I-B_{11}-RA_1)$ and $I-R$ have the following triangular structure

$$U = \begin{bmatrix} u_0 & u_1 & u_2 & u_3 & \cdots \\ & u_0 & u_1 & u_2 & \cdots \\ & & u_0 & u_1 & \cdots \\ & & & u_0 & \ddots \\ & & & & \ddots \end{bmatrix}.$$

Their inverses also have the same structure and can be written as

$$V = \begin{bmatrix} v_0 & v_1 & v_2 & v_3 & \cdots \\ & v_0 & v_1 & v_2 & \cdots \\ & & v_0 & v_1 & \cdots \\ & & & \ddots & \\ & & & v_0 & \ddots \\ & & & & \ddots \end{bmatrix},$$

where

$$v_0 = u_0^{-1}, \text{ and } v_n = -u_0^{-1} \sum_{j=0}^{n-1} u_{n-j} v_j, \; n \geq 1.$$

Let $y_i(k)$ be the probability that there are i type k packets in the system, $k = 1, 2, \; i \geq 0$. We have

$$y_i(1) = \boldsymbol{x}_i \mathbf{1}, \text{ and } y_i(2) = x_{0,i} + \boldsymbol{x}_1 (I - R)^{-1} \mathbf{e}_{i+1},$$

where $\mathbf{e}_j$ is the jth column of an infinite identity matrix.

The mean number of type 1, i.e. HP, in the system $E[X(1)]$ is given as

$$E[X(1)] = \sum_{j=1}^{\infty} j \boldsymbol{x}_j \mathbf{1} = \boldsymbol{x}_1 (I - R)^{-1} \mathbf{1}, \tag{9.16}$$

and for type 2, i.e. LP, in the system $E[X(2)]$ is given as

$$E[X(2)] = \sum_{i=0}^{\infty} \sum_{j=1}^{\infty} j x_{i,j} = \boldsymbol{x}_0 \phi + \boldsymbol{x}_1 (I - R)^{-1} \phi, \tag{9.17}$$

where $\phi = [1, \; 2, \; 3, \; \cdots]^T$.

The waiting time for class 1, is simply obtained in the same manner as the waiting time in the Geo/Geo/1 system since a class 2 packet has no impact on the waiting time of type 1 packet. Let $W_q(1)$ be the waiting time in the queue for a type 1 packet and let $w_i^{(q)}(1) = Pr\{W_q(1) = i\}$ we have

$$w_0^{(q)}(1) = \boldsymbol{x}_0 \mathbf{1}, \tag{9.18}$$

and

$$w_i^{(q)}(1) = \sum_{v=1}^{i} \boldsymbol{x}_v \mathbf{1} \binom{i-1}{v-1} b_1^v (1 - b_1)^{i-v}, \; i \geq 1. \tag{9.19}$$

Let $W(1)$ be the waiting time in the system (including service time) for class 1 packet, with $w_i(1) = Pr\{W(1) = i\}$, then we have

$$w_i(1) = \sum_{j=0}^{i-1} w_j^{(q)}(1)(1-b_1)^{i-j-1} b_1, \quad i \geq 1. \tag{9.20}$$

The expected waiting time in the system for class 1 packets $E[W(1)]$ is obtained using the Little's Law as

$$E[W(1)] = a_1^{-1} E[X(1)], \tag{9.21}$$

and $E[W_q(1)]$ can be similarly obtained by first computing the expected number in the queue for the class 1 packets as

$$E[X_q(1)] = \sum_{j=1}^{\infty} (j-1)\boldsymbol{x}_j \mathbf{1} = \boldsymbol{x}_1 R(I-R)^{-2}, \tag{9.22}$$

hence

$$E[W_q(1)] = a_1^{-1} E[X_q(1)]. \tag{9.23}$$

The waiting time in the queue for class 2 packets is a bit more involved to obtain. Keep in mind that even when a type 2 packet is receiving service it can still be preempted if a class 1 arrives during the service. Hence we can only determine a class 2 packet's waiting until it becomes the leading waiting packet in the queue for the first time, i.e. no class 1 or 2 ahead of it waiting. However we have to keep in mind that it could become the leading class 2 packet but with class 1 packets that arrive during its wait in the queue becoming the leading packets in the system. We can also study its waiting time in the system until it leaves the system. The later is a function of the former. We present an absorbing DTMC for studying the waiting time of a class 2 packet until it becomes the leading packet in the system for the first time.

Consider an absorbing DTMC with the following state space $\Delta = \{(i,j)\}, i \geq 0, j \geq 0$, with state $(0,0)$ as the absorbing state, where the first element refers to the number of HP packets in the system and the second element refers to the number of LP packets ahead of the target LP packet. The transition matrix of the absorbing DTMC $\tilde{P}$ is given as

$$\tilde{P} = \begin{bmatrix} \tilde{B}_{0,0} & \tilde{B}_{0,1} & & & & \\ \tilde{B}_{1,0} & \tilde{B}_{1,1} & \tilde{A}_0 & & & \\ & \tilde{A}_2 & \tilde{A}_1 & \tilde{A}_0 & & \\ & & \tilde{A}_2 & \tilde{A}_1 & \tilde{A}_0 & \\ & & & \tilde{A}_2 & \tilde{A}_1 & \tilde{A}_0 \\ & & & & \ddots & \ddots & \ddots \end{bmatrix},$$

where

$$\tilde{B}_{0,0} = \begin{bmatrix} 1 & & & & & \\ \tilde{B}_{00}^{(2)} & \tilde{B}_{00}^{(1)} & & & & \\ & \tilde{B}_{00}^{(2)} & \tilde{B}_{00}^{(1)} & & & \\ & & \tilde{B}_{00}^{(2)} & \tilde{B}_{00}^{(1)} & & \\ & & & \tilde{B}_{00}^{(2)} & \tilde{B}_{00}^{(1)} & \\ & & & & \ddots & \ddots \end{bmatrix},$$

with

$$\tilde{B}_{00}^{(1)} = (1-a_1)(1-b_2),\ \tilde{B}_{00}^{(2)} = (1-a_1)b_2,$$

$$\tilde{B}_{0,1} = \begin{bmatrix} 0 & 0 & & & & \\ \tilde{B}_{01}^{(2)} & \tilde{B}_{01}^{(1)} & & & & \\ & \tilde{B}_{01}^{(2)} & \tilde{B}_{01}^{(1)} & & & \\ & & \tilde{B}_{01}^{(2)} & \tilde{B}_{01}^{(1)} & & \\ & & & \tilde{B}_{01}^{(2)} & \tilde{B}_{01}^{(1)} & \\ & & & & \ddots & \ddots \end{bmatrix},$$

with

$$\tilde{B}_{01}^{(2)} = a_1 b_2,\ \tilde{B}_{01}^{(1)} = a_1(1-b_2),$$

$$\tilde{B}_{1,0} = \begin{bmatrix} \tilde{B}_{10}^{(00)} & & & & & \\ & \tilde{B}_{10}^{(1)} & & & & \\ & & \tilde{B}_{10}^{(1)} & & & \\ & & & \tilde{B}_{10}^{(1)} & & \\ & & & & \tilde{B}_{10}^{(1)} & \\ & & & & & \ddots \end{bmatrix},$$

with

$$\tilde{B}_{10}^{(00)} = (1-a_1)b_1,\ \tilde{B}_{10}^{(1)} = (1-a_1)b_1,$$

$$\tilde{A}_0 = Ia_1(1-b_1),\ \tilde{A}_1 = \tilde{B}_{11} = I((1-a_1)(1-b_1)+a_1b_1),\ \tilde{A}_2 = I(1-a_1)b_1.$$

Now we define $\mathbf{z}^{(0)} = [\mathbf{z}_0^{(0)}, \mathbf{z}_1^{(0)}, \mathbf{z}_2^{(0)}, \cdots]$ with $\mathbf{z}_i^{(0)} = [z_{i,0}^{(0)}, z_{i,1}^{(0)}, z_{i,2}^{(0)}, \cdots]$, $i \geq 0$, and let $\mathbf{z}^{(0)} = \boldsymbol{x}$. Define $W^*(2)$ as the waiting time in the queue for a type 2 packet until it becomes the leading packet waiting in the system for the first time, and $W_j^* = Pr\{W^*(2) \leq j\}$ then we have

$$W_i^* = z_{0,0}^{(i)},\ i \geq 1,\ \text{where}\ \mathbf{z}^{(i)} = \mathbf{z}^{(i-1)}\tilde{P}, \tag{9.24}$$

keeping in mind that

$$W_0^* = z_{0,0}^{(0)} = x_{0,0}. \tag{9.25}$$

9.3 Geo/Geo/1 Non-preemptive

In a non-preemptive system no LP packet's service is going to be started if there is an HP packet in the system. However if an LP packet's service is started when there is no HP packet in the system, this LP packet's service will not be interrupted, i.e. no preemption.

Now we consider a system with two classes of packets that arrive according to the Bernoulli process with parameters $a_k,\ k = 1,2$, where $k = 1$ refers to a high priority (HP) packet and $k = 2$ refers to a lower priority (LP) packet. Each packet k requires geometric service with parameter b_k. Since we are dealing in discrete time then during a time slot there is the possibility of having up to two packets, one of each class type, arriving. Hence we define $a_{i,j}$ as the probability of a i type HP packet and j of type LP packet arriving, $(i,j) = 0,1$. Hence we have

$$a_{0,0} = (1-a_1)(1-a_2),\ a_{0,1} = (1-a_1)a_2,\ a_{1,0} = a_1(1-a_2),\ a_{1,1} = a_1a_2.$$

We assume the system is stable. It is simple to show that provided $\frac{a_1}{b_1} + \frac{a_2}{b_2} < 1$ the system will be stable. For a stable system let $x_{i,j,k}$ be the probability that we find i number of type HP and j number of type LP packets in the system, including the one receiving system (if there is one receiving service) and the one receiving service is type k. Here we notice that we have to keep track of which type of packet is receiving service since it is a non-preemptive system. It is immediately clear that the steady state equations for this non-preemptive system is going to be more involved. We skip this step and use the matrix-geometric method directly. First let us write

$$\boldsymbol{x}_i = [\boldsymbol{x}_{i,0},\ \boldsymbol{x}_{i,1},\ \boldsymbol{x}_{i,2},\ \boldsymbol{x}_{i,3},\ \cdots],\ \boldsymbol{x} = [\boldsymbol{x}_0,\ \boldsymbol{x}_1,\ \boldsymbol{x}_2,\ \cdots],$$

and write $\boldsymbol{x}_{i,j} = [x_{i,j,1},\ x_{i,j+1,2}], i \geq 1, j \geq 0$. The state space for this system can be written as

$$\Delta = \Delta_1 \cup \Delta_2 \cup \Delta_3,$$

where

- $\Delta_1 = \{(0,0)\}$,
- $\Delta_2 = \{(0,j),\ j \geq 1,\}$,
- $\Delta_3 = \{(i,j+k-1,k),\ i \geq 1, j \geq 0, k = 1,2.\}$

The transition matrix P associated with this DTMC is given as

$$P = \begin{bmatrix} B_{0,0} & B_{0,1} & & & & \\ B_{1,0} & A_1 & A_0 & & & \\ & A_2 & A_1 & A_0 & & \\ & & A_2 & A_1 & A_0 & \\ & & & A_2 & A_1 & A_0 \\ & & & & \ddots & \ddots & \ddots \end{bmatrix},$$

where

$$B_{0,0} = \begin{bmatrix} B_{00}^{(00)} & B_{00}^{(01)} & & & & & \\ B_{00}^{(10)} & B_{00}^{(1)} & B_{00}^{(0)} & & & & \\ & B_{00}^{(2)} & B_{00}^{(1)} & B_{00}^{(0)} & & & \\ & & B_{00}^{(2)} & B_{00}^{(1)} & B_{00}^{(0)} & & \\ & & & B_{00}^{(2)} & B_{00}^{(1)} & B_{00}^{(0)} & \\ & & & & \ddots & \ddots & \ddots \end{bmatrix},$$

with

$$B_{00}^{(00)} = a_{0,0},\ B_{00}^{(01)} = a_{0,1},\ B_{00}^{(10)} = a_{0,0}b_2,\ B_{00}^{(0)} = a_{0,1}(1-b_2),$$

$$B_{00}^{(1)} = a_{0,0}(1-b_2) + a_{0,1}b_2,\ B_{00}^{(2)} = a_{0,0}b_2,$$

$$B_{0,1} = \begin{bmatrix} B_{01}^{(00)} & B_{01}^{(01)} & & & & \\ B_{01}^{(2)} & B_{01}^{(1)} & & & & \\ & B_{01}^{(2)} & B_{01}^{(1)} & & & \\ & & B_{01}^{(2)} & B_{01}^{(1)} & & \\ & & & B_{01}^{(2)} & B_{01}^{(1)} & \\ & & & & \ddots & \ddots \end{bmatrix},$$

with

$$B_{01}^{(00)} = [a_{1,0},\ 0],\ B_{01}^{(01)} = [a_{1,1},\ 0],\ B_{01}^{(2)} = [a_{1,0}b_2,\ a_{1,0}(1-b_2)],$$

$$B_{01}^{(1)} = [a_{1,1}b_2,\ a_{1,1}(1-b_2)],$$

$$B_{1,0} = \begin{bmatrix} B_{10}^{(00)} & B_{10}^{(0)} & & & & & \\ & B_{10}^{(1)} & B_{10}^{(0)} & & & & \\ & & B_{10}^{(1)} & B_{10}^{(0)} & & & \\ & & & B_{10}^{(1)} & B_{10}^{(0)} & & \\ & & & & B_{10}^{(1)} & B_{10}^{(0)} & \\ & & & & & \ddots & \ddots \end{bmatrix},$$

with

$$B_{10}^{(00)} = \begin{bmatrix} a_{0,0}b_1 \\ 0 \end{bmatrix}, \; B_{10}^{(0)} = \begin{bmatrix} a_{0,1}b_1 \\ 0 \end{bmatrix}, \; B_{10}^{(1)} = \begin{bmatrix} a_{0,0}b_1 \\ 0 \end{bmatrix},$$

$$A_0 = \begin{bmatrix} A_0^{(1)} & A_0^{(0)} & & \\ & A_0^{(1)} & A_0^{(0)} & \\ & & \ddots & \ddots \end{bmatrix}, \; A_0^{(1)} = \begin{bmatrix} a_{1,0}(1-b_1) & 0 \\ a_{1,0}b_2 & a_{1,0}(1-b_2) \end{bmatrix},$$

$$A_0^{(0)} = \begin{bmatrix} a_{1,1}(1-b_1) & 0 \\ a_{1,1}b_2 & a_{1,1}(1-b_2) \end{bmatrix},$$

$$A_2 = \begin{bmatrix} A_2^{(1)} & A_2^{(0)} & & \\ & A_2^{(1)} & A_2^{(0)} & \\ & & \ddots & \ddots \end{bmatrix}, \; A_2^{(1)} = \begin{bmatrix} a_{0,0}b_1 & 0 \\ 0 & 0 \end{bmatrix}, \; A_2^{(0)} = \begin{bmatrix} a_{0,1}b_1 & 0 \\ 0 & 0 \end{bmatrix},$$

$$A_1 = \begin{bmatrix} A_1^{(1)} & A_1^{(0)} & & \\ & A_1^{(1)} & A_1^{(0)} & \\ & & \ddots & \ddots \end{bmatrix}, \; A_1^{(0)} = \begin{bmatrix} a_{0,1}(1-b_1)+a_{1,1}b_1 & 0 \\ a_{0,1}b_2 & a_{0,1}(1-b_2) \end{bmatrix},$$

$$A_1^{(1)} = \begin{bmatrix} a_{0,0}(1-b_1)+a_{1,0}b_1 & 0 \\ a_{0,0}b_2 & a_{0,0}(1-b_2) \end{bmatrix}.$$

9.3.1 Stationary Distribution

If the system is stable we know that there is a relationship between the vector $\boldsymbol{x}$ and the matrix P of the form

$$\boldsymbol{x} = \boldsymbol{x}P, \;\; \boldsymbol{x}\mathbf{1} = 1,$$

and there exists a matrix R for which

$$\boldsymbol{x}_{i+1} = \boldsymbol{x}_i R,$$

where R is the minimal nonnegative solution to the matrix quadratic equation

$$R = A_0 + RA_1 + R^2 A_2.$$

It was shown by Alfa (1998) and earlier by Miller (1981) for the continuous time case that the matrix R has the following structure

$$R = \begin{bmatrix} R_0 & R_1 & R_2 & R_3 & \cdots \\ & R_0 & R_1 & R_2 & \cdots \\ & & R_0 & R_1 & \cdots \\ & & & R_0 & \ddots \\ & & & & \ddots \end{bmatrix}. \tag{9.26}$$

The arguments leading to this result are as follow. The matrix R is the expected number of visits to the HP state $i+1$ before first returning to the HP state i, given that the process started from state i. By this definition of the matrix R it is straightforward to see how the upper triangular pattern arises, i.e. because the LP number in the system cannot reduce when there is an HP in the system. Since the number of LP arrivals is independent of the state of the number of the HP in the system, the rows of R will be repeating. This structure makes it easier to study the matrix R. However, the R_i, $i \geq 0$ also has an additional structure. We have

$$R_i = \begin{bmatrix} r_i(1,1) & 0 \\ r_i(2,1) & r_i(2,2) \end{bmatrix}. \tag{9.27}$$

The matrix R can be computed to a suitable truncation level and then used to compute the stationary vector after the boundary equation has been solved. We have

$$\boldsymbol{x}_{i+1} = \boldsymbol{x}_i R, \; i \geq 1, \tag{9.28}$$

and specifically we have

$$x_{i+1,j,1} = \sum_{v=0}^{j} [x_{i,j-v,1} r_v(1,1) + x_{i,j-v+1,2} r_v(2,1)], \; i \geq 1, \; j \geq 0, \tag{9.29}$$

and

$$x_{i+1,j+1,2} = \sum_{v=0}^{j} x_{i,j-v+1,2} r_{j-v}(2,2), \; i \geq 1, \; j \geq 0. \tag{9.30}$$

9.4 MAP/PH/1 Preemptive

Consider a system with two classes of packets that arrive according to the Markovian arrival process represented by four matrices $D_{0,0}$, $D_{1,0}$, $D_{0,1}$, and $D_{1,1}$ all of dimension $n_t \times n_t$, where D_{i_1,i_2} refers to i_1 arrivals of type 1 (HP) and i_2 arrivals of type 2 (LP). For example, $D_{0,1}$ refers to no arrival of type 1 (the HP class) packet and 1 arrival of type 2 (the LP class) packet. The matrix $D = D_{0,0} + D_{1,0} + D_{0,1} + D_{1,1}$ is stochastic. Let us assume that it is irreducible and let its stationary distribution be $\boldsymbol{\pi} = \boldsymbol{\pi} D, \quad \boldsymbol{\pi}\mathbf{1} = 1$, then $a_1 = \boldsymbol{\pi}(D_{1,0} + D_{1,1})\mathbf{1}$ and $a_2 = \boldsymbol{\pi}(D_{0,1} + D_{1,1})\mathbf{1}$, where a_k is the probability of a type k arrival in a time slot. Packets of type k class have PH service times with parameters $(\boldsymbol{\beta}_k, S_k)$ of order $n_s(k)$, $k = 1,2$. Let $(b_k)^{-1} = \boldsymbol{\beta}_k(I - S_k)^{-1}\mathbf{1}$. We consider the preemptive discipline. The service discipline is as follows. No LP packet can start receiving service if there is an HP packet in the system and if an LP packet is receiving service (when there is no HP packet in the system), this LP packet service will be interrupted if an HP packet arrives before its service is completed. An interrupted service in phase j_1 is resumed in phase j_2 with probability Q_{j_1,j_2} which is an element of the stochastic matrix Q. If for example, $Q = I$ then an interrupted service starts from the phase where it was stopped, i.e. preemptive resume whereas if $Q = \mathbf{1}\boldsymbol{\beta}_1$ then the service is started from beginning again, i.e. preemptive repeat. We know that a_k, and b_k are the mean arrival rates and service rates, respectively of class k packet. It is simple to show that provided that $\frac{a_1}{b_1} + \frac{a_2}{b_2} < 1$ the system will be stable. This system is easier to study using the matrix-analytic method.

Consider the following state space

$$\Delta = \{(0,0,\ell) \cup (0,i_2,\ell,j_2) \cup (i_1,i_2-1,\ell,j_1,j_2^{'}),\ \ell = 1,2,\cdots,n_t; i_1 \geq 1; i_2 \geq 1;$$

$$j_1 = 1,2,\cdots,n_s(1);\ j_2 = 1,2,\cdots,n_s(2); j_2^{'} = 1,2,\cdots,n_s(2)\ \}.$$

The elements of the state space are ℓ referring to the phase of arrival, i_v, $v = 1,2$, referring to the number of type v packets in the system, and j_v, $v = 1,2$ the phase of service of type v in service, and $j_2^{'}$ is the phase at which the type 2 packet's service resumes next time after preemption. The first tuple $(0,0,\ell)$ refers to an empty system, the second tuple $(0,i_2,\ell,j_2)$ refers to the system with only type 2 packets in the system and the third tuple (i_1,i_2,ℓ,j_1) refers to the case where there is at least one type 1 packet in the system, which is receiving service. This state space describes the DTMC that represents this preemptive priority system and the associated transition matrix is of the form

$$P = \begin{bmatrix} B_{0,0} & B_{0,1} & & & \\ B_{1,0} & B_{1,1} & A_0 & & \\ & A_2 & A_1 & A_0 & \\ & & A_2 & A_1 & A_0 \\ & & & A_2 & A_1 & A_0 \\ & & & & \ddots & \ddots & \ddots \end{bmatrix},$$

where

$$B_{0,0} = \begin{bmatrix} B_{00}^{(00)} & B_{00}^{(01)} & & & & \\ B_{00}^{(10)} & B_{00}^{(1)} & B_{00}^{(0)} & & & \\ & B_{00}^{(2)} & B_{00}^{(1)} & B_{00}^{(0)} & & \\ & & B_{00}^{(2)} & B_{00}^{(1)} & B_{00}^{(0)} & \\ & & & B_{00}^{(2)} & B_{00}^{(1)} & B_{00}^{(0)} \\ & & & & \ddots & \ddots & \ddots \end{bmatrix},$$

with

$$B_{00}^{(00)} = D_{0,0},\ B_{00}^{(01)} = D_{0,1} \otimes \boldsymbol{\beta}_2,\ B_{00}^{(10)} = D_{0,0}\mathbf{s}_2,\ B_{00}^{0} = D_{0,1} \otimes S_2,$$

$$B_{00}^{(1)} = D_{0,0} \otimes S_2 + D_{0,1} \otimes (\mathbf{s}_2\boldsymbol{\beta}_2),\ B_{00}^{2} = D_{0,0} \otimes (\mathbf{s}_2\boldsymbol{\beta}_2),$$

$$B_{0,1} = \begin{bmatrix} B_{01}^{(00)} & B_{01}^{(01)} & & & & \\ B_{01}^{(2)} & B_{01}^{(1)} & B_{01}^{(0)} & & & \\ & B_{01}^{(2)} & B_{01}^{(1)} & B_{01}^{(0)} & & \\ & & B_{01}^{(2)} & B_{01}^{(1)} & B_{01}^{(0)} & \\ & & & B_{01}^{(2)} & B_{01}^{(1)} & B_{01}^{(0)} \\ & & & & \ddots & \ddots & \ddots \end{bmatrix},$$

with

$$B_{01}^{(00)}=D_{1,0} \otimes \boldsymbol{\beta}_1,\ B_{01}^{(01)}=D_{1,1} \otimes \boldsymbol{\beta}_1,\ B_{01}^{(2)}=D_{1,0} \otimes (\mathbf{s}_2\boldsymbol{\beta}_1),\ B_{01}^{(0)}=D_{1,1} \otimes (S_2Q) \otimes \boldsymbol{\beta}_1,$$

$$B_{01}^{(1)} = D_{1,0} \otimes (S_2Q) \otimes \boldsymbol{\beta}_1 + D_{1,1}(\mathbf{s}_2\boldsymbol{\beta}_2)Q \otimes \boldsymbol{\beta}_1,$$

$$B_{1,0} = \begin{bmatrix} B_{10}^{(00)} & B_{10}^{(01)} & & & & \\ & B_{10}^{(1)} & B_{10}^{(0)} & & & \\ & & B_{10}^{(1)} & B_{10}^{(0)} & & \\ & & & B_{10}^{(1)} & B_{10}^{(0)} & \\ & & & & B_{10}^{(1)} & B_{10}^{(0)} \\ & & & & & \ddots & \ddots \end{bmatrix},$$

with

$$B_{10}^{(00)}=D_{0,0} \otimes \mathbf{s}_1,\ B_{10}^{(01)}=D_{0,1} \otimes (\mathbf{s}_1\boldsymbol{\beta}_1),\ B_{10}^{(0)}=D_{0,1} \otimes (\mathbf{s}_1\boldsymbol{\beta}_2),\ B_{10}^{(1)}=D_{0,0} \otimes (\mathbf{s}_1\boldsymbol{\beta}_2),$$

$$B_{1,1} = \begin{bmatrix} B_{11}^{(1)} & B_{11}^{(0)} & & & & & \\ & B_{11}^{(1)} & B_{11}^{(0)} & & & & \\ & & B_{11}^{(1)} & B_{11}^{(0)} & & & \\ & & & B_{11}^{(1)} & B_{11}^{(0)} & & \\ & & & & B_{11}^{(1)} & B_{11}^{(0)} & \\ & & & & & \ddots & \ddots \end{bmatrix},$$

with

$$B_{11}^{(1)} = (D_{0,0} \otimes S_1 + D_{1,0} \otimes (\mathbf{s}_1 \boldsymbol{\beta}_1)) \otimes I, \; B_{11}^{0} = (D_{1,1} \otimes (\mathbf{s}_1 \boldsymbol{\beta}_1) + D_{0,1} \otimes S_1) \otimes I,$$

$$A_0 = \begin{bmatrix} A_0^{(1)} & A_0^{(0)} & & \\ & A_0^{(1)} & A_0^{(0)} & \\ & & \ddots & \ddots \end{bmatrix}, \; A_0^{(1)} = D_{1,0} \otimes S_1 \otimes I, \; A_0^{(0)} = D_{1,1} \otimes S_1 \otimes I,$$

$$A_2 = \begin{bmatrix} A_2^{(1)} & A_2^{(0)} & & \\ & A_2^{(1)} & A_2^{(0)} & \\ & & \ddots & \ddots \end{bmatrix}, \; A_2^{(1)} = D_{0,0} \otimes (\mathbf{s}_1 \boldsymbol{\beta}_1) \otimes I, \; A_2^{(0)} = D_{0,1} \otimes (\mathbf{s}_1 \boldsymbol{\beta}_1) \otimes I,$$

$$A_1 = \begin{bmatrix} A_1^{(1)} & A_1^{(0)} & & \\ & A_1^{(1)} & A_1^{0} & \\ & & \ddots & \ddots \end{bmatrix},$$

$$A_1^{(1)} = (D_{0,0} \otimes S_1 + D_{1,0}(\mathbf{s}_1 \boldsymbol{\beta}_1)) \otimes I, \; A_1^{0} = (D_{0,1} \otimes S_1 + D_{1,1}(\mathbf{s}_1 \boldsymbol{\beta}_1)) \otimes I.$$

9.4.1 Stationary Distribution

If the system is stable we know that there is a relationship between the vector $\boldsymbol{x}$ and the matrix P of the form

$$\boldsymbol{x} = \boldsymbol{x}P, \; \boldsymbol{x}\mathbf{1} = 1,$$

and there exists a matrix R for which

$$\boldsymbol{x}_{i+1} = \boldsymbol{x}_i R,$$

where R is the minimal nonnegative solution to the matrix quadratic equation

$$R = A_0 + RA_1 + R^2 A_2,$$

and

$$\boldsymbol{x} = [\boldsymbol{x}_0,\ \boldsymbol{x}_1,\ \boldsymbol{x}_2,\ \cdots].$$

It was shown by Alfa (1998) that the matrix R has the following structure

$$R = \begin{bmatrix} R_0 & R_1 & R_2 & R_3 & \cdots \\ & R_0 & R_1 & R_2 & \cdots \\ & & R_0 & R_1 & \cdots \\ & & & R_0 & \ddots \\ & & & & \ddots \end{bmatrix}. \tag{9.31}$$

The arguments leading to this result are as follow. The matrix R is the expected number of visits to the HP state $i+1$ before first returning to the HP state i, given that the process started from state i. By this definition of the matrix R it is straightforward to see how the upper triangular pattern arises, i.e. because the LP number in the system can not reduce when there is an HP in the system. Since the number of LP arrivals is independent of the state of the number of the HP in the system, the rows of R will be repeating. This structure makes it easier to study the matrix R.

Solving for R involves a recursive scheme as follows:
First we write out the R equations in block form as

$$R_0 = A_0^{(1)} + R_0 A_1^{(1)} + (R_0)^2 A_2^{(1)}, \tag{9.32}$$

$$R_1 = A_0^{(0)} + R_0 A_1^{(0)} + R_1 A_1^{(1)} + (R_0 R_1 + R_1 R_0) A_2^{(1)}, \tag{9.33}$$

$$R_j = R_{j-1} A_1^{(0)} + R_j A_1^{(1)} + \sum_{v=0}^{j-1} R_{j-v-1} R_v A_2^{(0)} + \sum_{v=0}^{j} R_{j-v} R_v A_2^{(1)},\ j \geq 2. \tag{9.34}$$

Let us write

$$F_2 = A_2^{(1)},$$

$$F_0 = I - A_1^{(1)} - R_0 F_2,$$

and

$$F_{1,j} = \delta_{1,j} A_0^{(1)} + R_{j-1} A_1^{(0)} + \sum_{v=0}^{j-1} R_{j-v-1} R_v A_2^{(0)} + (1 - \delta_{1,j}) \sum_{v=0}^{j} R_{j-v} R_v F_2,\ j \geq 1,$$

where, δ_{ij} is the Kronecker delta written as $\delta = \begin{cases} 1, & i=j, \\ 0, & \text{otherwise} \end{cases}$, then we have the solution to the matrix sequence $R_j,\ j=1,2,\cdots$ given as

- R_0 is the minimal non-negative solution to the matrix quadratic equation

$$R_0 = A_0^{(1)} + R_0 A_1^{(1)} + (R_0)^2 A_2^{(1)}, \tag{9.35}$$

  which can be solved for using any of the special techniques for the QBD such as the Logarithmic Reduction or the Cyclic Reduction, and
- $R_j,\ j \geq 1$ are given as the solutions to the matrix linear equations

$$(F_0^T \otimes I - F_2^T \otimes R_0) vecR_j = vecF_{1,j}^T, \tag{9.36}$$

  where for an $M \times N$ matrix $Y = [Y_{.1},\ Y_{.2},\ \cdots,\ Y_{.j},\ \cdots,\ Y_{.N}]$, where $Y_{.j}$ is the jth column vector of the matrix Y, we define a $1 \times NM$ column vector

$$vecY = \begin{bmatrix} Y_{.1} \\ Y_{.2} \\ \vdots \\ Y_{.j} \\ \cdots \\ Y_{.N} \end{bmatrix}.$$

It was pointed out in Alfa (1998) that $(R_{k+1})_{i,j} \leq (R_k)_{i,j},\ k \geq 1$, hence we can truncate R at an appropriate point of desired accuracy.

We need to obtain the boundary results $[\boldsymbol{x}_0,\ \boldsymbol{x}_1]$ and this is carried out using the boundary equation that

$$[\boldsymbol{x}_0,\ \boldsymbol{x}_1] = [\boldsymbol{x}_0,\ \boldsymbol{x}_1] B[R],\ [\boldsymbol{x}_0,\ \boldsymbol{x}_1]\mathbf{1} = 1,$$

where

$$B[R] = \begin{bmatrix} B_{00} & B_{01} \\ B_{10} & B_{11} + RA_2 \end{bmatrix}.$$

This can be solved using censored Markov chain idea to obtain $\boldsymbol{x}_0$ and after routine algebraic operations we have

$$\boldsymbol{x}_1 = \boldsymbol{x}_0 B_{01} (I - B_{11} - RA_1)^{-1}.$$

Note that for most practical situations the matrix $(I - B_{11} - RA_1)$ is non-singular. The vector $\boldsymbol{x}_0$ is normalized as

$$\boldsymbol{x}_0 \mathbf{1} + \boldsymbol{x}_0 [B_{01} (I - B_{11} - RA_1)^{-1} (I - R)^{-1}] \mathbf{1} = 1.$$

Both the matrices $(I - B_{11} - RA_1)$ and $I - R$ have the following block triangular structures

$$U = \begin{bmatrix} U_0 & U_1 & U_2 & U_3 & \cdots \\ & U_0 & U_1 & U_2 & \cdots \\ & & U_0 & U_1 & \cdots \\ & & & U_0 & \ddots \\ & & & & \ddots \end{bmatrix}.$$

Their inverses also have the same structure and can be written as

$$V = \begin{bmatrix} V_0 & V_1 & V_2 & V_3 & \cdots \\ & V_0 & V_1 & V_2 & \cdots \\ & & V_0 & V_1 & \cdots \\ & & & V_0 & \ddots \\ & & & & \ddots \end{bmatrix},$$

where

$$V_0 = U_0^{-1}, \text{ and } V_\ell = -U_0^{-1} \sum_{j=0}^{\ell-1} U_{\ell-j} V_j, \ \ell \geq 1.$$

Given $\boldsymbol{x}$ the major performance measures such as the distribution, and the mean, of the number of type k, $k = 1,2$ packets in the system is straightforward to obtain. The waiting times for type 1 is straightforward to obtain also. It is the same idea as used for obtaining the waiting time of the one class MAP/PH/1 or PH/PH/1 model. However, the waiting for the type 2 class is more involved. For details of how to obtain all these performance measures see Alfa et al (2003). The details will be omitted from this book.

9.5 MAP/PH/1 Non-preemptive

Consider a system with two classes of packets that arrive according to the Markovian arrival process represented by four matrices $D_{0,0}$, $D_{1,0}$, $D_{0,1}$, and $D_{1,1}$ all of dimension $n_t \times n_t$, where D_{i_1,i_2} refers to i_k arrivals of type k, $k = 1,2$. For example, $D_{0,1}$ refers to no arrival of type 1 (the HP class) packet and 1 arrival of type 2 (the LP class) packet. The matrix $D = D_{0,0} + D_{1,0} + D_{0,1} + D_{1,1}$ is stochastic. Let us assume that it is irreducible and let its stationary distribution be $\boldsymbol{\pi} = \boldsymbol{\pi} D$, $\ \boldsymbol{\pi}\mathbf{1} = 1$, then $a_1 = \boldsymbol{\pi}(D_{1,0} + D_{1,1})\mathbf{1}$ and $a_2 = \boldsymbol{\pi}(D_{0,1} + D_{1,1})\mathbf{1}$, where again, a_k is the probability of a type k arrival during a time slot. Packets of type k class have PH service times with parameters $(\boldsymbol{\beta}_k, S_k)$ of order $n_s(k)$, $k = 1,2$. Let $(b_k)^{-1} = \boldsymbol{\beta}_k(I - S_k)^{-1}\mathbf{1}$. We consider the nonpreemptive discipline. The service discipline is as follows. No LP packet can start receiving service if there is an HP packet in the system. However if an LP packet starts to receive service when there is no HP packet in the system, this LP packet service will continue until completion even if an HP packet arrives before the service is completed. We know that a_k, b_k are the mean arrival rates and service rates, respectively of class k packet. It is simple to show that provided that $\frac{a_1}{b_1} + \frac{a_2}{b_2} < 1$ the system will be stable. This system is easier to study using the matrix-analytic method.

The state space for this system can be written as

$$\Delta = \Delta_1 \cup \Delta_2 \cup \Delta_3,$$

where

- $\Delta_1 = \{(0,0,\ell),\ \ell = 1,2,\cdots,n_t\}$,
- $\Delta_2 = \{(0,i_2,\ell,j_2),\ i_2 \geq 1,\ \ell = 1,2,\cdots,n_t; j_2 = 1,2,\cdots,n_s(2)\}$,
- $\Delta_3 = \{(i_1,i_2+v-1,v,\ell,j_v),\ i_1 \geq 1; i_2 \geq 0;\ v = 1,2;$
  $\ell = 1,2,\cdots,n_t;\ j_v = 1,2,\cdots,n_s(v)\}$.

Throughout this state space definition we have, i_v, $v = 1,2$ referring to the number of type v packets in the system, ℓ to the phase of arrival, and k_v to the phase of type v packet in service. The first tuple $(0,0,\ell)$ refers to an empty system, the second $(0,i_2,\ell,j_2)$ to having no type 1 and at least one type 2 packets in the system, and the third tuple (i_1,i_2+v-1,v,ℓ,j_v) refers to having i_1 and i_2+v-1 type 1 and type 2 packets in the system ,with a type v in service whose service is in phase j_v.

The transition matrix P associated with this DTMC is given as

$$P = \begin{bmatrix} B_{0,0} & B_{0,1} & & & & \\ B_{1,0} & A_1 & A_0 & & & \\ & A_2 & A_1 & A_0 & & \\ & & A_2 & A_1 & A_0 & \\ & & & A_2 & A_1 & A_0 \\ & & & & \ddots & \ddots & \ddots \end{bmatrix},$$

where

$$B_{0,0} = \begin{bmatrix} B_{00}^{(00)} & B_{00}^{(01)} & & & & \\ B_{00}^{(10)} & B_{00}^{(1)} & B_{00}^{(0)} & & & \\ & B_{00}^{(2)} & B_{00}^{(1)} & B_{00}^{(0)} & & \\ & & B_{00}^{(2)} & B_{00}^{(1)} & B_{00}^{(0)} & \\ & & & B_{00}^{(2)} & B_{00}^{(1)} & B_{00}^{(0)} \\ & & & & \ddots & \ddots & \ddots \end{bmatrix},$$

with

$$B_{00}^{(00)} = D_{0,0},\ B_{00}^{(01)} = D_{0,1} \otimes \boldsymbol{\beta}_2,\ B_{00}^{(10)} = D_{0,0} \otimes \mathbf{s}_2,\ B_{00}^{(0)} = D_{0,1} \otimes S_2,$$

$$B_{00}^{(1)} = D_{0,0} \otimes S_2 + D_{0,1} \otimes (\mathbf{s}_2\boldsymbol{\beta}_2),\ B_{00}^{2} = D_{0,0} \otimes (\mathbf{s}_2\boldsymbol{\beta}_2),$$

$$B_{0,1} = \begin{bmatrix} B_{01}^{(00)} & B_{01}^{(01)} & & & & \\ B_{01}^{(2)} & B_{01}^{(1)} & & & & \\ & B_{01}^{(2)} & B_{01}^{(1)} & & & \\ & & B_{01}^{(2)} & B_{01}^{(1)} & & \\ & & & B_{01}^{(2)} & B_{01}^{(1)} & \\ & & & & \ddots & \ddots \end{bmatrix},$$

with

$$B_{01}^{(00)} = [D_{1,0} \otimes \boldsymbol{\beta}_1,\ 0],\ B_{01}^{(01)} = [D_{1,1} \otimes \boldsymbol{\beta}_1,\ 0],\ B_{01}^{(2)} = [D_{1,0} \otimes (\mathbf{s}_2\boldsymbol{\beta}_1),\ D_{1,0} \otimes S_2],$$

$$B_{01}^{(1)} = [D_{1,1} \otimes (\mathbf{s}_2\boldsymbol{\beta}_1),\ D_{1,1} \otimes S_2],$$

$$B_{1,0} = \begin{bmatrix} B_{10}^{(00)} & B_{10}^{(0)} & & & & \\ & B_{10}^{(1)} & B_{10}^{(0)} & & & \\ & & B_{10}^{(1)} & B_{10}^{(0)} & & \\ & & & B_{10}^{(1)} & B_{10}^{(0)} & \\ & & & & B_{10}^{(1)} & B_{10}^{(0)} \\ & & & & \ddots & \ddots \end{bmatrix},$$

with

$$B_{10}^{(00)} = \begin{bmatrix} D_{0,0} \otimes \mathbf{s}_1 \\ 0 \end{bmatrix},\ B_{10}^{(0)} = \begin{bmatrix} D_{0,1} \otimes (\mathbf{s}_1\boldsymbol{\beta}_2) \\ 0 \end{bmatrix},\ B_{10}^{(1)} = \begin{bmatrix} D_{0,0} \otimes (\mathbf{s}_2\boldsymbol{\beta}_2) \\ 0 \end{bmatrix},$$

$$A_0 = \begin{bmatrix} A_0^{(1)} & A_0^{(0)} & & \\ & A_0^{(1)} & A_0^{(0)} & \\ & & \ddots & \ddots \end{bmatrix},\ A_0^{(1)} = \begin{bmatrix} D_{1,0} \otimes S_1 & 0 \\ D_{1,0} \otimes (\mathbf{s}_2\boldsymbol{\beta}_1) & D_{1,0} \otimes S_2 \end{bmatrix},$$

$$A_0^{(0)} = \begin{bmatrix} D_{1,1} \otimes S_1 & 0 \\ D_{1,1} \otimes (\mathbf{s}_2\boldsymbol{\beta}_1) & D_{1,1} \otimes S_2 \end{bmatrix},$$

$$A_2 = \begin{bmatrix} A_2^{(1)} & A_2^{(0)} & & \\ & A_2^{(1)} & A_2^{(0)} & \\ & & \ddots & \ddots \end{bmatrix}, \; A_2^{(1)} = \begin{bmatrix} D_{0,0} \otimes (\mathbf{s}_1 \boldsymbol{\beta}_1) & 0 \\ 0 & 0 \end{bmatrix}, \; A_2^{(0)} = \begin{bmatrix} D_{0,1} \otimes (\mathbf{s}_1 \boldsymbol{\beta}_1) & 0 \\ 0 & 0 \end{bmatrix},$$

$$A_1 = \begin{bmatrix} A_1^{(1)} & A_1^{(0)} & & \\ & A_1^{(1)} & A_1^{(0)} & \\ & & \ddots & \ddots \end{bmatrix}, \; A_1^{(0)} = \begin{bmatrix} D_{0,1} \otimes S_1 + D_{1,1} \otimes (\mathbf{s}_1 \boldsymbol{\beta}_1) & 0 \\ D_{0,1} \otimes (\mathbf{s}_2 \boldsymbol{\beta}_1) & D_{0,1} \otimes S_2 \end{bmatrix},$$

$$A_1^{(1)} = \begin{bmatrix} D_{0,0} \otimes S_1 + D_{1,0} \otimes (\mathbf{s}_1 \boldsymbol{\beta}_1) & 0 \\ D_{0,0} \otimes (\mathbf{s}_2 \boldsymbol{\beta}_1) & D_{0,0} \otimes S_2 \end{bmatrix}.$$

9.5.1 Stationary Distribution

If the system is stable we know that there is a relationship between the vector $\boldsymbol{x}$ and the matrix P of the form

$$\boldsymbol{x} = \boldsymbol{x}P, \;\; \boldsymbol{x}\mathbf{1} = 1,$$

and there exists a matrix R for which

$$\boldsymbol{x}_{i+1} = \boldsymbol{x}_i R,$$

where R is the minimal nonnegative solution to the matrix quadratic equation

$$R = A_0 + RA_1 + R^2 A_2.$$

It was shown by Alfa (1998) that the matrix R has the following structure

$$R = \begin{bmatrix} R_0 & R_1 & R_2 & R_3 & \cdots \\ & R_0 & R_1 & R_2 & \cdots \\ & & R_0 & R_1 & \cdots \\ & & & R_0 & \ddots \\ & & & & \ddots \end{bmatrix}. \tag{9.37}$$

The arguments leading to this result are as follow. The matrix R is the expected number of visits to the HP state $i+1$ before first returning to the HP state i, given that the process started from state i. By this definition of the matrix R it is straightforward to see how the upper triangular pattern arises, i.e. because the LP number in the system cannot reduce when there is an HP in the system. Since the number of LP arrivals is independent of the state of the number of the HP in the system, the rows of R will be repeating. This structure makes it easier to study the matrix R. However, the R_i, $i \geq 0$ also has an additional structure. We have

$$R_i = \begin{bmatrix} R_i(1,1) & 0 \\ R_i(2,1) & R_i(2,2) \end{bmatrix}. \tag{9.38}$$

The matrix R can be computed to a suitable truncation level and then use it to compute the stationary vector after the boundary equation has been solved. We have

$$\boldsymbol{x}_{i+1} = \boldsymbol{x}_i R, \; i \geq 1, \tag{9.39}$$

and specifically we have

$$\boldsymbol{x}_{i+1,j,1} = \sum_{v=0}^{j} [\boldsymbol{x}_{i,j-v,1} R_v(1,1) + \boldsymbol{x}_{i,j-v+1,2} R_v(2,1)], \; i \geq 1, \, j \geq 0, \tag{9.40}$$

and

$$\boldsymbol{x}_{i+1,j+1,2} = \sum_{v=0}^{j} \boldsymbol{x}_{i,j-v+1,2} R_{j-v}(2,2), \; i \geq 1, \, j \geq 0. \tag{9.41}$$

From here on it is straightforward (but laboriously involved) to obtain the stationary distributions and the expected values of the key performance measures. We will skip them in this book and refer the reader to Alfa (1998).

9.6 Problems

1. Question 1: A Geo/Geo/1 system has arrival an service parameters given as a and b, respectively. After a customer finishes service it may return to the queue, with probability θ, for another service, which is also geometric with parameter b. A separate queue (which we call Queue 2) is set up for customers that have returned for another service. There is no restriction as to how many times a customer may return for service. The serve only attends to Queue 2 only if the first queue is empty. Find the stability condition of this system and the stationary distribution of number in the system.
2. Question 2: Develop a DTMC for representing the number in the system for a preemptive Geo/Geo/1 system with three classes of priorities, each with different service parameters. The arrival process is represented by $a_{i,j,k}$, where $a_{i,j,k}$ represents the probability of arrival of i Type 1, j Type 2 and k Type 3 items,

with $(i,j,k) = 0,1$. A Type r item has a geometric service with parameter b_r. We $1,2,3$ be a decreasing order of priorities, i.e. Type 1 has a higher priority than Type 2 and Type 2 has a higher priority than Type 3.

3. Question 3: Consider a Geo/Geo/1 system with two classes of customers, Type 1 and Type 2. Arrival process is represented by $a_{i,j}$ with $a_{i,j}$ representing probability that there are i Type 1 and j Type 2 arrivals, with $(i,j) = 0,1$. Both Type 1 and Type 2 have the same service parameters b, and service preemptive.When the server is attending to Type 1 class (which is the higher priority) if the number of Type 2 items waiting is more than K, the server will transfer $L < K$ of the Type 2 items into the Type 1 class. Set this problem up as a DTMC and study how long it takes for a Type 2 item to be classified at a Type 1.
4. Question 4: Repeat Question 3 for the case of non-preemptive system.

References

Alfa AS (1998) Matrix-geometric solution of discrete time MAP/PH/1 priority queue. Naval Research Logistics 45:23–50

Alfa AS, Liu B, He QM (2003) Discrete-time analysis of MAP/PH/1 multiclass general preemptive priority queue. Naval Research Logistics 50:662–682

Miller DG (1981) Computation of steady-state probabilities for the M/M/1 priority queues. Operations Research 29:945–958

Chapter 10
Special Single Node Queueing Models

10.1 Introduction

In this chapter we present five special models;

1. Queues with multiclass packets
2. LCFS Geo/Geo/1 system
3. SIRO GeoGeo/1 system
4. Double-ended queues
5. Finite source model

10.2 Queues with Multiclass of Packets

In most queueing systems we find that the service disciplines are either FCFS or priority based when we have more than one class of packets. In the last chapter we dealt with the priority systems of queues already. However sometimes in telecommunications we have multiclass systems with FCFS or sometimes LCFS disciplines. In this chapter we deal with two classes of multiclass packets – the no priority case and the LCFS. Consider a queueing system with $K < \infty$ classes of packets, with no priority. We assume that the arrival process is governed by the marked Markovian arrival process as defined by He and Neuts (1998). This arrival process is described by $n_t \times n_t$, $K+1$ matrices D_k, $k = 0, 1, \cdots, K$, with the elements $(D_k)_{i,j}$ representing the probability of a transition from state i to state j with the arrival of type $k(k = 1, 2, \cdots, K)$ and $(D_0)_{i,j}$ represents no arrival. We assume that $0 \leq (D_k)_{i,j} \leq 1$ and $D = \sum_{k=0}^{K} D_k$ is an irreducible stochastic matrix. Let $\boldsymbol{\pi}$ be the stationary vector of the matrix D, then the arrival rate of class k is given by $\lambda_k = \boldsymbol{\pi} D_k \mathbf{1}$. Each class of packet is served according to a PH distribution represented by $(\boldsymbol{\beta}_k, S_k)$ of order $n_s(k)$, $k = 0, 1, 2, \cdots, K$, with mean service time of

A.S. Alfa, *Applied Discrete-Time Queues*, DOI 10.1007/978-1-4939-3420-1_10

class k given as $\mu_k^{-1} = \boldsymbol{\beta}_k(I - S_k)^{-1}\mathbf{1}$. There is a single server attending to these packets. We may have the service in a FCFS or LCFS order. In the next two sections we deal with these two types of service disciplines.

10.2.1 Multiclass systems – with FCFS - the MMAP [K]/PH[K]/1

This class of queues is difficult to analyze using traditional approaches. van Houdt and Blondia (2002) have extended the idea of age process to analyze this system's waiting time. One may also obtain the queue length information from this waiting time information. In this section we present a summary of the results by van Houdt and Blondia (2002).

Define a new matrix S and column vector $\mathbf{s}$, where

$$S = \begin{bmatrix} S_1 & & & \\ & S_2 & & \\ & & \ddots & \\ & & & S_K \end{bmatrix}, \quad \mathbf{s} = \begin{bmatrix} \mathbf{s}_1 \\ \mathbf{s}_2 \\ \vdots \\ \mathbf{s}_K \end{bmatrix}.$$

We now construct a DTMC that represents the age of the leading packet that is in service. We call this the Age process. This process captures how long the packet in service has been in the system up to a point in time. Consider the following state space

$$\Delta = \{(0, J_n^{(4)}) \cup (J_n^{(1)}, J_n^{(2)}, J_n^{(3)}, J_n^{(4)}),\ J_n^1 \geq 1;$$

$$J_n^{(2)} = 1, 2, \cdots, K; J_n^{(3)} = 1, 2, \cdots, n_s(J_n^{(2)}); J_n^{(4)} = 1, 2, \cdots, n_t\}.$$

Here at time n, $J_n^{(1)}$ refers to the age of the packet that is receiving service, $J_n^{(2)}$ the class of packet that is in service, $J_n^{(3)}$ the phase of service of the packet in service, and $J_n^{(4)}$ the arrival phase of the packets. This DTMC has the GI/M/1-type structure with the transition matrix P written as

$$P = \begin{bmatrix} B_0 & C & & & \\ B_1 & A_1 & A_0 & & \\ B_2 & A_2 & A_1 & A_0 & \\ \vdots & \vdots & \vdots & \ddots & \ddots \end{bmatrix}, \tag{10.1}$$

where

$$B_0 = D_0,\ B_j = \mathbf{s} \otimes (D_0)^j,\ j \geq 1,$$

$$C = [(\boldsymbol{\beta}_1 \otimes D_1),\ (\boldsymbol{\beta}_2 \otimes D_2),\ \cdots,\ (\boldsymbol{\beta}_K \otimes D_K)],$$

$$A_0 = S \otimes I_{n_t},\ A_j = \mathbf{s} \otimes ((D_0)^{j-1} C),\ j \geq 1.$$

Provided the system is stable, we can easily obtain its stationary distribution using the standard GI/M/1 results presented earlier in the book. van Houdt and Blondia (2002) did give the conditions under which the system is ergodic and as long as it is ergodic then it is stable. Note that ergodicity in this particular case can only be considered after the state $(0, J_n^4)$ is eliminated from the state space.

10.2.1.1 Stationary distribution and waiting times

In order to proceed with obtaining the stationary distribution we first have to obtain the R matrix which is given as the minimal nonnegative solution to the matrix polynomial equation

$$R = \sum_{j=0}^{\infty} R^j A_j.$$

Let the stationary distribution be $\boldsymbol{x}$ where

$$\boldsymbol{x} = \boldsymbol{x} P,\ \boldsymbol{x}\mathbf{1} = 1,$$

with

$$\boldsymbol{x}_0 = [x_{0,1},\ x_{0,2},\ \cdots,\ x_{0,n_t}],\ \boldsymbol{x}_i = [\boldsymbol{x}_{i,1},\ \boldsymbol{x}_{i,2},\ \cdots,\ \boldsymbol{x}_{i,K}],$$

$$\boldsymbol{x}_{i,j} = [\boldsymbol{x}_{i,j,1},\ \boldsymbol{x}_{i,j,2},\ \cdots,\ \boldsymbol{x}_{i,j,n_s(j)}],\ \boldsymbol{x}_{i,j,\ell} = [x_{i,j,\ell,1},\ x_{i,j,\ell,2},\ \cdots,\ x_{i,j,\ell,n_t}].$$

We follow the standard procedure to obtain the boundary vectors $\boldsymbol{x}_0$ and $\boldsymbol{x}_1$ and then normalize the stationary vector.

Let W_k be the time spent in the system by a class k packet before leaving, i.e. queueing time plus service time. Further define $w_j^{(k)} = Pr\{W_k = j\}$, then we have

$$w_j^{(k)} = \sum_{v=1}^{n_s(k)} (\lambda_k)^{-1} (\mathbf{t}_k)_v \sum_{u=1}^{n_t} x_{j,k,v,u},\ j \geq 1, \tag{10.2}$$

where $(\mathbf{t}_k)_v$ is the vth element of the column vector $\mathbf{t}_k$. Note that $w_0^{(k)} = 0$ since every packet has to receive a service of at least one unit of time.

In a later paper Houdt and Blondia (2004) presented a method that converts the GI/M/1 structure of this problem to a QBD type. The reader is referred to their paper for more details on this and other related results.

10.2.2 Multiclass systems – with LCFS - the MMAP [K]/PH[K]/1

In a LCFS system with non-geometric service times we need to identify from which phase a service starts again if it is interrupted. Hence we define the matrix $Q^{(k)}$ as in the case of the preemptive priority queue earlier, with $Q^{(k)}_{\ell_1,\ell_2}$ referring to the probability that given the phase of an ongoing service of a packet of class k is in phase ℓ_1, its future service phase is chosen to start from (after its service resumption) is phase ℓ_2. Note that this $Q^{(k)}$ is different from the Q in the previous chapters but they both lead to the same concept. However, this current definition makes obtaining some additional performance measures more convenient. The presentation in this section is based on the paper by He and Alfa (2000). It uses the idea and results from tree-structured Markov chains.

Consider a pair (k,j) that is associated with a packet of a type k that is in phase j of service. By being in phase j of service we imply that its service is in phase j if it is currently receiving service or phase j is where its service will start next, if it is currently waiting in the queue. At time n let there be v packets in the system. Associated with this situation let us define

$$q_n = (k_1,j_1)(k_2,j_2)\cdots(k_v,j_v).$$

Here q_n is defined as the queue string consisting of the status of all the packets in the system at the beginning of time n. Let v_n be the phase of the arrival process at time n. It is easy to see that (q_n, v_n) is an irreducible and aperiodic DTMC. Its state space is

$$\Delta = \Omega \times \{1,2,\cdots,n_t\},$$

where n_t is the phase of arrival and

$$\Omega = \{0\} \cup \{\mathbf{J} : \mathbf{J} = (k_1,j_1)(k_2,j_2)\cdots(k_v,j_v),\ 1 \le k_w \le K,$$

$$1 \le j_w \le n_s(k),\ 1 \le w \le n,\ n \ge 1\}.$$

In fact this DTMC is a tree-structured QBD in the sense that q_n can only increase or decrease by a maximum of 1. We define an addition operation "$+$" here for strings as follows: if we have $\mathbf{J} = (k_1,j_1)(k_2,j_2)\cdots(k_v,j_v)$ then $\mathbf{J} + (k,j) = (k_1,j_1)(k_2,j_2)\cdots(k_v,j_v)(k,j)$.

Let the DTMC be in state $\mathbf{J}+(k,j) \in \Omega$ at time n, i.e. $q_n = \mathbf{J}+(k,j)$, then we have the following transition blocks:

1. When a new packet arrives and there is no service completion, we define

$$A_0((k,j),(k,j')(k_e,j_e))$$

$$= Pr\{q_{n+1} = \mathbf{J}+(k,j')(k_e,j_e), v_{t+1} = i'|q_n = \mathbf{J}+(k,j), v_n = i\}$$

$$= (S_k Q^k)_{j,j'}(\boldsymbol{\beta}_{k_e})_{j_e} D_{k_e},\ 1 \le (k,k_e) \le K,\ 1 \le (j,j') \le n_s(k),\ 1 \le j_e \le n_s(k_e), \tag{10.3}$$

2. When a new packet arrives and a service is completed or there is no new packet arrival and there is no service completion, we define

$$A_1((k,j),(k_e,j_e))$$

$$= Pr\{q_{n+1} = \mathbf{J}+(k_e,j_e), v_{n+1} = i'|q_n = \mathbf{J}+(k,j), v_n = i\}$$

$$= \begin{cases} (S_k)_{j,j_e} D_0 + ((\mathbf{s}_k)_j \boldsymbol{\beta}_{k_e})_{j_e} D_{k_e}, & k = k_e,\ 1 \le (j,j') \le n_s(k), \\ ((\mathbf{s}_k)_j \boldsymbol{\beta}_{k_e})_{j_e} D_{k_e}, & k \ne k_e,\ 1 \le j \le n_s(k),\ 1 \le j_e \le n_s(k_e) \end{cases}$$

3. When a service is completed and there is no new arrival, we define

$$A_2(k,j) = Pr\{q_{n+1} = \mathbf{J}, v_{n+1} = i'|q_n = \mathbf{J}+(k,j), v_n = i\}$$

$$= (\mathbf{s}_k)_j D_0,\ 1 \le k \le K,\ 1 \le j \le n_s(k). \tag{10.4}$$

4. When there is no packet in the system, i.e. $q_n = \mathbf{J} = 0$ we define

$$A_0(0,(k,j)) = Pr\{q_{n+1} = (k_e,j_e), v_{n+1} = i'|q_n = 0, v_n = i\}$$

$$= (\boldsymbol{\beta}_{k_e})_{j_e} D_{k_e},\ 1 \le k_e \le K,\ 1 \le j_e \le n_s(k_e). \tag{10.5}$$

$$A_1(0,0) = Pr\{q_{n+1} = 0, v_{n+1} = i'|q_t = 0, v_n = i\} = D_0. \tag{10.6}$$

10.2.2.1 Stability conditions

We summarize the conditions under which the system is stable as given by He and Alfa (2000). Analogous to the classical M/G/1 system this current system has a set of matrices $G_{k,j}$, $1 \le k \le K$, $1 \le j \le n_s(k)$, which are minimal non-negative solutions to the matrix equation

$$G_{k,j} = A_2(k,j) + \sum_{k_e=1}^{K} \sum_{j_e=1}^{n_s(k_e)} A_1((k,j),(k_e,j_e))G_{k_e,j_e}$$

$$+ \sum_{j'=1}^{n_s(k)} \sum_{k_e=1}^{K} \sum_{j_e=1}^{n_s(k_e)} A_0((k,j),(k,j')(k_e,j_e))G_{k_e,j_e}G_{k,j'}, \tag{10.7}$$

and are known to be unique if the DTMC (q_n, v_n) is positive recurrent (see He He (1998)). Consider the matrices $p((k,j),(k_e,j_e))$ defined as

$$p((k,j),(k_e,j_e)) = \begin{cases} A_1((k,j),(k_e,j_e)) + \sum_{j_e=1}^{n_s(k_e)} A_0((k,j),(k,j')(k_e,j_e)), \; k \neq k_e \\ A_1((k,j),(k,j_e)) + \sum_{j_e=1}^{n_s(k_e)} A_0((k,j),(k,j')(k,j_e)) \\ \quad + A_0((k,j),(k,j_e)(k,j_e))G_{k,j_e} \\ \quad + \sum_{k'=1}^{K} \sum_{j'=1:(k',j')\neq(k,j_e)}^{n_s(k_e)} A_0((k,j),(k,j_e)(k',j'))G_{k',j'}, \\ \qquad k = k_e. \end{cases} \tag{10.8}$$

Associated with these matrices is another matrix P of the form

$$P = \begin{bmatrix} p((1,1),(1,1)) & p((1,1),(1,2)) & \cdots & p((1,1),(K,n_s(K))) \\ p((1,2),(1,1)) & p((1,2),(1,2)) & \cdots & p((1,2),(K,n_s(K))) \\ \vdots & \vdots & \vdots & \vdots \\ p((K,n_s(K)),(1,1)) & p((K,n_s(K)),(1,2)) & \cdots & p((K,n_s(K)),(K,n_s(K))) \end{bmatrix}. \tag{10.9}$$

Let $sp(P)$ be the spectral radius of the matrix P, i.e. its largest absolute eigenvalue. Then the Markov chain $\{q_n, v_n\}$ is positive recurrent if and only if $sp(P) < 1$.

10.2.2.2 Performance measures

Here we present an approach for obtaining information about the queue length. First we need to redefine the DTMC in order to make it easier to study the queue length.

We consider the DTMC $\{\tilde{q}_n, v_n, k_n, j_n\}$ where at time n

- $\tilde{q}_n$ is the string consisting of the states of packets waiting in the queue
- v_n is the phase of MMAP[K]
- k_n is the type of packets in service, if any
- j_n is the phase of the service time, if any.

We let the random variable $\tilde{q}_n$ assume the value -1 when there is no packet in the system and assumes the value 0 when there is only one packet in the system. Hence $\tilde{q}_n$ assumes values in the set $\{-1\} \cup \Omega$.

Now we consider the one step block transitions of the DTMC $\{\tilde{q}_n, v_n, k_n, j_n\}$ as follows:

1. $\tilde{q}_n$ going from -1 to -1 and from -1 to 0 as

$$\tilde{A}_1(-1,-1) = D_0, \text{ and } \tilde{A}_0(-1,0) = (D_1 \otimes \boldsymbol{\beta}_1,\ D_2 \otimes \boldsymbol{\beta}_2,\ \cdots,\ D_K \otimes \boldsymbol{\beta}_K),$$

2. $\tilde{q}_t$ going from 0 to -1 as

$$\tilde{A}_2(0,-1) = D_0 \otimes \begin{bmatrix} \mathbf{s}_1 \\ \mathbf{s}_2 \\ \vdots \\ \mathbf{s}_K \end{bmatrix},$$

3. $\tilde{q}_t$ going from $\mathbf{J} + (k,j)$ to $\mathbf{J} + (k,j)(k_e,j_e)$, or to $\mathbf{J} + (k,j)$, or to $\mathbf{J}$, $\mathbf{J} \in \Omega$ as

$$\tilde{A}_0(k_e,j_e) = \sum_{w=1}^{K} D_w \otimes \left[\begin{pmatrix} 0 \\ \vdots \\ (S_{k_e}Q_{k_e})_{j_e}\boldsymbol{\beta}_w \\ 0 \\ \vdots \\ 0 \end{pmatrix} (0,\ \cdots,\ 0,\ I_{n_s(w)},\ 0\cdots,\ 0) \right];$$

$$\tilde{A}_1(k,j) = D_0 \otimes \begin{pmatrix} S_1 & & & \\ & S_2 & & \\ & & \ddots & \\ & & & S_K \end{pmatrix}$$

$$+ \sum_{w=1}^{K} D_w \otimes \left[\begin{pmatrix} \mathbf{s}_1\boldsymbol{\beta}_w \\ \vdots \\ \mathbf{s}_K\boldsymbol{\beta}_w \end{pmatrix} (0,\ \cdots,\ 0,\ I_{n_s(w)},\ 0\cdots,\ 0) \right];$$

$$\tilde{A}_2(k,j) = D_0 \otimes \left[\begin{pmatrix} \mathbf{s}_1\mathbf{e}_j \\ \vdots \\ \mathbf{s}_K\mathbf{e}_j \end{pmatrix} (0,\ \cdots,\ 0,\ I_{n_s(w)},\ 0\cdots,\ 0) \right],$$

where $(S_{k_e}Q_{k_e})_{j_e}$ is the column j_e of the matrix $S_{k_e}Q_{k_e}$. One notices that matrix $\tilde{A}_1(k,j)$ is independent of (k,j).

Let

$$\boldsymbol{x}(\mathbf{J},i,k,j)) = Pr\{(\tilde{q}_n, v_n, k_n, j_n) = (\mathbf{J},i,k,j))\}|_{n\to\infty};$$

and

$$\boldsymbol{x}(-1,i) = Pr\{\{(\tilde{q}_n, v_n)) = (-1,i)\}|_{n\to\infty}.$$

Further define

1.

$$\boldsymbol{x}(\mathbf{J},i,k) = [\boldsymbol{x}(\mathbf{J},i,k,1),\ \boldsymbol{x}(\mathbf{J},i,k,2),\ \cdots,\ \boldsymbol{x}(\mathbf{J},i,k,n_s(k))],$$

2.

$$\boldsymbol{x}(\mathbf{J},i) = [\boldsymbol{x}(\mathbf{J},i,1),\ \boldsymbol{x}(\mathbf{J},i,2),\ \cdots,\ \boldsymbol{x}(\mathbf{J},i,K)],$$

3.

$$\boldsymbol{x}(\mathbf{J}) = [\boldsymbol{x}(\mathbf{J},1),\ \boldsymbol{x}(\mathbf{J},2),\ \cdots,\ \boldsymbol{x}(\mathbf{J},n)],\ \mathbf{J} \neq -1;$$

4.

$$\boldsymbol{x}(-1) = [\boldsymbol{x}(-1,1),\ \boldsymbol{x}(-1,2),\ \cdots,\ \boldsymbol{x}(-1,n_t)].$$

Let $R = \sum_{k=1}^{K} \sum_{j=1}^{n_s(k)} R_{k,j}$ where the matrix sequence $\{R_{k,j},\ 1 \leq j \leq n_s(k),\ 1 \leq k \leq K\}$ is the minimal non-negative solutions to the matrix equations

$$R_{k,j} = \tilde{A}_1(k,j) + R_{k,j}\tilde{A}_1 + R_{k,j} \sum_{k=1}^{K} \sum_{j=1}^{n_s(k)} R_{k_e,j_e}\tilde{A}_2(k_e,j_e). \tag{10.10}$$

The computations of $\{R_{k,j},\ 1 \leq j \leq n_s(k),\ 1 \leq k \leq K\}$ can be carried out using the algorithm presented in Yeung and Alfa (1999) or the one in Yeung and Sengupta (1994).

The probability distribution of the queue length and the associated auxiliary variables can be obtained from

1.

$$\boldsymbol{x}(\mathbf{J} + (k,j)) = \boldsymbol{x}(\mathbf{J})R_{k,j},\ \ \mathbf{J} \in \Omega,\ 1 \leq k \leq K,\ 1 \leq j \leq n_s(k); \tag{10.11}$$

2.

$$\boldsymbol{x}(0) = \boldsymbol{x}(-1)\tilde{A}_0(-1,0) + \boldsymbol{x}(0)\tilde{A}_1(0,0) + \sum_{k=1}^{K}\sum_{j=1}^{n_s(k)} \boldsymbol{x}(0)R_{k,j}\tilde{A}_2(k,j), \quad (10.12)$$

3.

$$\boldsymbol{x}(-1) = \boldsymbol{x}(-1)\tilde{A}_1(-1,-1) + \boldsymbol{x}(0)\tilde{A}_2(0,-1); \quad (10.13)$$

4.

$$\boldsymbol{x}(-1)\mathbf{1} + \boldsymbol{x}(0)(I-R)^{-1}\mathbf{1} = 1, \quad (10.14)$$

10.3 Waiting Time Distribution for the LCFS Geo/Geo/1 System

Consider the Geo/Geo/1 system discussed earlier in section 5.4, where arrival probability is a and service completion probability is b. We have the transition matrix P which captures the behaviour of the number in the system as

$$P = \begin{bmatrix} \bar{a} & a & & & \\ \bar{a}b & \bar{a}\bar{b}+ab & a\bar{b} & & \\ & \bar{a}b & \bar{a}\bar{b}+ab & a\bar{b} & \\ & & \ddots & \ddots & \ddots \end{bmatrix}. \quad (10.15)$$

We showed how to obtain the distribution $\boldsymbol{x} = \boldsymbol{x}P, \ \boldsymbol{x}\mathbf{1} = 1$, where $\boldsymbol{x} = [x_0, x_1, x_2, \cdots]$ and x_k is the probability that there are k items in the system, including the one receiving service. It clear that whether our service discipline is FCFS, LCFS (last come first served) or even SIRO, this distribution will be the same. What changes is the distribution of the waiting times.

Our approach for obtaining the distribution of waiting time in the queue for LCFS is based on an absorbing Markov chain, which is given as P^* given as:

$$P^* = \begin{bmatrix} 1 & & & & \\ \bar{a}b & \bar{a}\bar{b}+ab & a\bar{b} & & \\ & \bar{a}b & \bar{a}\bar{b}+ab & a\bar{b} & \\ & & \ddots & \ddots & \ddots \end{bmatrix} \quad (10.16)$$

Let the vector $\mathbf{y}^{(n)} = \left[y_0^{(n)}, y_1^{(n)}, \ldots\right]$ be the state vector of the system at time n, with

$$\mathbf{y}^{(0)} = [x_0 + b\sum_{j=1}^{\infty},\ \bar{b}\sum_{j=1}^{\infty} x_j,\ 0,\ 0,\ \cdots].$$

The first element of $y_0^{(0)} = x_0 + b\sum_{j=1}^{\infty}$ captures the probability that an arriving customer finds the system empty or at its arrivals the customer in service completes service. In both cases this customer will be the one to be served next. The term for $y_1^{(0)} = \bar{b}\sum_{j=1}^{\infty} x_j$ captures the probability that arriving customer is the last in the queue and the ongoing service is not completed yet, hence this customer will either be the next to be served after the current service is completed and no new arrivals before then or this customer may be overtaken by an arriving customer at a later time. Now consider the following recursion

$$\mathbf{y}^{(n+1)} = \mathbf{y}^{(n)} P^*. \tag{10.17}$$

The element $y_0^{(n)}$ is the probability that the waiting time in the queue lasts for n or less time units, i.e.

$$y_0^{(n)} = Pr\{W_q \leq n\}, \tag{10.18}$$

where W_q is the waiting time in the queue under the LCFS system.

10.4 Waiting Time Distribution for the Geo/Geo/1 System with SIRO

Consider the Geo/Geo/1 system discussed earlier in section 5.4, where arrival probability is a and service completion probability is b. We have the transition matrix P which captures the behaviour of the number in the system as

$$P = \begin{bmatrix} \bar{a} & a & & \\ \bar{a}b & \bar{a}\bar{b}+ab & a\bar{b} & \\ & \bar{a}b & \bar{a}\bar{b}+ab & a\bar{b} \\ & & \ddots & \ddots & \ddots \end{bmatrix}. \tag{10.19}$$

We showed how to obtain the distribution $\boldsymbol{x} = \boldsymbol{x}P,\ \boldsymbol{x}\mathbf{1} = 1$, where $\boldsymbol{x} = [x_0,\ x_1,\ x_2,\ \cdots]$ and x_k is the probability that there are k items in the system, including the one receiving service. As pointed out earlier whether our service discipline is FCFS, LCFS or even SIRO (service in random order), this distribution will be the same. What changes is the distribution of the waiting times.

Our approach for obtaining the distribution of waiting time in the queue for SIRO is based on an absorbing Markov chain, which is given as P^* given as:

$$P^* = \begin{bmatrix} 1 & & & & \\ c_1 & a_{1,1} & a_{1,2} & & \\ c_2 & a_{2,0} & a_{2,1} & a_{2,2} & \\ c_3 & & a_{3,0} & a_{3,1} & a_{3,2} \\ c_4 & & & a_{4,0} & a_{4,1} & a_{4,2} \\ \vdots & & & & \ddots & \ddots & \ddots \end{bmatrix}, \tag{10.20}$$

where

$$c_j = \frac{1}{j}\bar{a}b + \frac{1}{j+1}ab, j = 1,2,\cdots, \; a_{1,1} = \bar{a}\bar{b} + \frac{1}{2}ab, \; a_{1,2} = a\bar{b},$$

$$a_{i,0} = \frac{i}{i+1}\bar{a}b, \quad a_{i,1} = \bar{a}b + \frac{i}{i+1}\bar{a}b, \quad a_{i,2} = ab, \; i = 2,3,\cdots.$$

Define a vector $\mathbf{y} = [y_0, y_1, y_2, \cdots]$, with

$$y_0 = x_0 + \sum_{j=1}^{\infty} \frac{b}{j} x_j,$$

and

$$y_k = \frac{k}{k+1} x_{k+1} ab + x_k a\bar{b}, \; k = 1,2,\cdots.]$$

The elements of the vector $\mathbf{y}$ capture whether a customer goes immediately for service is the probability that of joining the queue and there are k customers in the queue.

Define the vector $\mathbf{z}^{(n)} = \left[z_0^{(n)}, z_1^{(n)}, \ldots \right]$, with

$$\mathbf{z}^{(0)} = \mathbf{y}^{(0)}.$$

Then we can write the following recursion equation

$$\mathbf{z}^{(n+1)} = \mathbf{z}^{(n)} P^*. \tag{10.21}$$

The element $z_0^{(n)}$ is the probability that the waiting time in the queue lasts for n or less time units, i.e.

$$z_0^{(n)} = Pr\{W_q \leq n\}, \tag{10.22}$$

where W_q is the waiting time in the queue under the SIRO system.

10.5 Pair Formation Queues (aka Double-ended Queues)

In a pair formation type of queue two different types of items arrive according to some known processes, and each item needs to be paired up with an item of the other type before it can leave the system. For example, let us say we have two processes A and B, when a type $A(B)$ arrives it pairs up with a waiting time $B(A)$, otherwise it waits in the system until a type $B(A)$ arrives if there is none waiting. This type of queue and its variants is very commonly observed in assembly lines (usually more than two products to match up), but most commonly in telecommunications is in token buckets. We present some of the models of the pair formation system in this section. For a general exposition to this class of problems see Alfa and Neuts (2004) and references therein.

Consider two processes A and B and let us assume they arrive according to the geometric distribution with parameters a and b, respectively. Let $X_n(A)$ and $X_n(B)$ be the total number of type A and type B, respectively that have arrived to the system by time n. Further let $X_n = X_n(A) - X_n(B)$ be the excess at time n. It is clear that if $X_n < 0$ then we have excess of type B and excess of type A if $X_n > 0$. In this case, $\{X_n,\ n \geq 0\}$ is a Markov chain with state space $\{\cdots, -2, \ -1,\ 0,\ 1,\ 2,\ \cdots\}$ and its transition matrix P can be written as

$$P = \begin{bmatrix} \ddots & \ddots & & & & & \\ \ddots & \ddots & \ddots & & & & \\ & A_2 & A_1 & A_0 & & & \\ & & A_2 & A_1 & A_0 & & \\ & & & A_2 & A_1 & A_0 & \\ & & & & \ddots & \ddots & \ddots \\ & & & & & \ddots & \ddots \end{bmatrix},$$

where $A_0 = (1-b)a$, $A_1 = (1-a)(1-b) + ab$, and $A_2 = b(1-a)$. One interesting thing about pair formation problem is that it is not a stable system. When $a < b$ the system drifts left towards $-\infty$, and also when $a > b$ it drifts right towards $+\infty$. However when $a = b$ the system is null recurrent and usually difficult to analyze. For most practical problems we don't usually come across this. What we normally come across are situations where the system is bounded at one end or both ends. First we consider the case where the system is bounded on the left. Such systems are observed in communications.

10.5.1 Pair formation bounded at one end

Here we consider the case where as soon as the number of type B waiting exceeds a value $K < -\infty$ no more type B is allowed into the system. This is he case with token buckets, where B are tokens and A are packets. Our new process is thus $\{Y_n = max(K, X_n(A) - X_n(B))\}$, so our state space is $\{-K, -K+1, \cdots, -2, -1, 0, 1, 2, \cdots\}$. The transition matrix for this Markov chain is given as

$$P_1 = \begin{bmatrix} A_1 + A_2 & A_0 & & & & \\ A_2 & A_1 & A_0 & & & \\ & A_2 & A_1 & A_0 & & \\ & & A_2 & A_1 & A_0 & \\ & & & A_2 & A_1 & A_0 \\ & & & & \ddots & \ddots & \ddots \\ & & & & & \ddots & \ddots \end{bmatrix}.$$

This system is stable provided that $a < b$. If this is true then we have

$$\boldsymbol{x} = \boldsymbol{x}P_2, \ \boldsymbol{x}\mathbf{1} = 1,$$

where $\boldsymbol{x} = [x_{-K}, x_{-K+1}, \cdots, x_{-1}, x_0, x_1, x_2, \cdots]$.

It is clear that is a special system of birth death process and there is a scalar r which is the minimal non-negative solution to the equation

$$r = (1-b)a + r[ab + (1-b)(1-a)] + r^2(1-a)b.$$

It is straightforward to show that the probability distribution of the number of items in the system is given as

$$x_i = (1-r)r^{-K+i}, \quad -\infty < -K \leq i \leq \infty. \tag{10.23}$$

Another very important and useful measure is the waiting time distribution of type A customers, which in telecommunication system is usually the packets waiting for tokens (type B customers). Let the waiting time be W and $w_j = Pr\{W = j\}$, $j = 0, 1, 2, \cdots$.

It is straightforward to show that

$$w_0 = x_0 b + \sum_{i=1}^{K} x_{-i}. \tag{10.24}$$

and

$$w_1 = x_0(1-b)b,$$

$$w_2 = [x_0(1-b)^2 + 2x_1[(1-b)b]b,$$

$$w_3 = [x_0(1-b)^3 + 3x_1(1-b)^2 b + 3x_2(1-b)b^2]b,$$

and the general result is

$$w_j = b\sum_{v=0}^{j-1} x_v \binom{j}{j-v} b^v(1-b)^{j-v}, \; j \geq 1. \tag{10.25}$$

10.6 Finite Source Queues

Most of the queueing systems discussed so far in the book assume that the source of customers is infinite. There are several queueing systems in which the source is finite and the size is small enough to make the assumption of infinite source unrealistic. In this section we present a model for a finite source of customers.

Consider a single server system with finite number of customers, $I < \infty$. The probability of any of the customers not already in the system yet arriving is Bernoulli with parameter a. Hence if we already have $\mathscr{K}$ customers in the system, then potential arrivals will be from the population $I - \mathscr{K}$. Let $\mathscr{A}$ be the number of arriving customers at any time and define

$$a_{v,k} = Pr\{\mathscr{A} = v | \mathscr{K} = k\} = \binom{I-k}{v} a^v(1-a)^{I-k-v}, \; 0 \leq v \leq I-k, \; 0 \leq k \leq I.$$

Services are geometric with parameter b, and only one customer can be served at a time.

Let X_n = number of customers in the system at time n and Y_n the number that has arrived by time n. Further let us define

$$x_{i,j}^{(n)} = Pr\{X_n = i, Y_n = j\},$$

and further let $x_{i,j} = x_{i,j}^{(n)}|_{n\to\infty}$, then we have

$$x_{0,0} = x_{0,0}a_{0,0}, \tag{10.26}$$

$$x_{0,j} = x_{0,0}a_{0,j} + x_{1,j}a_{0,j}b, \; j \geq 1, \tag{10.27}$$

$$x_{1,j} = x_{0,j-1}a_{1,j-1} + x_{1,j-1}a_{1,j-1}b + x_{1,j}a_{0,j}(1-b) + x_{i+1,j}a_{0,j}b, \; j \geq 1, \tag{10.28}$$

$$x_{i,j} = \sum_{v=0}^{I-j} x_{i-v,j-v}a_{v,j-v}(1-b) + \sum_{v=0}^{I-j} x_{i-v+1,j-v}a_{v,j-v}b, \; i \geq 1, \; 1 \leq j \leq I-j. \tag{10.29}$$

This set of equations can be written in matrix form as follow. Let $\boldsymbol{x} = [\boldsymbol{x}_0, \boldsymbol{x}_1, \cdots, \boldsymbol{x}_I]$, where $\boldsymbol{x}_i = [x_{i,i}, x_{i,i+1}, \cdots, x_{i,I}]$ and P the corresponding transition matrix, then we have

$$P = \begin{bmatrix} A_{0,0} & A_{0,1} & A_{0,2} & \cdots & A_{0,I} \\ A_{1,0} & A_{1,1} & A_{1,2} & \cdots & A_{1,I} \\ & A_{2,1} & A_{2,2} & \cdots & A_{2,I} \\ & & \ddots & \ddots & \ddots \\ & & & A_{I-1,I} & A_{I,I} \end{bmatrix}.$$

The we can write

$$\boldsymbol{x} = \boldsymbol{x}P, \;\; \boldsymbol{x}\mathbf{1} = 1,$$

and apply any of the tools for solving finite DTMC to obtain the stationary distribution $\boldsymbol{x}$.

10.7 Problems

1. Question 1: Consider a single server queue with PH service times represented by $(\boldsymbol{\beta}, S)$ of dimension m. There are only $I < \infty$ customers who use this system. Each customer may arrive with probability a.
   a. Develop the transition matrix of the number of customers in the system at any time n.
   b. How would you compute the the steady state distribution on the number in the system?
   c. Develop an algorithm for computing the waiting time in the system.
   d. What is the probability of the system being idle?
2. Question 2: A double-ended queue has two types of components arriving according to the Bernoulli process with arrival parameters a_1 and a_2 for of type 1 and type 2, respectively. A type 1 and a type 2 need to be assembled together, before they can leave the system . Let the assembly time follow a geometric distribution with parameter b. The matched pair which cannot go through the assembly immediately waits in an infinite buffer queue, whereas the items 1 and 2 waiting t be matched do so in finite buffers spaces of sizes K_1 and K_2, respectively.
   a. Develop a DTMC that captures the behaviour of the numbers in this system.
   b. How would you obtain the stationary distributions of these numbers assuming the system is stable?
   c. What are the stability conditions of this system?

References

Alfa AS, Neuts MF (2004) Pair formation in a MAP with two event labels. Journal of Applied Probability 41(4):1124–1137

He QM (1998) Classification of Markov processes of matrix m/g/1 type with a tree structure and its applications to the map[k]/ph[k]/1 queue. Internal report, Dalhousie University

He QM, Alfa AS (2000) The discrete time MMAP[K]/PH[K]/1/LCFS-GPR queue and its variants. In: latouche G, Taylor P (eds) Advances in algorithmic Methods for Stochastic Models,, Notable Pub. Inc, pp 167–190

He QM, Neuts MF (1998) Markov arrival process with marked transitions. Stochastic Processes and Applications 74:37–52

Houdt BV, Blondia C (2004) The waiting time distribution of a type k customer in a discrete-time MMAP[K]/PH[K]/c (c=1,2) queue using QBDs. Stochastic Models 20(1):55–69

van Houdt B, Blondia C (2002) The delay distribution of a type k customer in a first-come-first-served MMAP[K]/PH[K]/1 queue. Journal of Applied Probability 39:213–223

Yeung RW, Alfa AS (1999) The quasi-birth-and-death type Markov chain with a tree structure. Stochastic Models 15:639–659

Yeung RW, Sengupta B (1994) Matrix product-form solutions for Markov chains with a tree structure. Adv Appl Probab 26:965–987

Chapter 11
Queues with Time Varying Parameters

11.1 Introduction

Most queueing systems discussed in the literature are those with input parameters that are stationary or static. However as most people will find, these parameters in real life are usually time-varying, especially the arrival parameters. For example, the arrival probability of a packet may depend on the time of the day. Usually our interests in such cases involve the performance measures at different times. In this Chapter we introduce the readers to this class of queueing systems. We start by considering those with geometric inter-arrival times and geometric service times.

At time n, let X_n be the number of units in the system, including the one being processes. Note that X_n assumes the values in the state space $\mathscr{I} = \{0,1,2,3,\cdots\}$. Let D_n be the number of service completions at time n, where $D_n = 0,1$ and let A_n be the number of arrivals at time n, where $A_n = 0,1$. The stochastic process $\{X_n, n = 0,1,2,\cdots\}, X_n \in \mathscr{I}$, is a DTMC, with the relationship

$$X_{n+1} = (X_n - D_n)^+ + A_n, \tag{11.1}$$

where $(z)^+ = max\{0,z\}$.

The DTMC representing this BD process is now discussed in the next section.

11.2 Discrete Time Time-Varying B-D Process

Let a_n be the probability of an arrival of customers in the system at time n, with $\bar{a}_n = 1 - a_n$. Also, let b_n be the probability that a service completion occurs at time n, given there is a customer in the system, with $\bar{b}_n = 1 - b_n$. This system is the Geo$_n$/Geo$_n$/1 system.

A.S. Alfa, *Applied Discrete-Time Queues*, DOI 10.1007/978-1-4939-3420-1_11

If we now define a Markov chain with state space $\{X_n, n \geq 0\}$, where X_n is the number of customers in the system, then the transition matrix P_n of this chain is given as

$$P_n = \begin{bmatrix} \bar{a}_n & a_n & & & \\ \bar{a}_n b_n & \bar{a}_n \bar{b}_n + a_n b_n & a_n \bar{b}_n & & \\ & \bar{a}_n b_n & \bar{a}_n \bar{b}_n + a_n b_n & a_n \bar{b}_n & \\ & & \ddots & \ddots & \ddots \end{bmatrix} \quad (11.2)$$

If we define $x_i^{(n)} = Pr\{X_n = i\}$ as the probability that there are i customers in the system at time n, and letting $\boldsymbol{x}^{(n)} = \left[x_0^{(n)}, x_1^{(n)}, \ldots\right]$ then we have

$$\boldsymbol{x}^{(n+1)} = \boldsymbol{x}^{(n)} P_n, \text{ or } \boldsymbol{x}^{(n+1)} = \boldsymbol{x}^{(0)} \prod_{j=0}^{n} P_j.$$

Of interest to us is x_i^n, $i \geq 0$, $n \geq 0$. This can be obtained in several ways, after conditioning on the vector $\boldsymbol{x}^{(0)} = [x_0^{(0)}, x_1^{(0)}, \cdots, x_i^0, \cdots]$. For practical purposes we assume that, i.e.

$$x_0^{(0)} = 1.0, \text{ and } x_i^{(0)} = 0, \forall i \geq 1.$$

This is a special case, which is not unrealistic. We can apply the rectangular matrix approach or apply the $z-$transform approach from which we need to obtain the emptiness probabilities at all time points which can then be used to obtain the distributions.

Let us consider the first approach.

11.2.1 The Method of Rectangular Matrix

It is clear that

$$\mathbf{x}^{(0)} = [1, 0, 0, \cdots]$$

hence

$$\mathbf{x}^{(1)} = [x_0^{(1)}, x_1^{(1)}, 0, 0, \cdots]$$

with

$$\mathbf{x}^{(n)} = [x_0^{(n)}, x_1^{(n)}, x_2^{(n)}, \cdots, x_n^{(n)}, 0, \cdots].$$

Given that $x_j^{(n)} = 0,\ \forall j \geq n+1$, let us write

$$\tilde{\mathbf{x}}^{(n)} = [x_0^{(n)},\ x_1^{(n},\ \cdots,\ x_n^{(n)}].$$

For computational purposes all we need at each time step n is a modified matrix $\tilde{P}_n$ which is an $(n+1) \times (n+2)$ portion of the original matrix P_n, i.e.

$$\tilde{P}_n = \begin{bmatrix} \bar{a}_n & a_n & & & & \\ \bar{a}_n b_n & \bar{a}_n \bar{b}_n + a_n b_n & a_n \bar{b}_n & & & \\ & \bar{a}_n b_n & \bar{a}_n \bar{b}_n + a_n b_n & a_n \bar{b}_n & & \\ & & \ddots & \ddots & \ddots & \\ & & & \bar{a}_n b_n & \bar{a}_n \bar{b}_n + a_n b_n & a_n \bar{b}_n \end{bmatrix}.$$

From here we can obtain

$$\tilde{\mathbf{x}}^{(n+1)} = \tilde{\mathbf{x}}^{(n)} \tilde{P}_n,\ \ n = 0, 1, 2, 3, \cdots.$$

Remember this is for the case where $x_0^{(0)} = 1.0$. Other cases will have to be specifically developed.

11.2.2 The Use of z-Transforms for the Geo$_n$/Geo$_n$/1 system

First let us write out the detailed equations in scalar form.

$$x_0^{(n+1)} = x_0^{(n)} \bar{a}_n + x_1^{(n)} \bar{a}_n b_n \tag{11.3}$$

$$x_1^{(n+1)} = x_0^{(n)} a_n + x_1^{(n)} \left(\bar{a}_n \bar{b}_n + a_n b_n\right) + x_2^{(n)} \bar{a}_n b_n \tag{11.4}$$

$$x_i^{(n+1)} = x_{i-1}^{(n)} a_n \bar{b}_n + x_i^{(n)} \left(\bar{a}_n \bar{b}_n + a_n b_n\right) + x_{i+1}^{(n)} \bar{a}_n b_n,\ i \geq 2 \tag{11.5}$$

Now let us write

$$X^{(n)}(z) = \sum_{i=0}^{\infty} x_i^{(n)} z^i,\ |z| \leq 1,$$

and set as follows:

$$x_0^{(n+1)} = x_0^{(n)} \bar{a}_n + x_1^{(n)} \bar{a}_n b_n \tag{11.6}$$

$$z x_1^{(n+1)} = z x_0^{(n)} a_n + z x_1^{(n)} \left(\bar{a}_n \bar{b}_n + a_n b_n\right) + z x_2^{(n)} \bar{a}_n b_n \tag{11.7}$$

$$z^i x_i^{(n+1)} = z^i x_{i-1}^{(n)} a_n \bar{b}_n + z^i x_i^{(n)} \left(\bar{a}_n \bar{b}_n + a_n b_n\right) + z^i x_{i+1}^{(n)} \bar{a}_n b_n,\ i \geq 2, \tag{11.8}$$

and taking the sum over i from 0 to ∞ then we have

$$X^{(n+1)}(z) = X^{(n)}(z)[\bar{a}_n\bar{b}_n + a_nb_n + z^{-1}\bar{a}_nb_n + za_n\bar{b}_n] + x_0^{(n)}[\bar{a}_n + za_n - \bar{a}_n\bar{b}_n - a_nb_n - z^{-1}\bar{a}_nb_n - za_n\bar{b}_n].$$

$$= A_n(z)X^{(n)}(z) + B_n(z)x_0^{(n)},$$

where

$$A_n(z) = \bar{a}_n\bar{b}_n + a_nb_n + z^{-1}\bar{a}_nb_n + za_n\bar{b}_n,$$

and

$$B_(z) = \bar{a}_n + za_n - \bar{a}_n\bar{b}_n - a_nb_n - z^{-1}\bar{a}_nb_n - za_n\bar{b}_n = a_n(z) - A(z).$$

where $a_n(z) = \bar{a}_n + za_n$

It is clear that by applying this equation recursively we have

$$X^{(1)}(z) = A_0(z)X^{(0)}(z) + B_0(z)x_0^0,$$

$$X^{(2)}(z) = A_1(z)X^{(1)}(z) + B_1(z)x_0^{(1)} = A_1(z)[A_0(z)X^{(0)}(z) + B_0(z)x_0^{(0)}] + B_1(z)x_0^{(1)},$$

and continuing with this we have

$$X^{(n+1)}(z) = B_n(z)x_0^{(n)} + \left[\prod_{v=0}^{n} A_v(z)\right] X^{(0)}(z) + \sum_{v=1}^{n} \left[\prod_{k=v}^{n} A_k(z)\right] B_{v-1}(z)x_0^{(v-1)}.$$

We can easily write $x_i^{(n+1)}$ as a function of $x_0^{(m)}$, $m = 1,2,\cdots,n$ and $x_j^{(0)}$, $j = 0,1,2,\cdots$. If we condition the system as earlier, i.e. that $x_0^{(0)} = 1$, then we have

$$x_i^{(n+1)} = f(a_m, b_m, x_0^{(m)}), \; m = 0,1,2,\cdots,n, \; i \geq 0.$$

11.3 The Geo$_n$/Geo$_n$/k system

Consider a Geo$_n$/Geo$_n$/k queue in which we have packets arriving according to the Bernoulli process, i.e. the inter-arrival times have time-varying geometric distribution with parameter $a_n = Pr\{an\ arrival\ at\ any\ time\ epoch\ n\}$, with $\bar{a}_n = 1 - a_n$. There are $k < \infty$ identical servers in parallel each with a time-varying geometric service distribution with parameter b_n and $\bar{b}_n = 1 - b_n$. Since the severs are all identical and have geometric service process, we know that if there are j packets in service, then the probability of i ($i \leq j$) of those packets having their service completed in one time slot n, written as $b_n^{(j,i)}$ is given by a binomial distribution, i.e.

$$b_n^{(j,i)} = \binom{j}{i} b_n^i (1-b_n)^{j-i}, \quad 0 \le i \le j. \tag{11.9}$$

Consider the DTMC $\{X_n,\ n \ge 0\}$ which has the state space $\{0,1,2,3,\cdots\}$ where X_n is the number of packets in the system at time slot n. If we define $p_n^{(i,j)} = Pr\{X_{n+1} = j|X_n = i\}$, then we have

$$p_n^{(0,0)} = \bar{a}_n, \tag{11.10}$$

$$p_n^{(0,1)} = a_n, \tag{11.11}$$

$$p_n^{(0,j)} = 0,\ \forall j \ge 2, \tag{11.12}$$

and

$$p_n^{(i,j)} = \begin{cases} a_n b_n^{(i,0)}, & j = i+1,\ i < k \\ \bar{a}_n b_n^{(i,i)}, & j = 0, \quad i < k \\ a_n b_n^{(i,i-j+1)} + \bar{a}_n b_n^{(i,i-j)}, & 1 \le j \le i \\ 0, & j > i+1,\ i < k \end{cases}, \tag{11.13}$$

and

$$p_n^{(i,j)} = \begin{cases} c_n^{(0)} = a_n b_n^{(k,0)}, & j = i+1, & i \ge k \\ c_n^{(k+1)} = \bar{a}_n b_n^{(k,k)}, & j = i-k, & i \ge k \\ c_n^{(i-j+1)} = a_n b_n^{(k,i-j+1)} + \bar{a}_n b_n^{(k,i-j)}, & 1 \le j \le i,\ i \ge k, & \\ 0, & j > i+1, & i \ge k \\ 0, & i > k, & j \le i-k \end{cases} \tag{11.14}$$

We display the case of $k = 3$ first and then show the general case in a block matrix form. For $k = 3$ the transition matrix of this DTMC is given as

$$P_n = \begin{bmatrix} p_n^{(0,0)} & p_n^{(0,1)} & & & & & & \\ p_n^{(1,0)} & p_n^{(1,1)} & p_n^{(1,2)} & & & & & \\ p_n^{(2,0)} & p_n^{(2,1)} & p_n^{(2,2)} & p_n^{(2,3)} & & & & \\ c_n^{(4)} & c_n^{(3)} & c_n^{(2)} & c_n^{(1)} & c_n^{(0)} & & & \\ & c_n^{(4)} & c_n^{(3)} & c_n^{(2)} & c_n^{(1)} & c_n^{(0)} & & \\ & & c_n^{(4)} & c_n^{(3)} & c_n^{(2)} & c_n^{(1)} & c_n^{(0)} & \\ & & & \ddots & \ddots & \ddots & \ddots & \ddots \end{bmatrix}. \tag{11.15}$$

Again we know that

$$\boldsymbol{x}^{(n+1)} = \boldsymbol{x}^{(n)} P_n, \text{ or } \boldsymbol{x}^{(n+1)} = \boldsymbol{x}^{(0)} \prod_{j=0}^{n} P_j.$$

Of interest to us is $x_i^{(n)}$, $i \geq 0$, $n \geq 0$. This can be obtained in several ways, after conditioning on the vector $\boldsymbol{x}^{(0)} = [x_0^{(0)}, x_1^{(0)}, \cdots, x_i^0, \cdots]$. For practical purposes we assume that

$$x_0^{(0)} = 1.0, \text{ and } x_i^{(0)} = 0, \forall i \geq 1.$$

We can apply the rectangular matrix approach.

Given that $x_j^{(n)} = 0$, $\forall j \geq n+1$, let us write

$$\tilde{\mathbf{x}}^{(n)} = [x_0^{(n)}, x_1^{(n)}, \cdots, x_n^{(n)}].$$

As such all we need at each time step n is a modified matrix $\tilde{P}_n$ which is an $(n+1) \times (n+2)$ portion of the original matrix P_n, i.e.

$$\tilde{P}_n = \begin{bmatrix} p_n^{(0,0)} & p_n^{(0,1)} & & & & & & & \\ p_n^{(1,0)} & p_n^{(1,1)} & p_n^{(1,2)} & & & & & & \\ p_n^{(2,0)} & p_n^{(2,1)} & p_n^{(2,2)} & p_n^{(2,3)} & & & & & \\ c_n^{(4)} & c_n^{(3)} & c_n^{(2)} & c_n^{(1)} & c_n^{(0)} & & & & \\ & c_n^{(4)} & c_n^{(3)} & c_n^{(2)} & c_n^{(1)} & c_n^{(0)} & & & \\ & & c_n^{(4)} & c_n^{(3)} & c_n^{(2)} & c_n^{(1)} & c_n^{(0)} & & \\ & & & \ddots & \ddots & \ddots & \ddots & \ddots & \\ & & & & c_n^{(4)} & c_n^{(3)} & c_n^{(2)} & c_n^{(1)} & c_n^{(0)} \end{bmatrix}.$$

and

$$\tilde{\mathbf{x}}^{(n+1)} = \tilde{\mathbf{x}}^{(n)} \tilde{P}_n.$$

This idea is easily extended to the case of PH/PH/k system.

11.4 The Geo$_n$/Geo$_n^Y$/1 system

In some systems packets are served in batches, rather than individually. Here we consider a case where arrival is time-varying Bernoulli with $a_n = \Pr\{\text{an arrival in a time slot n}\}$ and $\bar{a}_n$ its complement. Service is time-varying geometric with parameter b_n and also in batches of size $\mathscr{B}_n$, with $\theta_n^{(j)} = Pr\{\mathscr{B}_n = i,\ i \geq 1\}$, with $\sum_{j=1}^{\infty} \theta_n^{(j)} = 1$. We can write $b_n^{(0)} = 1 - b$ and $b_n^{(i)} = b_n \theta_n^{(i)}$. We therefore have a vector $\mathbf{b}_{(n)} = [b_n^{(0)},\ b_n^{(1)},\ \cdots]$ which represents the number of packets served in a time slot. The mean service rate is thus given as $\mu_n = b_n \bar{\theta}_n$, where $\bar{\theta}_n = \sum_{j=1}^{\infty} j\theta_n^{(j)}$. Consider the DTMC $\{X_n,\ n \geq 0\}$ with state space $\{0, 1, 2, \cdots\}$, where X_n is the number of items in the system at time n. It is straightforward to show that the transition matrix of this DTMC is

$$P_n = \begin{bmatrix} c_n^{(0)} & d_n^{(1)} & & & & \\ c_n^{(1)} & d_n^{(1)} & d_n^{(0)} & & & \\ c_n^{(2)} & d_n^{(2)} & d_n^{(1)} & d_n^{(0)} & & \\ c_n^{(3)} & d_n^{(3)} & d_n^{(2)} & d_n^{(1)} & d_n^{(0)} & \\ \vdots & \vdots & \vdots & \vdots & \vdots & \ddots \end{bmatrix},$$

where

$$c_n^{(0)} = \bar{a}_n,\ d_n^0 = a_n,\ d_n^{(j)} = a_n b_n^{(j)} + \bar{a}_n b_n^{(j-1)},\ j \geq 1,\ c_n^{(j)} = 1 - \sum_{k=0}^{j} d_n^{(k)},\ j \geq 1.$$

Again we know that

$$\boldsymbol{x}^{(n+1)} = \boldsymbol{x}^{(n)} P_n, \text{ or } \boldsymbol{x}^{(n+1)} = \boldsymbol{x}^{(0)} \prod_{j=0}^{n} P_j.$$

Of interest to us is $x_i^{(n)},\ i \geq 0,\ n \geq 0$. This can be obtained in several ways, after conditioning on the vector $\boldsymbol{x}^{(0)} = [x_0^{(0)},\ x_1^{(0)},\ \cdots,\ x_i^{(0)},\ \cdots]$. For practical purposes we assume that

$$x_0^{(0)} = 1.0, \text{ and } x_i^{(0)} = 0,\ \forall i \geq 1.$$

We can apply the rectangular matrix approach.

Given that $x_j^{(n)} = 0,\ \forall j \geq n+1$, let us write

$$\tilde{\mathbf{x}}^{(n)} = [x_0^{(n)},\ x_1^{(n)},\ \cdots,\ x_n^{(n)}].$$

As such all we need at each time step n is a modified matrix $\tilde{P}_n$ which is an $(n+1) \times (n+2)$ portion of the original matrix P_n, i.e.

$$\tilde{P}_n = \begin{bmatrix} c_n^{(0)} & d_n^{(1)} & & & & & \\ c_n^{(1)} & d_n^{(1)} & d_n^{(0)} & & & & \\ c_n^{(2)} & d_n^{(2)} & d_n^{(1)} & d_n^{(0)} & & & \\ c_n^{(3)} & d_n^{(3)} & d_n^{(2)} & d_n^{(1)} & d_n^{(0)} & & \\ \vdots & \vdots & \vdots & \vdots & \vdots & \ddots & \\ c_n^{(n-1)} & d_n^{(n-1)} & d_n^{(n-2)} & d_n^{(n-3)} & \cdots & d_n^{(1)} & d_n^{(0)} \end{bmatrix},$$

and

$$\tilde{\mathbf{x}}^{(n+1)} = \tilde{\mathbf{x}}^{(n)} \tilde{P}_n.$$

This idea is easily extended to the case of PH/PH^Y/1 system.

11.5 The PH$_n$/PH$_n$/1 System

Let the time-varying arrival process be represented by the time-varying phase type distribution at time n with representation $(\boldsymbol{\alpha}_n, T_n)$ of order u with $\mathbf{t}_n = \mathbf{1} - T_n\mathbf{1}$, and the time-varying service by a time-varying phase type $(\boldsymbol{\beta}_n, S_n)$ of order v with $\mathbf{s}_n = \mathbf{1} - S_n\mathbf{1}$.

If we now define a Markov chain with state space $\{X_n, Y_n, Z_n\} n \geq 0$, where at time n, X_n is the number of customers in the system, Y_n is the phase of arrival and Z_n is the phase of service then the transition matrix P_n of this chain is given as

$$P_n = \begin{bmatrix} B_n & C_n & & & \\ E_n & A_n & U_n & & \\ & D_n & A_n & U_n & \\ & & \ddots & \ddots & \ddots \end{bmatrix},$$

where,

$$B_n = T_n,\; C_n = (\mathbf{t}_n\boldsymbol{\alpha}_n) \otimes \boldsymbol{\beta}_n,\; E_n = T_n \otimes \mathbf{s}_n,\; U_n = (\mathbf{t}_n\boldsymbol{\alpha}_n) \otimes S_n,$$

$$A_n = (\mathbf{t}_n\boldsymbol{\alpha}_n) \otimes (\mathbf{s}_n\boldsymbol{\beta}_n) + T_n \otimes S_n,\; D_n = T_n \otimes (\mathbf{s}_n\boldsymbol{\beta}_n).$$

Let $\boldsymbol{x}^{(n)} = [\boldsymbol{x}_0^{(n)}, \boldsymbol{x}_1^{(n)}, \boldsymbol{x}_2^{(n)}, \cdots]$, with $\boldsymbol{x}_0^{(n)} = [x_{0,1}^{(n)}, x_{0,2}^{(n)}, \cdots, x_{0,u}^{(n)}]$, and for $i \geq 1$, $\boldsymbol{x}_i^{(n)} = [\boldsymbol{x}_{i,1}^{(n)}, \boldsymbol{x}_{i,2}^{(n)}, \cdots, \boldsymbol{x}_{i,u}^{(n)}]$, where $\boldsymbol{x}_{i,j}^{(n)} = [x^{(n)}i,j,1, x_{i,j,2}^{(n)}, \cdots, x_{i,j,v}^{(n)}]$. We have

$$\boldsymbol{x}^{(n+1)} = \boldsymbol{x}^{(n)} P_n, \text{ or } \boldsymbol{x}^{(n+1)} = \boldsymbol{x}^{(0)} \prod_{j=0}^{n} P_j.$$

Of interest to us is $\boldsymbol{x}_i^{(n)}$, $i \geq 0$, $n \geq 0$. This can be obtained in several ways, after conditioning on the vector $\boldsymbol{x}^{(0)}\mathbf{1} = 1$. For practical purposes we assume that We can apply the rectangular matrix approach as follows

11.5.0.1 The Method of Rectangular Matrix for the PH$_n$/PH$_n$/1 System

It is clear that if

$$\boldsymbol{x}_0^{(0)}\mathbf{1} = 1,$$

hence

$$\boldsymbol{x}^{(1)} = [\boldsymbol{x}_0^{(1)}, \boldsymbol{x}_1^{(1)}, 0, 0, \cdots]$$

with

$$\boldsymbol{x}^{(n)} = [\boldsymbol{x}_0^{(n)}, \boldsymbol{x}_1^{(n)}, \boldsymbol{x}_2^{(n)}, \cdots, x_n^{(n)}, 0, \cdots].$$

Given that $x_j^{(n)} = 0$, $\forall j \geq n+1$, let us write

$$\tilde{\mathbf{x}}^{(n)} = [\boldsymbol{x}_0^{(n)}, \boldsymbol{x}_1^{(n)}, \cdots, \boldsymbol{x}_n^{(n)}].$$

As such all we need at each time step n is a modified matrix

$$\boldsymbol{x}^{(n)} = [x_0^{(n)}, x_1^{(n)}, x_2^{(n)}, \cdots, x_n^{(n)}, 0, \cdots].$$

Given that $x_j^{(n)} = 0$, $\forall j \geq n+1$, let us write

$$\tilde{\boldsymbol{x}}^{(n)} = [x_0^{(n)}, x_1^{(n)}, \cdots, x_n^{(n)}].$$

For computational purpose all we need at each time step n is a modified matrix $\tilde{P}_n$ which is an $(n+1)$ blocks $\times(n+2)$ blocks portion of the original matrix P_n, i.e.

$$\tilde{P}_n = \begin{bmatrix} B_n & C_n & & & & \\ E_n & A_n & U_n & & & \\ & D_n & A_n & U_n & & \\ & & \ddots & \ddots & \ddots & \\ & & & D_n & A_n & U_n \end{bmatrix}.$$

From here we can obtain

$$\tilde{\boldsymbol{x}}^{(n+1)} = \tilde{\boldsymbol{x}}^{(n)}\tilde{P}_n, \quad n = 0, 1, 2, 3, \cdots.$$

Remember this is for the case where $\boldsymbol{x}_0^{(0)}\mathbf{1} = 1.0$. We thus have

$$\boldsymbol{x}^{(0)} = [\boldsymbol{x}_0^{(0)},\ \mathbf{0},\ \mathbf{0},\ \cdots],$$

$$\boldsymbol{x}^{(1)} = [\boldsymbol{x}_0^{(1)},\ \boldsymbol{x}_1^{(1)},\ \mathbf{0},\ \mathbf{0},\ \cdots] = [\boldsymbol{x}_0^{(0)}B_0,\ \boldsymbol{x}_0^{(0)}E_0,\ \mathbf{0},\ \mathbf{0},\ \cdots]$$

$$\boldsymbol{x}^{(2)} = [\boldsymbol{x}_0^{(2)},\ \boldsymbol{x}_1^{(2)},\ \boldsymbol{x}_2^{(2)},\ \mathbf{0},\ \mathbf{0},\ \cdots] = [\boldsymbol{x}_0^{(0)}B_0B_1 + \boldsymbol{x}_0^{(0)}E_0E_1,\ \boldsymbol{x}_0^{(0)}B_0C_1 + \boldsymbol{x}_0^{(0)}E_0A_1,\ \boldsymbol{x}_0^{(0)}E_0U_1,\ \mathbf{0},\ \mathbf{0},\ \cdots]$$

and this continues.

11.6 The Geo$_n$/G/1 System

This is a case where the arrival process has a time-varying geometric distribution, with parameter $a_n,\ n = 0,1,2,\cdots$. The service has a general distribution. and described as follows. Let S be the service time. We let $Pr\{S=j\} = b_j,\ j = 1,2,\cdots$. At time n, let X_n be the number in the system and Y_n the remaining service time of the customer in service, and define $x_{i,j}^{n} = Pr\{X_n = i, Y_n = j\},\ \ i = 1,2,\cdots; j = 1,2,\cdots$, with $x_0^{(n)} = Pr\{X_n = 0\}$, since when there is no customer in the system there is no service going on we do not need the second index in this case. Then we have the following Chapman-Kolmogorov equations

$$x_0^{(n+1)} = x_0^{(n)}\bar{a}_n + x_{1,1}^{(n)}\bar{a}_n,$$

$$x_{1,j}^{(n+1)} = x_0^{(n)}a_nb_j + x_{1,j+1}^{(n)}\bar{a}_n + x_{2,1}^{(n)}\bar{a}_nb_j + x_{1,1}^{(n)}a_nb_j,\ j = 1,2,\cdots,$$

$$x_{i,j}^{(n+1)} = x_{i-1,j+1}^{(n)}a_n + x_{i,j+1}^{(n)}\bar{a}_n + x_{i+1,1}^{(n)}\bar{a}_nb_j + x_{i,1}^{(n)}a_nb_j,\ i = 2,3,\cdots;\ j = 1,2,\cdots.$$

For $|z| \le 1$ and $|y| \le 1$, let

$$X^{(n)}(z,y) = x_0^{(n)} + \sum_{i=1}^{\infty}\sum_{j=1}^{\infty} x_{i,j}^{(n)} z^i y^j,$$

$$A_n(z) = a_nz + 1 - a_n,\ \text{and}\ B(y) = \sum_{j=1}^{\infty} b_jy^j.$$

Define

$$q_i^{(n)} = Pr\{X_n = i\} = \sum_{j=1}^{\infty} x_{i,j}^{(n)},\ i \ge 1,\ \text{and}\ q_0^{(n)} = x_0^{(n)}.$$

This is usually one of the most important measures in a queueing system, and most important is $E[X_n]$, the mean number in the system at time n, which is given as

$$E[X_n] = \frac{dX^{(n)}(z,1)}{dz}|_{z\to 1}.$$

By defining

- $\beta_m^{(i)}$ as the coefficient of y^m in $(B(y))^i$
- $\alpha_i^{(m,n)}$ as the coefficient of z^i in $\prod_{v=m}^{n} A_v(z)$
- $\lambda_{m,n} = \sum_{v=}^{n} a_v$,

and after routine calculations with the three Chapman-Kolmogorov equations it was shown in Alfa and Chen (1991) that

$$E[X_n] = \lambda_{0,n-1} - \sum_{m=0}^{n-1} [1 - q_0^{(m)}] \sum_{v=1}^{n-m} \beta_{n-m}^{(v)},$$

where $q_0^{(n)}$ us obtained recursively from

$$q_0^{(n+1)} = \sum_{i=0}^{n+1} \alpha_i^{(0,n)} \beta_{n+1}^{(i)} + \sum_{m=0}^{n} \sum_{i=0}^{n-m+1} [\beta_{n-m}^{(i)} - \beta_{n-m+1}^{(i)}] q_0^{(m)}.$$

Note that β_j^n can be computed recursively as

$$\beta_j^{(n)} = \sum_{m=1}^{j-n+1} \beta_{j-m}^{(n-1)} b_m,\ j \geq n,$$

where

$$\beta_j^{(1)} = b_j, \text{ and } \beta_n^{(n)} = (b_1)^n.$$

11.7 The $\text{Geo}_n^{(X_n)}$/G/1 system

We consider the case of time-varying batch arrivals and general service. Let $\mathbf{A}_n$ be the number of arrivals at epoch n with $a_n^{(i)} = Pr\{A_n = i\}$, $i = 0, 1, 2, \cdots$. The service time S has a dis distribution given by $P\{S = j\} = b_j$, $j = 1, 2, 3, \cdots$. In this analysis we include the study of the departure process as due to Minh (1978). We are able to use the results in Minh to study this case.

At time n, let X_n be the number in the system, R_n the remaining service time of the customer in service and C_n the number of customers that have arrive during the time interval $[0,n]$. It is clear that the process $\{X_n, R_n, C_n;\ n \geq 0\}$ is a discrete time Markov chain. Let

$$x_{i,j,k}^{(n)} = Pr\{X_n = i, R_n = j, C_n = k\}.$$

The random variable C_n is not necessary to have a Markovian system but was introduced as a surplus variable to allow the study of departure process. The Chapman-Kolmogorov equations associated with are

$$x_{0,0,0}^{(n+1)} = x_{0,0,0}^{(n)} a_n^0;$$

$$x_{0,0,k}^{(n+1)} = x_{0,0,k}^{(n)} a_n^0 + x_{1,2,k}^{(n)} a_n^0, \quad 1 \leq k;$$

$$x_{i,j,k}^{(n+1)} = \sum_{v=0}^{i-1} x_{i-v,j+1,k-v}^{(n)} a_n^v + \sum_{v=0}^{i} x_{i-v+1,1,k-v}^{(n)} a_n^v b_j$$

$$+ x_{0,0,k-i}^{(n)} a_n^i b_j, \quad 1 \leq i \leq k-1; 2‘k; 1 \leq j;$$

$$x_{k,j,k}^{(n+1)} = \sum_{v=0}^{k-1} x_{k-v,j+1,k-v}^{(n)} a_n^{(v)} + x_{0,0,0}^{(n)} a_n^{(k)} b_j, \; 1 \leq k; 1 \leq j.$$

A detailed analysis of this system can be found in Minh (1978).

For most real life situations the time varying aspects are periodic. For example in the case of road traffic, it seems to repeat every 24 hours and this example applies to many telecommunication systems. Next we consider the case of such period traffic.

Consider a Geo$_n$/Geo$_n$/1 system with parameters a_n and b_n, as discussed earlier in this chapter. If this system is periodic for both the arrival and service times have periodicity, i.e. of τ_a and τ_b, respectively we can write that $a_n = a_{n+\tau_a} = a$ and $b_n = b_{n+\tau_b},\ \ \tau_a \geq 1,\ \tau_b \geq 1$. Throughout this section we assume that the arrival parameters and service parameters have the same periodicity, i.e. that $\tau_a = \tau_b = \tau$, for simplicity. For example, let us consider the case of $\tau = 2$. In that case we have

$$P_n = \begin{bmatrix} \bar{a}_n & a_n & & & \\ \bar{a}_n b_n & \bar{a}_n \bar{b}_n + a_n b_n & a_n \bar{b}_n & & \\ & \bar{a}_n b_n & \bar{a}_n \bar{b}_n + a_n b_n & a_n \bar{b}_n & \\ & & \ddots & \ddots & \ddots \end{bmatrix}, \tag{11.16}$$

and

$$P_1 = P_{k\tau+1}, \text{ and } P_2 = P_{k\tau+2},\ k = 0, 1, 2, 3 \cdots.$$

Let us define

$$P_{n,m} = P_n P_{n+1} \cdots P_m, \quad \text{where} \quad m \geq n.$$

For this small example of $\tau = 2$ we have

$$P_{1,1} = P_1, \; P_{1,2} = P_1 P_2, \; P_{1,3} = P_1 P_2 P_1, \; P_{2,2} = P_2, \; P_{2,3} = P_2 P_1, \; \cdots.$$

It is clear that any matrix $P_{n,n+1}$ will be a 2−banded matrix of the form

$$P_{n,n+1} = \begin{bmatrix} p_{0,0} & p_{0,1} & p_{0,2} & & & & \\ p_{1,0} & p_{1,1} & p_{1,2} & p_{1,3} & & & \\ p_{2,0} & p_{2,1} & p_{2,2} & p_{2,3} & p_{2,4} & & \\ & p_{3,1} & p_{3,2} & p_{3,3} & p_{3,4} & p_{3,5} & \\ & & \ddots & \ddots & \ddots & \ddots & \ddots \end{bmatrix}.$$

We assume that all the $p_{i,j}$ elements are time varying and depend on n and $n+1$, but to avoid a messy notation we did not include the time-varying parameters in the elements.

This can be re-blocked to blocks of size 2×2 and written as

$$P_{n,n+1} = \begin{bmatrix} B_{n,n+1} & C_{n,n+1} & & \\ E_{n,n+1} & A_{n,n+1} & U_{n,n+1} & \\ & D_{n,n+1} & A_{n,n+1} & U_{n,n+1} \\ & & \ddots & \ddots & \ddots \end{bmatrix}.$$

For the general case of $\tau = k \geq 1$, we see that the matrix $P_{n,n+\tau}$ has the same structure except that the block matrices are now of order $\tau \times \tau$.

Now we discuss the general case, where $1 \leq \tau < \infty$. Let us define $\mathscr{P}_m = P_{m,m+\tau-1}$. The we can write

$$\mathscr{P}_m = P_{m,m+\tau-1} = P_{j\tau+m,j\tau+m+\tau-1}, \; j = 0,1,2,,3,\cdots.$$

For simplicity we write

$$\mathscr{P}_m = \begin{bmatrix} B^{(m)} & C^{(m)} & & \\ E^{(m)} & A^{(m)} & U^{(m)} & \\ & D^{(m)} & A^{(m)} & U^{(m)} \\ & & \ddots & \ddots & \ddots \end{bmatrix}.$$

If we assume that there exists a stationary distribution $\mathbf{y}_m$, then it is given as

$$\mathbf{y}_m = \boldsymbol{x}^{(m+k\tau)} = \boldsymbol{x}^{(m+k\tau)} \mathscr{P}_m, \; k = 0,1,2,\cdots, \; \text{with} \; \mathbf{y}_m \mathbf{1} = 1.$$

The conditions for stability are discussed in Alfa and Margolius (2008). For this case of Geo$_n$/Geo$_n$/1 it is simply that

$$\sum_{j=0}^{\tau-1} a_{k\tau+j} < \sum_{j=0}^{\tau-1} b_{k\tau+j}, \ \forall k \geq 0.$$

This problem can now be cast in a matrix form and matrix-analytic methods used for studying the system. We know that there exists a matrix R_m which is the minimal non-negative solution to the equation

$$R_m = U^{(m)} + R_m A^{(m)} + R_m^2 D^{(m)},$$

and the stationary distribution $\mathbf{y}_m = [\mathbf{y}_m(0),\ \mathbf{y}_m(1),\ \mathbf{y}_m(2),\ \cdots]$ is given as

$$\mathbf{y}_m(i+1) = \mathbf{y}_m(i) R_m, \ \ i = 1,2,3,\cdots.$$

After applying the boundary conditions we can easily obtain the whole distribution.

Because of the QBD nature of the matrix $\mathscr{P}_m$ for the Geo$_n$/Geo$_n$/1 case and the scalar components of the elements in the block matrices we find that actually $D^{(m)}$, $A^{(m)}$, and $U^{(m)}$ are independent of m and hence we have

$$\mathscr{P}_m = \begin{bmatrix} B^{(m)} & C^{(m)} & & \\ E^{(m)} & A & U & \\ & D & A & U \\ & & \ddots & \ddots & \ddots \end{bmatrix}.$$

So the matrix R_m is independent of m and it is simply

$$R = U + RA + R^2 D,$$

and the stationary distribution is given as

$$\mathbf{y}_m(i+1) = \mathbf{y}_m(i) R, \ \ i = 1,2,3,\cdots.$$

The parameter m comes into play mainly in the boundary condition. For the boundary information we solve

$$[\mathbf{y}_m(0),\ \mathbf{y}_m(1)] = [\mathbf{y}_m(0),\ \mathbf{y}_m(1)] \begin{bmatrix} B^{(m)} & C^{(m)} \\ E^{(m)} & A + RD^{(m)} \end{bmatrix},$$

and normalize by

$$\mathbf{y}_m(0)\mathbf{1} + \mathbf{y}_m(1)[I - R]^{-1}\mathbf{1} = 1.$$

11.8 Problems

1. Question 1: Consider a single server queue with time-varying geometric service times with parameter b_n at time n. There are only $I < \infty$ customers who use this system. Each customer may arrive with probability a_n at time n.

   a. Develop the time-varying transition matrix of the number of customers in the system at any time n.
   b. How would you compute the the distribution on the number in the system at time n, $n = 1, 2, 3 \cdots.$?

2. Question 2: Consider Question 1. Suppose the arrivals and service are periodic, i.e. there is a period k such that $a_{n+jk} = a_n$ and $b_{n+jk} = b_n$, $k = 1, 2, \cdots$, $j \geq 0$, $\forall n$.

   a. Develop the transition matrix of the number of customers in the system at any time $n + jk$.
   b. How would you compute the the distribution on the number in the system at time $n + jk$, $n = 1, 2, 3 \cdots$. over the period k?
   c. Discuss the behaviour of this system as $j \to \infty$, keeping in mind that $x_i^{(n+jk)} = x_i^{(n+(j+m)k)}$ as $\to \infty$, where $x_i^{(n)}$ is the probability that the number in the system at time n is i.

References

Alfa AS, Chen MY (1991) Approximating queue lengths in M(t)/G/1 queue using the maximum entropy principle. ACTA Informatica 28:801–815

Alfa AS, Margolius BH (2008) Two classes of time-inhomogeneous Markov chains: Analysis of the periodic case. Annals of Oper Res 160:121–137

Minh DL (1978) The discrete-time single-server queue with time-inhomogeneous compound Poisson input. Journal of Applied Probability 15:590–601

Chapter 12
Tandem Queues and Queueing Networks

12.1 Introduction

In this Chapter we consider tandem queues in discrete time. We focus mainly on queues where arrivals and service are not state dependent. The cases of state dependent tandem queues in discrete time have been adequately dealt with in the literature, see (Hsu and Burke (1976), Daduna (2001)). This chapter also discusses several open problems for discrete time tandem queues. Queueing networks are briefly discussed in this chapter.

12.2 Two Queues in Tandem – The Geometric Case

Throughout this section we let arrivals and services to be geometric and the buffer at the first queue is assumed to be unlimited. Later we show the extensions to cases with finite buffers and also to cases with phase type distributions.

Consider two queues in tandem (as shown in Fig 12.1) with arrivals to the first queue happening according to the geometric distribution with parameter a, i.e. the probability of an arrival at any time epoch being given as a. The service at the first queue is according to geometric distribution with parameter b_1 and at the second queue is also according to geometric distribution with parameter b_2. We let $\bar{a} = 1 - a$, $\bar{b}_j = 1 - b_j$, $j = 1,2$. A packet that arrives at the first queue is processed as soon as the server is available and then proceeds to the second queue for service which is also rendered as soon as the second server is available.

A.S. Alfa, *Applied Discrete-Time Queues*, DOI 10.1007/978-1-4939-3420-1_12

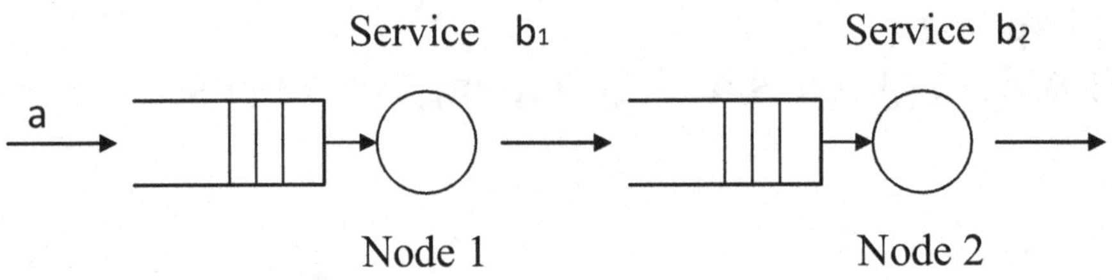

Fig. 12.1 Two Queues in Tandem

12.2.1 Second Queue with Infinite Buffer

For the queue considered let us assume that the buffer at the second queue is unlimited, so a packet that completes service at the first queue always has a space to wait for the service in the second queue, if the second server is busy. Define a bivariate DTMC $\{(X_n^{(1)}, X_n^{(2)}),\ n = 1,2,\cdots\},\ X_n^{(1)} = 0,1,2,\cdots; X_n^{(2)} = 0,1,2,\cdots$, where $X_n^{(k)}$ is the number of packets waiting in the kth queue, including the ones receiving service at time n. Let $A(n)$ be the number of arrivals to the first queue at time n. Note that $A(n)$ can only assume the values of 0 or 1. Further let $D(n)^{(k)}$ be the number of service completions at time n by server $k = 1,2$. It should be noted that each server can only complete at most one service at any time and when there are no customers in the system no service can be completed. So we can write the random variable representing this system as

$$(X_{n+1}^{(1)}, X_{n+1}^{(2)}) = ((X_n^{(1)} - D(n)^{(1)})^+ + A(n), (X_n^{(2)} - D(n)^{(2)})^+ + D(n)^{(1)}). \quad (12.1)$$

It is immediately clear that $(X_n^{(1)}, X_n^{(2)})$ is a bivariate DTMC and its transition matrix is of the QBD type. The transition matrix associated with this DTMC is P and is given as

$$P = \begin{bmatrix} B & C & & & \\ E & A_1 & A_0 & & \\ & A_2 & A_1 & A_0 & \\ & & A_2 & A_1 & A_0 \\ & & & \ddots & \ddots & \ddots \end{bmatrix}. \quad (12.2)$$

Matrix P in a block form is what we refer to as the outer matrix and it is a QBD. The details of the block matrices depend on several factors. In this subsection we consider two kinds of subsystems, 1) with no external arrivals to the second queue and 2) with external arrivals to the second queue. First we consider the case where there are no external arrivals into the second queue.

12.2.1.1 No External Arrivals into Second Queue

By saying no external arrivals to the second queue we imply that it is fed by the first queue only. In that case both $X_n^{(1)}$ and $X_n^{(2)}$ can either increase (decrease) by one only at maximum. Hence there are two QBDs involved here. The inner matrix is a QBD. Before we give the details of the entries in the transition matrix, first let us define two other matrices. Let the matrix $P_1^{(1)}$ and $P_1^{(2)}$ be written as

$$P_1^{(1)} = \begin{bmatrix} 1 & 0 & & & \\ b_2 & \bar{b}_2 & & & \\ & b_2 & \bar{b}_2 & & \\ & & b_2 & \bar{b}_2 & \\ & & & \ddots & \ddots \end{bmatrix}, \text{ and } P_1^{(2)} = \begin{bmatrix} 0 & 1 & 0 & & & \\ 0 & b_2 & \bar{b}_2 & & & \\ & 0 & b_2 & \bar{b}_2 & & \\ & & 0 & b_2 & \bar{b}_2 & \\ & & & \ddots & \ddots & \ddots \end{bmatrix},$$

respectively. The matrices $P_1^{(1)}$ and $P_1^{(2)}$ are the ones we refer to as the inner matrices, and they are also some form of special QBDs, just missing either the super-diagonal or the sub-diagonal components. The first matrix, $P_1^{(1)}$ should not be seen as a kind of an absorbing Markov chain, but as a behaviour of the second queue as seen by the first server. The 1 in the northwest corner refers to no change in the second queue as seen by the first server. This will be interpreted slightly differently depending on the context in which this matrix used. The second matrix $P_1^{(2)}$ is related to the first one. It is simply a shifted version of matrix $P_1^{(1)}$. Now the block matrices of the matrix P can be written as

$$B = \bar{a}P_1^{(1)},\ C = aP_1^{(1)},\ E = \bar{a}b_1P_1^{(2)}$$

$$A_0 = a\bar{b}_1P_1^{(1)},\ A_1 = \bar{a}\bar{b}_1P_1^{(1)} + ab_1P_1^{(2)},\ A_2 = \bar{a}b_1P_1^{(2)}.$$

The stability conditions for this Markov chain can be obtained through the operator geometric approach for the GI/G/1 system by studying the matrix $A = A_0 + A_1 + A_2$. Alternatively we can simply use an intuitive argument as follow.

Proposition: Since the input into the second queue is mainly the output of the first queue and there is no blocking, then stability conditions are that we require

$$\frac{a\bar{b}_1}{\bar{a}b_1} < 1, \text{ and } \frac{a\bar{b}_2}{\bar{a}b_2} < 1 \rightarrow \frac{a}{b_1} < 1, \text{ and } \frac{a}{b_2} < 1.$$

or simply,

$$max\left\{\frac{a}{b_1}, \frac{a}{b_2}\right\} < 1.$$

Let $x_{i,j} = Pr\{X_n^{(1)} = i, X_n^{(2)} = j\}|_{n\to\infty}$, with $\boldsymbol{x}_i = [x_{i,0},\ x_{i,1},\ x_{i,2},\ \cdots]$, $i = 0,1,2,\cdots$, and $\boldsymbol{x} = [\boldsymbol{x}_0,\ \boldsymbol{x}_1,\ \boldsymbol{x}_2,\ \cdots]$, then we have

$$\boldsymbol{x} = \boldsymbol{x}P, \quad \boldsymbol{x}\mathbf{1} = \mathbf{1}. \tag{12.3}$$

We present two different approaches for obtaining this stationary behaviour.

Algebraic Approach for Solution

Applying the relationship $\boldsymbol{x} = \boldsymbol{x}P$ in scalar form we obtain

$$x_{0,0} = x_{0,0}\bar{a} + x_{0,1}\bar{a}b_2 \tag{12.4}$$

$$x_{0,1} = x_{0,1}\bar{a}\bar{b}_2 + x_{1,0}\bar{a}b_1 + x_{0,2}\bar{a}b_2 + x_{1,1}\bar{a}b_1b_2, \tag{12.5}$$

$$x_{0,j} = x_{0,j}\bar{a}\bar{b}_2 + x_{1,j-1}\bar{a}b_1\bar{b}_2 + x_{0,j+1}\bar{a}b_2 + x_{1,j}\bar{a}b_1b_2,\ j = 2,3,\cdots \tag{12.6}$$

$$x_{1,0} = x_{1,0}\bar{a}\bar{b}_1 + x_{1,1}\bar{a}\bar{b}_1b_2 + x_{0,0}a + x_{0,1}ab_2, \tag{12.7}$$

$$x_{i,0} = x_{i,0}\bar{a}\bar{b}_1 + x_{i,1}\bar{a}\bar{b}_1b_2 + x_{i-1,0}a\bar{b}_1 + x_{i-1,1}a\bar{b}_1b_2,\ i = 2,3,\cdots. \tag{12.8}$$

Defining

$$\hat{b}_k(i) = \begin{cases} 1, & i = 0 \\ \bar{b}_k, & i \geq 1, \end{cases},\ k = 1,2,$$

and then for $i \geq 1$ and $j \geq 1$ we have

$$x_{i,j} = x_{i,j}\bar{a}\bar{b}_1\bar{b}_2 + x_{i,j}ab_1b_2 + x_{i+1,j}\bar{a}b_1b_2 + x_{i+1,j-1}\bar{a}b_1\hat{b}_2(j-1)$$

$$+ x_{i-1,j}a\hat{b}_1(i-1)\bar{b}_2 + x_{i-1,j+1}a\hat{b}_1(i-1)b_2,\ i \geq 1,\ j \geq 1. \tag{12.9}$$

Hsu and Burke (1976) using reversibility and Hunter (1983) using probabilistic and induction arguments, both showed that the output of a stable Geo/Geo/1 system with infinite buffer is also a geometric distribution with the same parameter as the input. Using that result Hsu and Burke (1976) showed that for discrete time tandem queues, the independence of simultaneous states of two systems also applies. Hence the discrete case and well known continuous case by Jackson (1954) have the same independence structures. That means

$$Pr\{X_n^{(1)} = i, X_n^{(2)} = j\}|_{n\to\infty} = Pr\{X_n^{(1)} = i\}Pr\{X_n^{(2)} = j\}|_{n\to\infty}.$$

Based on those results of Hsu and Burke (1976) and the Geo/Geo/1 results of Chapter 5 of this book, we can write

$$x_{0,0} = (b_1 - a)(b_1\bar{b}_1)^{-1}(b_2 - a)(b_2\bar{b}_2)^{-1},$$

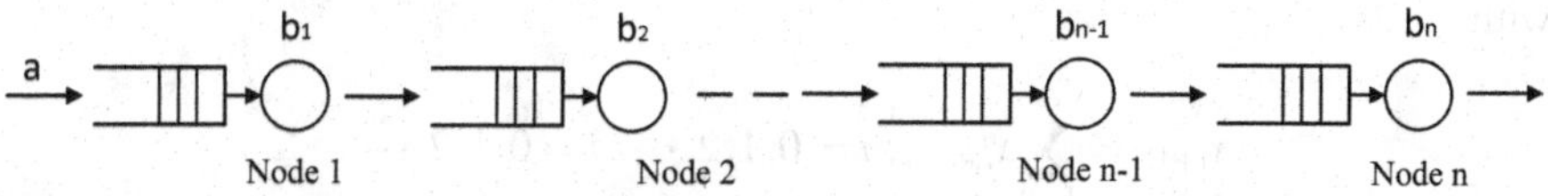

Fig. 12.2 A System of N Queues in Tandem

and

$$x_{i,j} = x_{0,0}\theta_1^i\theta_2^j,\ i = 1,2,\cdots,j = 1,2,\cdots,$$

where $\theta_k = (a\bar{b}_k)(\bar{a}b_k)^{-1}$, $k = 1,2$.

Extension to tandem queues with multiple nodes

Consider a system of N queues in tandem as shown in Fig 12.2. The result for the two queues in tandem can also be extended to the case of this $N-$tandem queues, with $N > 1$, provided that

$$\theta_k < 1,\ \forall k,\ 1 \le k \le N,$$

where b_k is the probability of service completion at server in node k.

Let $x_{i_1,i_2,\cdots,i_N} = Pr\{X_n^{(1)} = i_1, X_n^{(2)} = i_2,\cdots,X_n^N = i_N\}|_{n\to\infty}$, then we have

$$x_{i_1,i_2,\cdots,i_N} = \prod_{k=1}^{N}[(b_k - a)(b_1\bar{b}_k)^{-1}\theta_k^{i_k}]. \tag{12.10}$$

Using Matrix-geometric Approach

Considering the matrix P (for $N = 2$) in block matrix form we know that there exists a matrix R that is the minimal non-negative solution to the matrix equation

$$R = A_0 + RA_1 + R^2A_2.$$

Let us write R as

$$R = \begin{bmatrix} r_{0,0} & r_{0,1} & r_{0,2} & \cdots \\ r_{1,0} & r_{1,1} & r_{1,2} & \cdots \\ r_{2,0} & r_{2,1} & r_{2,2} & \cdots \\ \vdots & \vdots & \vdots & \cdots \end{bmatrix}.$$

Given the Rmatrix we can then write

$$\boldsymbol{x}_{i+1} = \boldsymbol{x}_i R,\ i = 1,2,\cdots,$$

with

$$x_{i+1,j} = \sum_{k=0}^{\infty} x_{i,k} r_{k,j},\ i = 0,1,2,\cdots; j = 0,1,2,\cdots,$$

and the boundary conditions obtained from

$$[\boldsymbol{x}_0,\ \boldsymbol{x}_1] = [\boldsymbol{x}_0,\ \boldsymbol{x}_1] \begin{bmatrix} B & C \\ E & A_1 + RA_2 \end{bmatrix},$$

and normalized by

$$\boldsymbol{x}_0 \mathbf{1} + \boldsymbol{x}_1 [I - R]^{-1} \mathbf{1} = 1.$$

Nevertheless, some theoretical analysis regarding the matrix R will be in order. We know now that for any node i

$$Pr\{X_n^i = 0\}|_{n\to\infty} = (b_i - a)(b_i \bar{b}_i)^{-1}, \text{ and } Pr\{X_n^i = k\}|_{n\to\infty} = (b_i - a)(b_i \bar{b}_i)^{-1} \theta_i^k,\ i \geq 0.$$

We know that the joint distribution of the number in the system at all nodes follow a product form solution given in Equation (12.10). From matrix-geometric results we also have $x_{i+1,j} = \sum_{v=0}^{\infty} x_{i,v} r_{v,j}$. Then combining these two results for the case of two queues in tandem, and after some algebraic operations, we end up with

$$\sum_{v=0}^{\infty} \theta_2^v r_{v,k} = \theta_1 \theta_2^k.$$

Letting $R_k(z) = \sum_{v=0}^{\infty} z^v r_{v,k}$ then we have

$$R_k(\theta_2) = \theta_1 \theta_2^k,\ k = 0,1,2,\cdots.$$

However, because it is of infinite dimension and does not have any specially known structure we may have to resort to its truncation for computational purposes, thereby making it an approximate method.

12.2.1.2 With External Arrivals into Second Queue and direct Departures from First Queue

By saying we have external arrivals to the second queue we imply that it is fed by both the first queue and an external source, as shown in Fig 12.3. Also by direct departures from the first queue we imply that some of the services completed in the first queue do not necessarily proceed to the second queue. In that case $X_n^{(1)}$ can

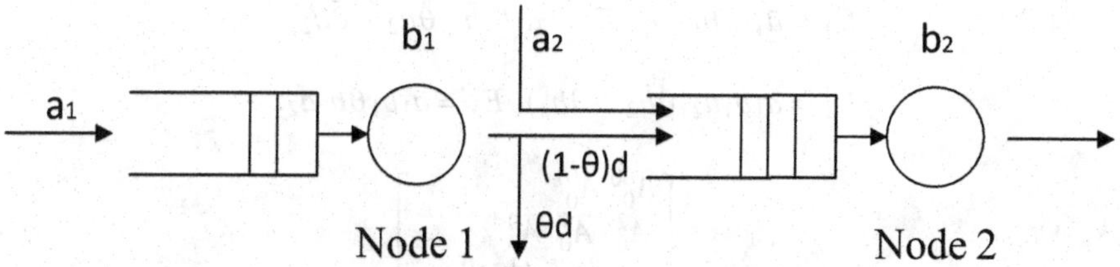

Fig. 12.3 A General System of Two Queues in Tandem

either increase (decrease) by one only at maximum, but $X_n^{(2)}$ may increase by more than one (up to 2). Let θ be the probability that a service that is completed in the first queue does not proceed to the second queue, with $\bar{\theta} = 1 - \theta$. Let a_k be the arrival probability to the kth, queue, $k = 1, 2,$.

The transition matrix is still the same QBD, and the blocks are given by

$$B = \begin{bmatrix} B_{00} & B_{01} & & & \\ & B_2 & B_1 & B_0 & \\ & & B_2 & B_1 & B_0 \\ & & & \ddots & \ddots & \ddots \end{bmatrix},$$

where

$$B_{00} = \bar{a}_1\bar{a}_2,\ B_{01} = \bar{a}_1 a_2,\ B_0 = \bar{a}_1 a_2 \bar{b}_2,\ B_1 = \bar{a}_1\bar{a}_2\bar{b}_2 + \bar{a}_1 a_2 b_2,\ B_2 = \bar{a}_1\bar{a}_2 b_2.$$

$$C = \begin{bmatrix} C_{00} & C_{01} & & & \\ & C_2 & C_1 & C_0 & \\ & & C_2 & C_1 & C_0 \\ & & & \ddots & \ddots & \ddots \end{bmatrix},$$

where

$$C_{00} = a_1\bar{a}_2,\ C_{01} = a_1 a_2,\ C_0 = a_1 a_2 \bar{b}_2,\ C_1 = a_1\bar{a}_2\bar{b}_2 + a_1 a_2 b_2,\ C_2 = a_1\bar{a}_2 b_2.$$

$$E = \begin{bmatrix} E_{00} & E_{01} & E_{02} & & & \\ & E_3 & E_2 & E_1 & E_0 & \\ & & E_3 & E_2 & E_1 & E_0 \\ & & & \ddots & \ddots & \ddots & \ddots \end{bmatrix},$$

where

$$E_{00} = \bar{a}_1 b_1 \theta \bar{a}_2,\ E_{01} = \bar{a}_1 b_1(\theta a_2 + \bar{\theta}\bar{a}_2),\ E_{02} = \bar{a}_1 b_1 \bar{\theta} a_2,$$

$$E_0 = \bar{a}_1 b_1 \bar{\theta} a_2 \bar{b}_2,\ E_1 = \bar{a}_1 b_1 \bar{b}_2(\bar{\theta} a_2 + \bar{\theta} \bar{a}_2),$$

$$E_2 = \bar{a}_1 b_1 \bar{a}_2(\theta \bar{b}_2 + \bar{\theta} b_2),\ E_3 = \bar{a}_1 b_1 \theta b_2 \bar{a}_2.$$

$$A_0 = \begin{bmatrix} A_0^{00} & A_0^{01} & & & \\ A_0^2 & A_0^1 & A_0^0 & & \\ & A_0^2 & A_0^1 & A_0^0 & \\ & & \ddots & \ddots & \ddots \end{bmatrix},$$

where

$$A_0^{00} = a_1 \bar{b}_1 \bar{a}_2,\ A_0^{01} = a_1 \bar{b}_1 a_2,\ A_0^0 = a_1 \bar{b}_1 a_2 \bar{b}_2,\ A_0^1 = a_1 \bar{b}_1(\bar{a}_2 \bar{b}_2 + a_2 b_2),\ A_0^2 = a_1 \bar{b}_1 \bar{a}_2 b_2.$$

$$A_1 = \begin{bmatrix} A_1^{00} & A_1^{01} & A_1^{02} & & & \\ A_1^3 & A_1^2 & A_1^1 & A_1^0 & & \\ & A_1^3 & A_1^2 & A_1^1 & A_1^0 & \\ & & \ddots & \ddots & \ddots & \ddots \end{bmatrix},$$

where

$$A_1^{00} = \bar{a}_1 \bar{b}_1 \bar{a}_2 + a_1 b_1 \theta \bar{a}_2,\ A_1^{01} = \bar{a}_1 \bar{b}_1 a_2 + a_1 b_1(\theta a_2 + \bar{\theta} \bar{a}_2),\ A_1^{02} = a_1 b_1 \bar{\theta} a_2,$$

$$A_1^0 = a_1 b_1 \bar{\theta} a_2 \bar{b}_2,\ A_1^1 = \bar{a}_1 \bar{b}_1 a_2 \bar{b}_2 + a_1 b_1 \bar{b}_2(\theta a_2 + \bar{\theta} \bar{a}_2),$$

$$A_1^2 = a_1 b_1(\theta \bar{a}_2 \bar{b}_2 + \theta a_2 b_2 + \bar{\theta} \bar{a}_2 b_2) + \bar{a}_1 \bar{b}_1(a_2 b_2 + \bar{a}_2 \bar{b}_2),\ A_1^3 = \bar{a}_1 b_2 \bar{a}_2 \bar{b}_1 + a_1 b_1 \theta \bar{a}_2 b_2.$$

$$A_2 = \begin{bmatrix} A_2^{00} & A_2^{01} & A_2^{02} & & & \\ A_2^3 & A_2^2 & A_2^1 & A_2^0 & & \\ & A_2^3 & A_2^2 & A_2^1 & A_2^0 & \\ & & \ddots & \ddots & \ddots & \ddots \end{bmatrix},$$

where

$$A_2^{00} = \bar{a}_1 b_1 \theta,\ A_2^{01} = \bar{a}_1 b_1(\theta a_2 + \bar{\theta} \bar{a}_2),\ A_2^{02} = \bar{a}_1 b_1 \bar{\theta} a_2,\ A_2^0 = \bar{a}_1 b_1 \bar{\theta} a_2 \bar{b}_2,$$

$$A_2^1 = \bar{a}_1 b_1(\bar{\theta}(\bar{a}_2 \bar{b}_2 + a_2 b_2) + \theta a_2 \bar{b}_2),\ A_2^2 = \bar{a}_1 b_1(\theta \bar{a}_2 \bar{b}_2 + \theta a_2 b_2 + \bar{\theta} \bar{a}_2 b_2),\ A_2^3 = \bar{a}_1 b_1 \theta \bar{a}_2 b_2.$$

Note that even though the block elements are scalar, uppercase letters have been used to avoid confusion.

This problem is now in the form of a QBD, which in theory could be analyzed by the matrix-geometric approach. However, because the block matrices are of the

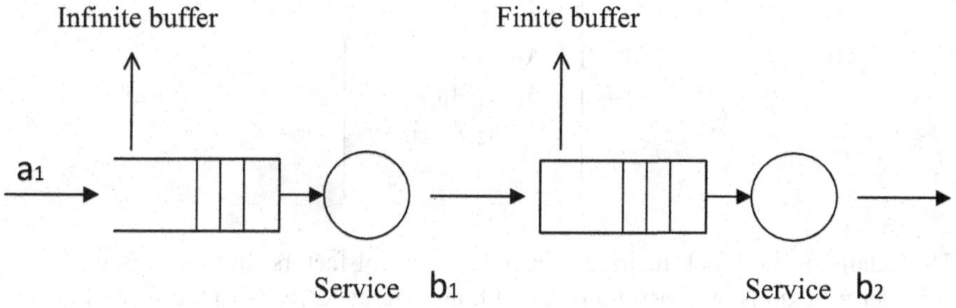

Fig. 12.4 Second Queue with Finite Buffer

infinite dimensions, and there are no known simple insights at the moment that can help us in analyzing this problem. It is still an open and challenging problem.

OPEN PROBLEM 01: We leave this as an open problem for those who might be interested in solving it

12.2.2 Second Queue with Finite Buffer

For the queue considered let us assume that the buffer at the second queue is now limited to a finite size $K < \infty$, as shown in Fig. 12.4.

Some of the packets that complete service at the first queue may find the second full, in which case they either stay in the first queue and because they are blocked they will hold up the server or they will be lost. In either case, the state space is the same. Define a bivariate DTMC $\{(X_n^{(1)}, X_n^{(2)}),\ n = 1, 2, \cdots, \}$ $X_n^{(1)} = 0, 1, 2, \cdots; X_n^{(2)} = 0, 1, 2, \cdots, K+1$, where at time n, X_n^k is the number of packets waiting in the kth queue, including the one receiving service. Let $A(n)$ be the number of arrivals to the first queue at time n. Note that $A(n)$ can only assume the values of 0 or 1. Further let $D(n)^{(k)}$ be the number of service completions at time n by server $k = 1, 2$. It should be noted that each server can only complete at most one service at any time and when there are no customers in the system no service can be completed. So we can write the random variable representing this system as

$$\{X_{n+1}^{(1)}, X_{n+1}^{(2)}\} = \{(X_n^{(1)} - D(n)^{(1)})^+ + A(n),\ min((X_n^{(2)} - D(n)^{(2)})^+ + D(n)^{(1)}, K+1)\}. \tag{12.11}$$

It is immediately clear that $(X_n^{(1)}, X_n^{(2)})$ is a bivariate DTMC and its transition matrix is of the QBD type. The transition matrix associated with this DTMC is P and is given as

$$P = \begin{bmatrix} B & C & & & \\ E & A_1 & A_0 & & \\ & A_2 & A_1 & A_0 & \\ & & A_2 & A_1 & A_0 \\ & & & \ddots & \ddots & \ddots \end{bmatrix}.$$

The details of the block matrices depend on several factors. In this subsection we consider two kinds of subsystems, 1) with no external arrivals to the second queue and 2) with external arrivals to the second queue. First we consider the case where there are no external arrivals into the second queue.

12.2.2.1 No External Arrivals into Second Queue

By saying no external arrivals to the second queue we imply that it is fed by the first queue. In that case both $X_n^{(1)}$ and $X_n^{(2)}$ can either increase (decrease) by only one at maximum. Hence there are two QBDs involved here. The inner matrix is a QBD.

The System with Loss:

Before we give the details of the entries in transition matrix, first let us define two other matrices. Consider the two $(K+2)\times(K+2)$ matrices $P_1^{(1)}$ and $P_1^{(2)}$ which are written as

$$P_1^{(1)} = \begin{bmatrix} 1 & 0 & & & \\ b_2 & \bar{b}_2 & & & \\ & b_2 & \bar{b}_2 & & \\ & & b_2 & \bar{b}_2 & \\ & & & \ddots & \ddots \\ & & & & b_2 & \bar{b}_2 \end{bmatrix}, \quad \text{and} \quad P_1^{(2)} = \begin{bmatrix} 0 & 1 & 0 & & & \\ 0 & b_2 & \bar{b}_2 & & & \\ & 0 & b_2 & \bar{b}_2 & & \\ & & 0 & b_2 & \bar{b}_2 & \\ & & & \ddots & \ddots & \ddots \\ & & & & 0 & 1 \end{bmatrix},$$

respectively. The matrices $P_1^{(1)}$ and $P_1^{(2)}$ are the ones we refer to as the inner matrices, and they are also some form of QBDs as presented earlier. Now the block matrices of the matrix P can be written as

$$B = \bar{a}P_1^{(1)},\ C = aP_1^{(1)},\ E = \bar{a}b_1P_1^{(2)}$$

$$A_0 = a\bar{b}_1P_1^{(1)},\ A_1 = \bar{a}\bar{b}_1P_1^{(1)} + ab_1P_1^{(2)},\ A_2 = \bar{a}b_1P_1^{(2)}.$$

The stability conditions for this Markov chain can be obtained by studying the matrix $A = A_0 + A_1 + A_2$. The matrix A is irreducible for this class of problem. Let its stationary distribution vector be $\boldsymbol{\pi}$, where $\boldsymbol{\pi} = \boldsymbol{\pi}A,\ \boldsymbol{\pi}\mathbf{1} = 1$.

Proposition: The stability conditions are given as follow; that we require

$$\pi A_0 \mathbf{1} < \pi A_2 \mathbf{1}.$$

Let

$$x_{i,j} = Pr\{X_n^{(1)} = i, X_n^{(2)} = j\}|_{n\to\infty},$$

with $\boldsymbol{x}_i = [x_{i,0},\ x_{i,1},\ x_{i,2},\ \cdots,\ x_{i,K+1}]$, $i = 0,1,2,\cdots$, and $\boldsymbol{x} = [\boldsymbol{x}_0,\ \boldsymbol{x}_1,\ \boldsymbol{x}_2,\ \cdots]$, then we have

$$\boldsymbol{x} = \boldsymbol{x}P,\ \ \boldsymbol{x}\mathbf{1} = \mathbf{1}. \tag{12.12}$$

From here it is straightforward to obtain all the key performance measures of this system using the standard matrix-analytic methods.

The System with No Loss but Blocking: When we deal with blocking, there are basically two types; one in which as soon as the second buffer is full, the first server stops working and the other one is where the first server still works on an item even after the second buffer is full, but stops after that item has been served and the second buffer is still full. We will consider both cases.

Case One: Whenever we have K in the second buffer,i.e. $K+1$ in the second queue, the first server is blocked and can no longer serve until at least one item leaves the second buffer. Now we define the two finite matrices of order $(K+2) \times (K+2)$, given as follows

$$P_1^{(1)} = \begin{bmatrix} 1 & 0 & & & \\ b_2 & \bar{b}_2 & & & \\ & b_2 & \bar{b}_2 & & \\ & & b_2 & \bar{b}_2 & \\ & & & \ddots & \ddots \\ & & & & b_2 & \bar{b}_2 \end{bmatrix}, \quad P_1^{(2)} = \begin{bmatrix} 0 & 1 & 0 & & & \\ 0 & b_2 & \bar{b}_2 & & & \\ & 0 & b_2 & \bar{b}_2 & & \\ & & 0 & b_2 & \bar{b}_2 & \\ & & & \ddots & \ddots & \ddots \\ & & & & b_2 & \bar{b}_2 \end{bmatrix}.$$

Once again we just need to apply the standard matrix-analytic results to this system to obtain solutions. It is straightforward.

Also the results can be easily extended to the case of $K > 1$ queues in tandem.

Case Two: First server keeps serving the Head of Line packet in its queue when the buffer of the second queue is full. However it temporarily suspends service after that service is completed, if the second server buffer is still full, and resumes service after the buffer of the second server queue is reduced to $K-1$. For this case we introduce an additional state of the second buffer. We let the state $(K+1)^*$ be the state when the second queue has $K+1$ in its system and the first server is blocked. So we have state space of the second queue as $\{0,1,2,\cdots,K,K+1,(K+1)^*\}$. Now we define the two finite matrices of order $(K+3) \times (K+3)$, given as follows

$$P_1^{(1)} = \begin{bmatrix} 1 & 0 & & & & \\ b_2 & \bar{b}_2 & & & & \\ & b_2 & \bar{b}_2 & & & \\ & & b_2 & \bar{b}_2 & & \\ & & & \ddots & \ddots & \\ & & & b_2 & \bar{b}_2 & 0 \\ & & & b_2 & & \bar{b}_2 \end{bmatrix}, \quad P_1^{(2)} = \begin{bmatrix} 0 & 1 & 0 & & & \\ 0 & b_2 & \bar{b}_2 & & & \\ & 0 & b_2 & \bar{b}_2 & & \\ & & 0 & b_2 & \bar{b}_2 & \\ & & & \ddots & \ddots & \ddots \\ & & & 0 & b_2 & \bar{b}_2 \\ & & & b_2 & 0 & \bar{b}_2 \end{bmatrix}.$$

12.2.2.2 With External Arrivals into Second Queue

The case of when we have external arrivals to the second queue and departures from first queue can be easily obtained by using the combined knowledge of the case with infinite second buffer and the external arrivals into second queue and departures from first queue. The only issue is that it is painstakingly laborious. But after the transition matrix is built we simply follow the standard matrix-analytic method again.

OPEN PROBLEM 02: We leave this as an exercise for the reader.

12.3 Two Queues in Tandem – The Phase Type Case

Throughout this section we let arrivals and services be phase types and the buffer at the first queue to be unlimited.

Consider two queues in tandem with arrivals to the first queue happening according to the phase type distribution represented by (α, T) of dimension r, with arrival rate given as a, where $a^{-1} = \alpha(I-T)^{-1}\mathbf{1}$. The service at the first queue is according to phase type distribution represented by (β_1, S_1) of dimension m_1 and at the second queue is also according to phase type (β_2, S_2) of dimension m_2. We let $\mathbf{s}_k = \mathbf{1} - S_k\mathbf{1}$, $k = 1, 2$, with the service rates given as b_k, where $b_k^{-1} = \beta_k(I-S_k)^{-1}\mathbf{1}$. A packet that arrives at the first queue is processed as soon as the server is available and then proceeds to the second queue for service which is also rendered as soon as the second server is available.

12.3.1 Second Queue with Infinite Buffer

For the queue considered let us assume that the buffer at the second queue is unlimited, so a packet that completes service at the first queue always has a space to wait for the service in the second queue, if the second server is busy.

Define a multivariate DTMC $\{(X_n^{(1)}, X_n^{(2)}, A_n, S_n^{(1)}, S_n^{(2)}),\ n = 1, 2, \cdots,\}$ $X_n^{(1)} = 0, 1, 2, \cdots; X_n^{(2)} = 0, 1, 2, \cdots$, where $X_n^{(k)}$ is the number of packets waiting in the kth queue, including the one receiving service at time n, A_n is the phase of arrival at time n, $S_n^{(1)}$ is the phase of service at the first queue, if there is an item in service and $S_n^{(2)}$ is the phase of service at the second queue is there is an item in service. It is immediately clear that $(X_n^{(1)}, X_n^{(2)} A_n, S_n^{(1)}, S_n^{(2)})$ is a DTMC and its transition matrix is of the QBD type. The transition matrix associated with this DTMC is P and is given as

$$P = \begin{bmatrix} B & C & & & \\ E & A_1 & A_0 & & \\ & A_2 & A_1 & A_0 & \\ & & A_2 & A_1 & A_0 \\ & & & \ddots & \ddots & \ddots \end{bmatrix}.$$

The details of the block matrices depend on several factors. In this subsection we consider two kinds of subsystems, 1) with no external arrivals to the second queue and 2) with external arrivals to the second queue. We only present the case where there are no external arrivals into the second queue, and leave the first case as an exercise for the reader.

12.3.1.1 No External Arrivals into Second Queue

By saying no external arrivals to the second queue we imply that it is fed by the first queue only. In that case both $X_n^{(1)}$ and $X_n^{(2)}$ can either increase (decrease) by one only at maximum. Hence there are two QBDs involved here. The inner matrix is some form of a QBD, with either the super-diagonal or sub-diagonal elements missing. Before we give the details of the entries in transition matrix, first let us define two other matrices. Let the matrix $P_1^{(1)}$ and $P_1^{(2)}$ be written as

$$P_1^{(1)} = \begin{bmatrix} 1 & 0 & & \\ \mathbf{s}_2 & S_2 & & \\ & \mathbf{s}_2\boldsymbol{\beta}_2 & S_2 & \\ & & \mathbf{s}_2\boldsymbol{\beta}_2 & S_2 \\ & & & \ddots & \ddots \end{bmatrix}, \text{ and } P_1^{(2)} = \begin{bmatrix} 0 & \boldsymbol{\beta}_2 & 0 & & \\ 0 & \mathbf{s}_2\boldsymbol{\beta}_2 & S_2 & & \\ & 0 & \mathbf{s}_2\boldsymbol{\beta}_2 & S_2 & \\ & & 0 & \mathbf{s}_2\boldsymbol{\beta}_2 & S_2 \\ & & & \ddots & \ddots & \ddots \end{bmatrix},$$

respectively. The matrices $P_1^{(1)}$ and $P_1^{(2)}$ are the ones we refer to as the inner matrices, and they are some form of QBDs. The first matrix, $P_1^{(1)}$ should not be seen as a kind of an absorbing Markov chain, but as a behaviour of the second queue as seen by the first server. The 1 in the left hand corner refers to no change in the second queue as seen by the first server. This will be interpreted slightly differently depending on the context in which this matrix used. The second matrix $P_1^{(2)}$ is related to the first one, and their relationship should be obvious by now. It is simply a shifted version of matrix $P_1^{(1)}$. Next we define a matrix $\mathscr{I}(A)$, where A is any square matrix as

$$\mathscr{I}(A) = \begin{bmatrix} A \otimes I & & & \\ & A \otimes I & & \\ & & A \otimes I & \\ & & & \ddots \end{bmatrix},$$

where I in this case is of same order as A. Now the block matrices of the matrix P can be written as

$$B = \mathscr{I}(T)P_1^{(1)},\ C = \mathscr{I}(\mathbf{t}\boldsymbol{\alpha})P_1^{(1)},\ E = \mathscr{I}(T \otimes \mathbf{s}_1)P_1^{(2)}$$

$$A_0 = \mathscr{I}((\mathbf{t}\boldsymbol{\alpha}) \otimes S_1))P_1^{(1)},\ A_1 = \mathscr{I}(T \otimes S_1)P_1^{(1)} + \mathscr{I}((\mathbf{t}\boldsymbol{\alpha}) \otimes (\mathbf{s}_1\boldsymbol{\beta}_1))P_1^{(2)},\ A_2 = \mathscr{I}(T \otimes (\mathbf{s}_1\boldsymbol{\beta}_1))P_1^{(2)}.$$

The stability conditions for this Markov chain can be obtained through the operator geometric approach for the GI/G/1 system by studying the matrix $A = A_0 + A_1 + A_2$. Alternatively we can simply use an intuitive argument as follow.

Proposition: Since the input into the second queue is mainly the output of the first queue and there is no blocking, then stability conditions are that we require

$$a < b_1 \text{ and } a < b_2,$$

or

$$min\{b_1, b_2\} > a.$$

12.3.2 Second Queue with Finite Buffer

For the queue considered let us assume that the buffer at the second queue is now limited to a finite size $K < \infty$. Some of the packets that complete service at the first queue may find the second buffer full, in which case they either stay in the first queue and because they are blocked they will hold up the server or they will be lost. In either case, the state space is the same. Define a multivariate DTMC $\{X_n^{(1)}, X_n^{(2)}, A_n, S_n^{(1)}, S_n^{(2)}\}$, $n = 1, 2, \cdots$, $X_n^{(1)} = 0, 1, 2, \cdots; X_n^{(2)} = 0, 1, 2, \cdots, K+1$, where X_n^k is the number of packets waiting in the kth queue, including the one receiving service at time n. Let $A(n)$ be the number of arrivals to the first queue at time n. Note that $A(n)$ can only assume the values of 0 or 1. It is immediately clear that $(X_n^{(1)}, X_n^{(2)}, A_n, S_n^{(1)}, S_n^{(2)})$ is a multivariate DTMC and its transition matrix is of the QBD type. The transition matrix associated with this DTMC is P and is given as

$$P = \begin{bmatrix} B & C & & & \\ E & A_1 & A_0 & & \\ & A_2 & A_1 & A_0 & \\ & & A_2 & A_1 & A_0 \\ & & & \ddots & \ddots & \ddots \end{bmatrix}.$$

Matrix P in a block form is what we refer to as the outer matrix and it is a QBD. The details of the block matrices depend on several factors. In this subsection we consider only the case where there are no external arrivals into the second queue.

12.3.2.1 No External Arrivals into Second Queue

By saying no external arrivals to the second queue we imply that it is fed by the first queue only. In that case both $X_n^{(1)}$ and $X_n^{(2)}$ can either increase (decrease) by one only at maximum. Hence there are two QBDs involved here. The inner matrix is a QBD.

<u>The System with Loss:</u>

Before we give the details of the entries in the transition matrix, first let us define two other matrices. Consider the two $(K+2) \times (K+2)$ matrices $P_1^{(1)}$ and $P_1^{(2)}$ which are written as

$$P_1^{(1)} = \begin{bmatrix} 1 & 0 & & & \\ \mathbf{s}_2 & S_2 & & & \\ & \mathbf{s}_2\boldsymbol{\beta}_2 & S_2 & & \\ & & \mathbf{s}_2\boldsymbol{\beta}_2 & S_2 & \\ & & & \ddots & \ddots \\ & & & & \mathbf{s}_2\boldsymbol{\beta}_2 & S_2 \end{bmatrix}, \text{ and } P_1^{(2)} = \begin{bmatrix} 0 & \boldsymbol{\beta}_2 & 0 & & & \\ 0 & \mathbf{s}_2\boldsymbol{\beta}_2 & S_2 & & & \\ & 0 & \mathbf{s}_2\boldsymbol{\beta}_2 & S_2 & & \\ & & 0 & \mathbf{s}_2\boldsymbol{\beta}_2 & S_2 & \\ & & & \ddots & \ddots & \ddots \\ & & & & 0 & S_2+\mathbf{s}_2\boldsymbol{\beta}_2 \end{bmatrix},$$

respectively. The matrices $P_1^{(1)}$ and $P_1^{(2)}$ are the ones we refer to as the inner matrices, and they are also QBDs. Once again the first matrix, $P_1^{(1)}$ should not be seen as a kind of an absorbing Markov chain, but as a behaviour of the second queue as seen by the first server. The 1 in the left hand corner refers to no change in the second queue as seen by the first server. This will be interpreted slightly differently depending on the context in which this matrix used. The second matrix $P_1^{(2)}$ is related to the first one, and their relationship should be obvious by now. It is simply a shifted version of matrix $P_1^{(1)}$. Now the block matrices of the matrix P can be written as

$$B = \mathscr{I}(T)P_1^{(1)},\ C = \mathscr{I}(\mathbf{t}\boldsymbol{\alpha})P_1^{(1)},\ E = \mathscr{I}(T \otimes \mathbf{s}_1)P_1^{(2)}$$

$$A_0 = \mathscr{I}((\mathbf{t}\boldsymbol{\alpha}) \otimes S_1)P_1^{(1)},\ A_1 = \mathscr{I}(T \otimes S_1)P_1^{(1)} + \mathscr{I}((\mathbf{t}\boldsymbol{\alpha}) \otimes (\mathbf{s}_1\boldsymbol{\beta}_1))P_1^{(2)},\ A_2 = \mathscr{I}(T \otimes (\mathbf{s}_1\boldsymbol{\beta}_1))P_1^{(2)}.$$

The stability conditions for this Markov chain can be obtained by studying the matrix $A = A_0 + A_1 + A_2$. The matrix A is irreducible for this class of problem. Let its stationary distribution vector be $\boldsymbol{\pi}$, where $\boldsymbol{\pi} = \boldsymbol{\pi}A,\ \boldsymbol{\pi}\mathbf{1} = 1$.

Proposition: The stability conditions are given as follow; that we require

$$\boldsymbol{\pi}A_0\mathbf{1} < \boldsymbol{\pi}A_2\mathbf{1}.$$

From here it is straightforward to obtain all the key performance measures of this system using the standard matrix-analytic methods.

The System with No Loss but with Blocking: When we deal with blocking, there are basically two types; one in which as soon as the second buffer is full, the first server stops working and the other one is where the first server still works on an item even after the second buffer is full, but stops after that item has been served and the second buffer is still full. We will consider both cases.

Case One: Whenever we have K in the second buffer,i.e. $K+1$ in the second queue, the first server is blocked and can no longer serve until at least one item leaves the second buffer. Now we define the two finite matrices of order $(K+2) \times (K+2)$, given as follows

$$P_1^{(1)} = \begin{bmatrix} 1 & 0 & & & & \\ \mathbf{s}_2 & S_2 & & & & \\ & (\mathbf{s}_2\boldsymbol{\beta}_2) & S_2 & & & \\ & & \ddots & \ddots & & \\ & & & (\mathbf{s}_2\boldsymbol{\beta}_2) & S_2 & \\ & & & & (\mathbf{s}_2\boldsymbol{\beta}_2) & S_2 \end{bmatrix},$$

$$P_1^{(2)} = \begin{bmatrix} 0 & \boldsymbol{\beta}_2 & 0 & & & \\ 0 & (\mathbf{s}_2\boldsymbol{\beta}_2) & S_2 & & & \\ & 0 & (\mathbf{s}_2\boldsymbol{\beta}_2) & S_2 & & \\ & & \ddots & \ddots & \ddots & \\ & & & 0 & (\mathbf{s}_2\boldsymbol{\beta}_2) & S_2 \\ & & & & (\mathbf{s}_2\boldsymbol{\beta}_2) & S_2 \end{bmatrix}.$$

Now the block matrices of the matrix P can be written as

$$B = \mathcal{J}(T)P_1^{(1)},\ C = \mathcal{J}(\mathbf{t}\boldsymbol{\alpha})P_1^{(1)},\ E = \mathcal{J}(T\otimes \mathbf{s}_1)P_1^{(2)}$$

$$A_0 = \mathcal{J}((\mathbf{t}\boldsymbol{\alpha})\otimes S_1)P_1^{(1)},\ A_1 = \mathcal{J}(T\otimes S_1)P_1^{(1)} + \mathcal{J}((\mathbf{t}\boldsymbol{\alpha})\otimes(\mathbf{s}_1\boldsymbol{\beta}_1))P_1^{(2)},\ A_2 = \mathcal{J}(T\otimes(\mathbf{s}_1\boldsymbol{\beta}_1))P_1^{(2)}.$$

The stability conditions for this Markov chain can be obtained by studying the matrix $A = A_0 + A_1 + A_2$. The matrix A is irreducible for this class of problem. Let its stationary distribution vector be $\boldsymbol{\pi}$, where $\boldsymbol{\pi} = \boldsymbol{\pi}A,\ \boldsymbol{\pi}\mathbf{1} = 1$.

Proposition: The stability conditions are given as follow; that we require

$$\boldsymbol{\pi}A_0\mathbf{1} < \boldsymbol{\pi}A_2\mathbf{1}.$$

Once again we just need to apply the standard matrix-analytic results to this system to obtain solutions. It is straightforward.

Also the results can be easily extended to the case of $K > 1$ queues in tandem.

Case Two: First server keeps serving the Head of Line packet in its queue when the buffer of the second queue is full. However it temporarily suspends service after that service is completed, if the second server buffer is still full, and resumes service after the buffer of the second server queue is reduced to $K-1$. For this case we introduce an additional state of the second buffer. We let the state $(K+1)^*$ be the state when the second queue has $K+1$ in its system and the first server is blocked. So we have state space of the second queue as $\{0,1,2,\cdots,K,K+1,(K+1)^*\}$. We can then define the two finite matrices of order $(K+3)\times(K+3)$ and use the same technique as for the geometric case. We leave that as an exercise for the reader.

$$P_1^{(1)} = \begin{bmatrix} 1 & 0 & & & \\ \mathbf{s}_2 & S_2 & & & \\ & (\mathbf{s}_2\boldsymbol{\beta}_2) & S_2 & & \\ & & \ddots & \ddots & \\ & & & (\mathbf{s}_2\boldsymbol{\beta}_2) & S_2 \\ & & & (\mathbf{s}_2\boldsymbol{\beta}_2) & S_2 \end{bmatrix},$$

$$P_1^{(2)} = \begin{bmatrix} 0 & \boldsymbol{\beta}_2 & 0 & & & \\ 0 & (\mathbf{s}_2\boldsymbol{\beta}_2) & S_2 & & & \\ & 0 & (\mathbf{s}_2\boldsymbol{\beta}_2) & S_2 & & \\ & & \ddots & \ddots & \ddots & \\ & & & 0 & (\mathbf{s}_2\boldsymbol{\beta}_2) & S_2 \\ & & & & (\mathbf{s}_2\boldsymbol{\beta}_2) & S_2 \end{bmatrix}.$$

the block matrices of the matrix P can be written as

$$B = \mathscr{I}(T)P_1^{(1)},\ C = \mathscr{I}(\mathbf{t}\boldsymbol{\alpha})P_1^{(1)},\ E = \mathscr{I}(T \otimes \mathbf{s}_1)P_1^{(2)}$$

$$A_0 = \mathscr{I}((\mathbf{t}\boldsymbol{\alpha}) \otimes S_1)P_1^{(1)},\ A_1 = \mathscr{I}(T \otimes S_1)P_1^{(1)} + \mathscr{I}((\mathbf{t}\boldsymbol{\alpha}) \otimes (\mathbf{s}_1\boldsymbol{\beta}_1))P_1^{(2)},\ A_2 = \mathscr{I}(T \otimes (\mathbf{s}_1\boldsymbol{\beta}_1))P_1^{(2)}.$$

The stability conditions for this Markov chain can be obtained by studying the matrix $A = A_0 + A_1 + A_2$. The matrix A is irreducible for this class of problem. Let its stationary distribution vector be $\boldsymbol{\pi}$, where $\boldsymbol{\pi} = \boldsymbol{\pi}A,\ \boldsymbol{\pi}\mathbf{1} = 1$. Then we have

Proposition: The stability conditions are given as follow; that we require

$$\boldsymbol{\pi}A_0\mathbf{1} < \boldsymbol{\pi}A_2\mathbf{1}.$$

Once again we just need to apply the standard matrix-analytic results to this system to obtain solutions. It is straightforward.

Also the results can be easily extended to the case of $K > 1$ queues in tandem.

12.3.3 Both First and Second Queues with Finite Buffers

Consider the case where both buffers are finite as shown in Fig. 12.5.

Let $K_i < \infty$ be the buffers size at node $i = 1,2$. This system has possible losses in the first queue because of the limited buffer size. Define a multivariate DTMC $\{X_n^{(1)}, X_n^{(2)}, A_n, S_n^{(1)}, S_n^{(2)}\}$, $n = 1,2,\cdots$, $X_n^{(1)} = 0,1,2,\cdots,K_1+1$; $X_n^{(2)} = 0,1,2,\cdots,K_2+1$, where X_n^k is the number of packets waiting in the kth queue, including the one receiving service at time n. Let $A(n)$ be the number of arrivals to the first queue at time n. Note that $A(n)$ can only assume the values of 0 or 1. It is immediately clear that $(X_n^{(1)}, X_n^{(2)}, A_n, S_n^{(1)}, S_n^{(2)})$ is a multivariate DTMC and its transition matrix is of the QBD type. The transition matrix associated with this DTMC is P and is given as

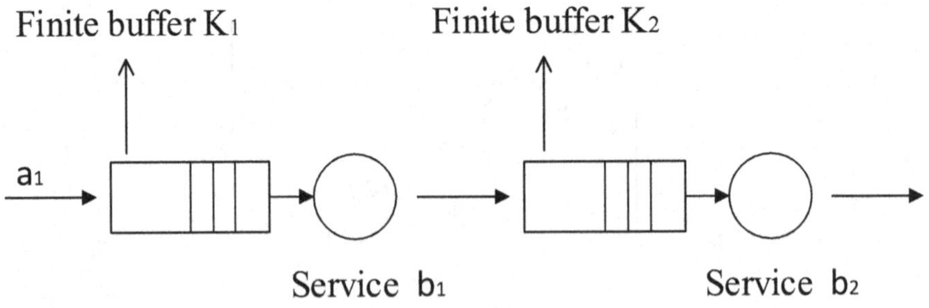

Fig. 12.5 Both First and Second Queues with Finite Buffers

$$P = \begin{bmatrix} B & C & & & & \\ E & A_1 & A_0 & & & \\ & A_2 & A_1 & A_0 & & \\ & & A_2 & A_1 & A_0 & \\ & & & \ddots & \ddots & \ddots \\ & & & & A_2 & A_0 + A_1 \end{bmatrix},$$

where all the block matrices are the same as in Section 12.2.2. All the techniques proposed earlier in Section 4.3 can be used to analyze this finite DTMC. They are straightforward and the matrices could be of huge dimensions – the problem.

12.4 Approximating ($N > 2$) Tandem Queues Using Above Results

At the beginning of this book we pointed out that some tandem queues can be approximated by a set of fewer nodes queues. Most of the tandem queues that we have studied so far can all be approximated by a series of fewer nodes queues. To avoid too much repetition of the same information we proceed to discuss tandem queues with phase type arrivals and services, as in Section 12.3.

12.4.1 Decomposition to N single queues

Consider a system of N queues in tandem. We can approximate it by a set of N single node queues which are assumed to be independent, as shown in Fig 12.6.

The assumption of independence is a key aspect of this approximation, since we know that they are not independent. We first analyze the first node queue as a simple single server queue. The output from this queue can then be approximated accordingly and used as the input into the second queue which can also be analyzed as a single node queue. This process is continued until all the N nodes have been analyzed and the behaviour of the N tandem queue, in terms of the joint probability distribution of the number in each queue is assumed to be the product of the distributions of number each independent queue. We first consider the case where the intermediate buffers are unlimited.

12.4.1.1 Infinite Intermediate Buffers

Consider each node of the tandem system as isolated nodes, and study one as a MAP/PH/1 system. For the first node, the MAP is given by two matrices $D_0 = T$ and $D_1 = \mathbf{t}\boldsymbol{\alpha}$. Using the MAP/PH/1 result of Chapter 5 we can decompose its transition

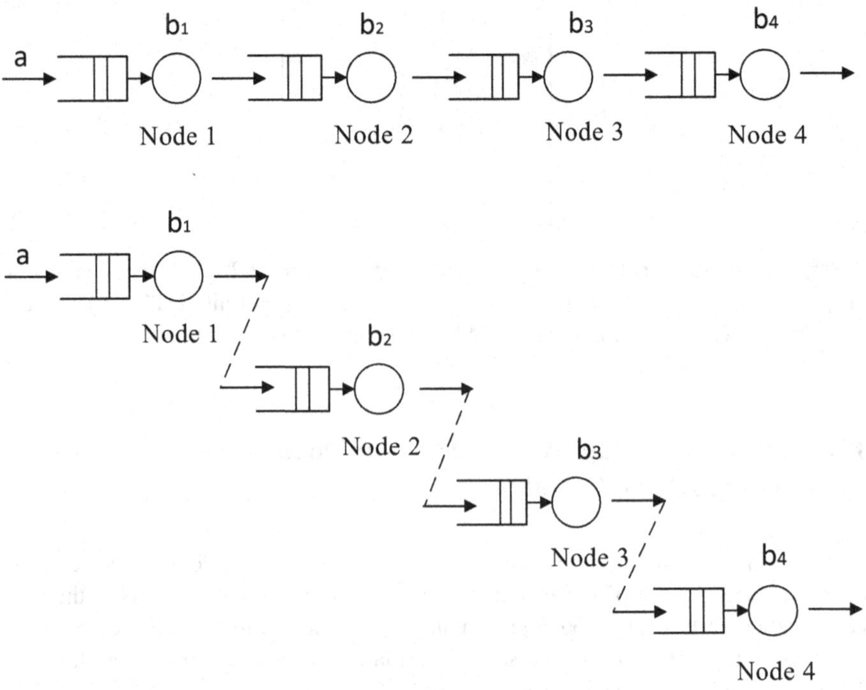

Fig. 12.6 Single Nodes Decomposition of System of N Queues in Tandem

matrix into two MAP matrices that capture the output process of the first queue. Let the two matrices D_0 and D_1 that capture the output which now becomes the input to the next node be

$$D_0 = \begin{bmatrix} B & C & & & \\ E & A_1^{(1)} & & & \\ & A_2 & A_1^{(1)} & & \\ & & A_2 & A_1^{(1)} & \\ & & & \ddots & \ddots \end{bmatrix}, \quad D_1 = \begin{bmatrix} 0 & 0 & & & \\ & A_1^{(2)} & A_0 & & \\ & & A_1^{(2)} & A_0 & \\ & & & A_1^{(2)} & A_0 \\ & & & \ddots & \ddots \end{bmatrix},$$

where matrices $A_1^{(1)}$ and $A_1^{(2)}$ are given later in the approximation procedure to follow. It is immediately clear that these two matrices form a an infinite dimension MAP and can be used to capture the output from the first queue and used as the input to the second queue studied in isolation.

So what we can do to approximate a two queue tandem system will be as follows.

The Approximation:

- *Step 1:* Study the first queue as a single node queue in isolation, with arrivals given as phase type $(\boldsymbol{\alpha}, T)$ which is a special MAP with $D_0 = T$ and $D_1 = \mathbf{t}\boldsymbol{\alpha}$, and phase service $(\boldsymbol{\beta}_1, S_1)$.
- *Step 2:* For the first queue, obtain the R matrix and all the needed performance measures as shown in Chapter 5.
- *Step 3:* Consider an infinite MAP given by (D_0, D_1) with entries given as $B = T$, $C = \mathbf{t}\alpha$, $E = T \otimes \mathbf{s}_1$, $A_0 = (\mathbf{t}\boldsymbol{\alpha}) \otimes S_1$, $A_1^{(1)} = T \otimes S_1$, $A_1^{(2)} = (\mathbf{t}\boldsymbol{\alpha}) \otimes \mathbf{s}_1\boldsymbol{\beta}_1)$, $A_2 = T \otimes (\mathbf{s}_1\boldsymbol{\beta}_1)$
- *Step 4:* Use this infinite MAP as it is or approximate it by a finite MAP given $(\tilde{D}_0, \tilde{D}_1)$ of some dimension $q < \infty$ to represent the input to the second queue.
- *Step 5:* Study the second queue as an isolated MAP/PH/1 system with arrivals given as $(\tilde{D}_0, \tilde{D}_1)$, service as $(\boldsymbol{\beta}_2, S_2)$.
- *Step 6:* Obtain the needed performance measures for this second queue as shown in Chapter 5.
- *Step 7:* By assuming that the two queues are practically independent we can obtain the joint performance measures as products of individual performance measures.

12.4.1.2 Finite Intermediate Buffers

When the intermediate buffers sizes are finite, then we have to take into account the blocking probabilities, unless they are loss systems. Estimating the blocking probabilities is not a simple procedure. A method for approximating them are presented in Schamber (1997). We now present a summary of that approach.

Consider a two-queue tandem queue in which the first queue has an infinite buffer size and the second server has a finite buffer queue size. We assume that it is a blocking system. The idea is as follows. We initially consider the first queue as a single server isolated queue in which the server is never blocked. Obtain the departure process from this queue or at least approximate it by a finite MAP. Using this departure process analyze the second queue as a finite buffer queue with loss. Obtain the loss probability. Use this loss probability as an estimate of the blocking probability. Let us call this probability θ. We then re-visit the first queue and analyze it as before, except that now we assume that the server may be blocked from serving with probability θ. Again we obtain the departure process based on this new analysis and repeat the process until the loss probability at the nth iteration is close to that of the $(n+1)^{st}$ iteration.

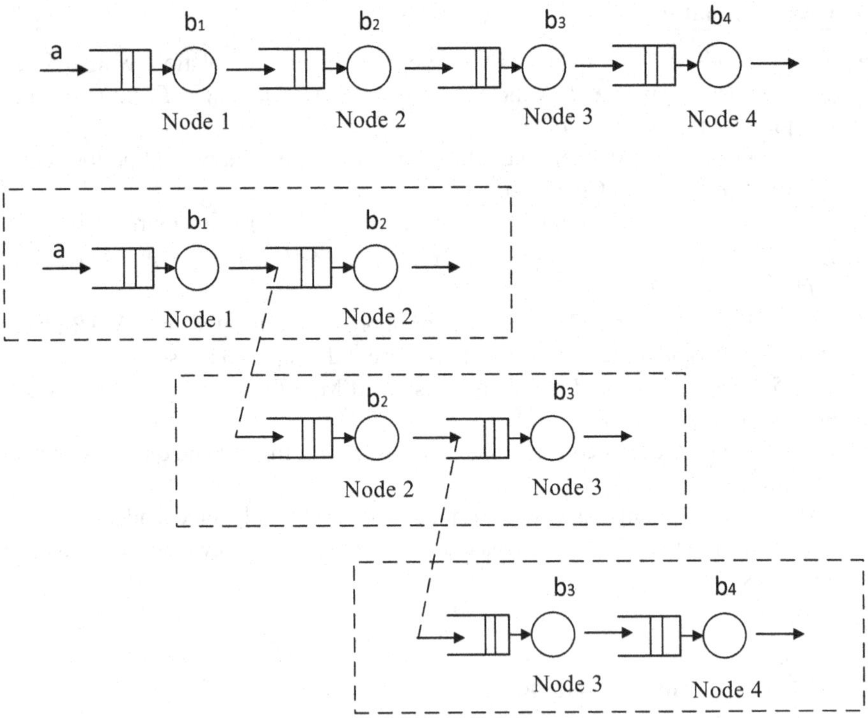

Fig. 12.7 Pair Nodes Decomposition of N Queues in Tandem

12.4.2 *Decomposition to* $N-1$ *two tandem queues*

When $N > 2$ we can use the decomposition approximations discussed in the last section. However, we can use a more improved approximation that decomposes the system into $N-1$ paired tandem queues with overlapping nodes, as shown in Fig 12.7.

The overlapping of nodes allows us to capture a better estimation of the intermediate arrival processes. This idea was made popular by Brandwajn (1985) and used for continuous time cases. For ease of presentation we consider a tandem queue with three queues, i.e. $N = 3$. If we analyze queues 1 and 2 as a two-node tandem queue we can estimate the input into the second queue more accurately. This input is then used for analyzing queue 2 and 3 as a two node tandem queue. This was shown to give a better estimation than decomposing this three queue system into three single node queues. Even though we are trying to focus on the case of three nodes in tandem, the idea applies to $N \geq 3$ nodes in tandem. Secondly, we will stay with geometric system since extension to the Phase type is quite straightforward.

We consider both cases where the intermediate buffers have unlimited and also limited capacities. The ideas are about the same in both cases, with very slight differences.

12.4.2.1 Infinite Intermediate Buffers

Consider three queues in tandem, with arrival into the first queue according to the geometric distribution with parameter a. Service at the nodes are also geometric with parameters b_i, $i = 1,2,3$. Let $\{(X_n^{(1)}, X_n^{(2)}, X_n^{(3)}), n \geq 0\}$ be the trivariate process that represents the number of items in queues 1, 2 and 3, respectively. It is immediately clear that this is a DTMC with QBD structure, given as

$$P = \begin{bmatrix} B & C & & & \\ E & A_1 & A_0 & & \\ & A_2 & A_1 & A_0 & \\ & & A_2 & A_1 & A_0 \\ & & & \ddots & \ddots & \ddots \end{bmatrix}.$$

Looking at each of the blocks in this matrix P we find that each of them is also a QBD type. For example,

$$A_1 = \begin{bmatrix} A_1(00) & A_1(01) & & & \\ A_1(10) & A_1(11) & A_1(12) & & \\ & A_1(21) & A_1(22) & A_1(23) & \\ & & A_1(32) & A_1(33) & A_1(34) \\ & & & \ddots & \ddots & \ddots \end{bmatrix}.$$

Once again if we consider each of the blocks in A_1 they are also QBD types.

The approximation is as follows. First we study a system of two queues in tandem, i.e. Queue 1 and Queue 2. From this we can obtain the flow from Queue 1 to Queue 2. Let us call this flow F. Next we study Queue 2 and Queue 3 as a system of two queues in tandem with F as the input to Queue 2 in this second system. This way queues 1,2, and 3 are studied by decomposition into two sets of tow queues in tandem, i.e. Queue 1 and Queue 2 and also Queue 2 and Queue 3.

12.5 Queueing Networks

Classical queueing networks in discrete time has not received much attention compared to its continuous time counterpart. This is partly because reversibility has not been properly established for it thereby making it not clear whether the Jackson type results do apply for most cases. In a queueing network, especially the open type we have possible external arrivals into nodes and this makes input into some nodes to be batch arrival process. Secondly at each queueing node we have both split and merge features. The merge feature also contributes to the possible batch arrivals at a node. All we present in this book is the idea behind decomposition of queueing networks for the cases where all the buffers are of infinite sizes. For more comprehensive study of discrete time queueing networks refer to Daduna (2001) and Schamber (1997).

The simplest of all queueing networks is the tandem network, for which we have already covered some basic ideas in this chapter. The key difference between a tandem queue and a queueing network is that in tandem queues there are no splits of outputs sent to other nodes, and there are no merges from other nodes (except in the case of external inputs into intermediate nodes). Generally, for discrete queueing networks one is left with making approximations because of the challenges regarding the possibility of multiple events occurring at the same time. First we discuss the most basic true queueing network in order to show the challenges involved.

12.5.1 *The Challenges of Discrete Time Queueing Networks*

Consider a three node queueing network, with the nodes numbered 1,2,3; see Fig 12.8.

Packets from external sources arrive to node i according to the Bernoulli process with parameter a_i and service is Geometric with parameter b_i. When a service is completed at node i a packet proceeds to node j with probability $r_{i,j}$, with $i = 1,2,3$ and $j = 0,1,2,3$. Here node 0 refers to the outside the system. For example $r_{i,0}$ is the probability that a packet that completes service at node i leaves the system completely, hence we have $\sum_{j=0}^{3} r_{i,j} = 1, \ \forall i$, with $r_{i,i} = 0$.

Let $X_i^{(n)}$ be the number of packets at node i at time n. Then we know that $\{X_1^{(n)}, X_2^{(n)}, X_3^{(n)}\}$ is a DTMC and its transition matrix can be written as

$$P = \begin{bmatrix} P_{0,0} & P_{0,1} & P_{0,2} & P_{0,3} & & \\ P_{1,0} & P_{1,1} & P_{1,2} & P_{1,3} & P_{1,4} & \\ & P_{2,1} & P_{2,2} & P_{2,3} & P_{2,4} & P_{2,5} \\ & & \ddots & \ddots & \ddots & \ddots & \ddots \end{bmatrix},$$

where the matrix $P_{i,j}$ captures the set of transitions $(i,*,*) \rightarrow (j,*,*)$ and $*$ here simply refers to all possible states of node 1 and node 2. The matrices $P_{i,j}$ are made up of another set of matrices $P_{i,j}^{(k,\ell)}$ which capture the set of transitions $(i,k,*) \rightarrow (j,\ell,*)$. Finally we have the elements of matrices $P_{i,j}^{(k,\ell)}$ written as $P_{i,j}^{(k,\ell)}(r,s)$ which represents the following:

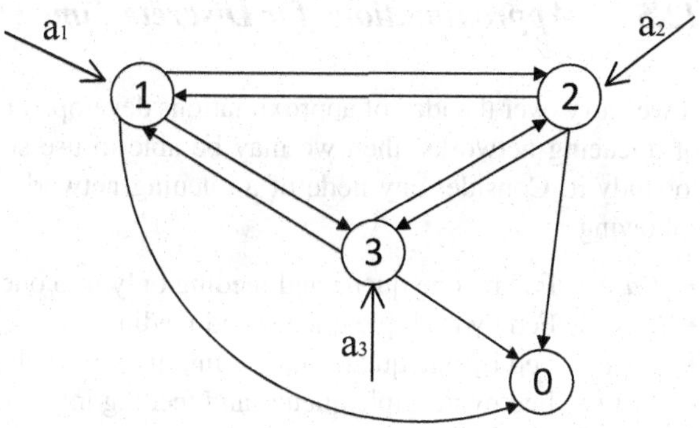

Fig. 12.8 Three Node QueueingNetwork

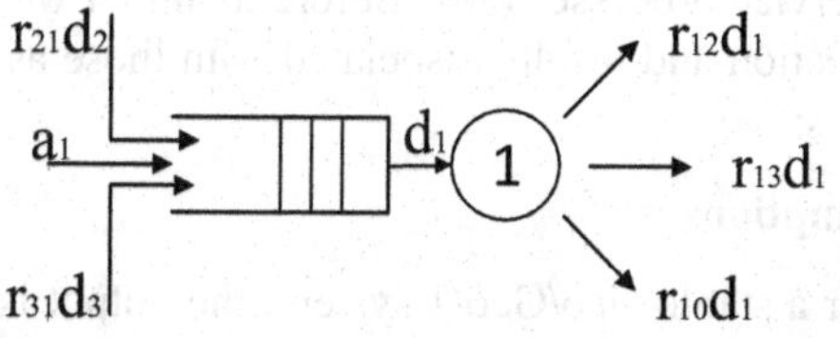

$$P_{i,j}^{(k,\ell)}(r,s) = Pr\{(X_1^{(n+1)} = j, X_2^{(n+1)} = \ell, X_3^{(n+1)} = s) | (X_1^{(n)} = i, X_2^{(n)} = k, X_3^{(n)} = r)\}.$$

First let us consider how many types of transitions are possible in this situation.

For $i \geq 1, j \geq 1, k \geq 1, \ell \geq 1, r \geq 1, s \geq 1$ we have up to $5^3 = 125$ different possible transition elements to consider. This is because we have $j = i-1, i, i+1, i+2, i+3$ possible values that j can assume. For example

$$Pr\{(X_1^{n+1} = i-1, X_2^{n+1} = j, X_3^{n+1} = k+1) | (X_1^n = i, X_2^n = j, X_3^n = k)\}$$

$$= \bar{a}_1\bar{a}_2a_3b_1\bar{b}_2\bar{b}_3r_{1,0} + \bar{a}_1\bar{a}_2\bar{a}_3b_1\bar{b}_2\bar{b}_3r_{1,3}$$

$$+ \bar{a}_1a_2\bar{a}_3b_1b_2\bar{b}_3(r_{2,0}r_{1,3} + r_{2,3}r_{1,0}).$$

We would need to obtain all the 125 entries for this case. Then for the cases where we have

$$\{0,*,*),\ (0,0,*),\ (0,0,0),\ (*,0,*),\ (*,*,0),\ (*,0,0),\ (0,*,0)\},$$

we need to know all the possible states that the transitions could go into and find all the transition probabilities. These are the boundary cases.

It is immediately clear that even for this simple 3 node queueing network the problem is cumbersome. So, we approach the study of discrete time queueing networks from an approximation point of view.

12.5.2 Approximations for Discrete Time Queueing Networks

If we carry over the idea of approximations developed for tandem queues to the case of queueing networks, then we may be able to use some form of approximations to study it. Consider any node of a queuing network. It will belong to one of the following

- *Case 1:* Fed by one queue and feeding only into one queue – **Simple Tandem**
- *Case 2:* Fed by multiple queues and feeding into one queue–**Merge**
- *Case 3:* Fed by one queue and feeding into multiple queue – **Split**
- *Case 4:* Fed by multiple queues and feeding into multiple queues – **Merge-Split**

We decompose a queueing network and study it as a decomposed system. We consider each of the above scenarios and limit ourselves to the geometric arrival and service processes only. Before then first we present some assumptions made in this section and results associated with those assumptions. The assumptions are as follows:

Assumptions

1. For a stable Geo/Geo/1 system, the output is a Bernoulli process with a known probability given as d
2. For a stable Geo^X/Geo/1 system, the output is a Bernoulli process with a known probability given as d
3. If an output from a node is d and it is destined to K destinations and that it is split according to the ratio θ_k between the K destinations, i.e. $\sum_{k=1}^{K}\theta_k = 1$, then the input to a destination k is geometric with parameter $\theta_k d$

12.5.3 Simple Tandem

This has already been studied in the previous sections.

12.5.4 Merge:

In a merge system we have K nodes labelled $1,2,\cdots,K$ feeding into a node we labelled 0. If each of the K nodes is generating geometric traffic with parameters $d_k,\ k=1,2,\cdots,K$, then we can obtain the input to node $i \neq 1,2,\cdots,K$, as a batch arrival process with $c_k,\ k=0,1,2,\cdots,K$, where c_k is the probability that the input to queue i is $k,\ k=0,1,2\cdots,K$. We can easily estimate the queue length for this node i using previous results obtained in Section 5.14.

12.5.5 *Split:*

In a split system we have a node 0 feeding into K nodes labelled $1,2,\cdots,K$. Let the output from node 0 be geometric with parameter d. Suppose node $i \neq 1,2,\cdots,K$, sends θ_k proportion of the output to node k, $k = 1,2,\cdots,K$, then the input into node k is geometric with parameter $\theta_k d$. We can easily estimate the queue length for this node i using previous results obtained in Chapter 5.

12.5.6 *Merge-Split:*

The merge-split case is simply a combination of the merge and split presented earlier.

References

Brandwajn A (1985) Equivalence and decomposition in queueing systems - a unified approach. Performance Evaluation 5:175–186

Daduna H (2001) Queueing networks with discrete time scale: Explicit expressions for the steady state behaviour of discrete time stochastic networks. In: Springer Lecture Notes in Computer Science, Springer

Hsu J, Burke PJ (1976) Behavior of tandem buffers with geometric input and Markovian output. IEEE Trans Communications 24:358–361

Hunter JJ (1983) Mathematical Techniques of Applied Probability - Volume 2, Discrete Time Model: Techniques and Applications. Academic Press, New York

Jackson RRP (1954) Queueing systems with phase type service. Operations Research 9:109–120

Schamber M (1997) Decomposition methods for finite queue networks with a non-renewal arrival process in discrete time. M.Sc thesis, University of Manitoba

Chapter 13
Optimization and Control of Discrete-Time Queues

13.1 Introduction

This chapter is short and is meant to be an introductory material to help point out how little there is in terms of results for the optimization of discrete time queues. It is meant to help encourage more research on the topic.

The key parameters that go into the analysis include; arrival process, service process, queue configuration, service discipline, and several other minor ones. As would be apparent to the reader by now, there are several performance measures used to determine how well a queueing system performs and these include, number in the system, queue length, waiting time, durations of busy and idle periods, number served during the busy period, etc. It is vey common that the performance measure of concern to a customer is not necessarily the same as the one of interest to the system operator. For example, a customer wants to have as small a waiting time as possible, ideally a zero waiting time in the queue! A system operator wants to make the idle time as small as possible, ideally zero idle times! These two, for example, are conflicting. Optimally selecting which parameters minimize waiting time while ensuring that the duration idle time is also minimum under the circumstances, is an example of trying to optimize the queueing system operations and also controlling the parameters.

By controlling a queue, we may decide to limit the arrival rate of customers in order to achieve a particular performance measure for those that are allowed to join the queue. Alternatively we may control the service process by increasing it appropriately to achieve the same goal but at some additional costs. Finally, we may even change the queue service policy to give preferential service to some customers and ensure they achieve the target performance measure; this leads to the other customers receiving poorer service!. The focus of this chapter is how to control a queueing system in order to achieve some of the performance measures of interest, given some constraints.

A.S. Alfa, *Applied Discrete-Time Queues*, DOI 10.1007/978-1-4939-3420-1_13

There are several good books that have dealt with the control and optimization of queueing system operations very thoroughly. Among them are Bertsekas (1995), Sennott (1999), Kitaev and Rykov (1995), (Puterman (2005)), and several others. These books cover optimization and control of queueing system from a very dynamic and multiple decision point aspects, hence they apply some versions of Markov decision process (MDP). For example, one problem discussed by Kitaev and Rykov (1995) in their introductory chapter is as follows: a single server system has M different possible rates of service it could offer. There is a reward of R units received for each customer served, a holding cost H for customers waiting, a cost C_m, $m = 1, 2, \cdots, M$ for using the mth service rate, and also a cost $\hat{C}_{i,j}$ for switching from a service rate of type i to a type j. The problem is to find when to use what service rate in order to minimize the total cost. This problem can be set up using the MDP idea. Our focus in this chapter is a set of simpler problems. We want to present cases that have less number of decision making parameters. For example, customers in a single server queue wish to keep their waiting times less than a particular amount. What percentage of the customers should be denied service and after what threshold of queue length should the rule be applied? We could decide on the percentage and then find the threshold using simple techniques, or we could set the threshold and then determine the percentage. Essentially what we will be doing in this case is controlling the arrival process. There will be other examples also where we control the service process, such as the following. Consider the same single server queue. Suppose we want to serve at a particular rate initially and admit all arriving customers. However we change the service rate after a particular queue length threshold in order to maintain the upper bound on the waiting time. This may be seen as an alternative way of achieving the same goal as in the previous case. The difference is that in the first case we lose some customers and limit the total number served per unit time and in the second case we serve all customers but put in additional effort in the service aspect. We will consider these types of problems in this chapter.

We classify our control into three categories; controlling arrivals, controlling service, and controlling the service discipline.

13.2 Controlling the Arrival Process

13.2.1 One Queue Length Threshold Problem

13.2.1.1 Finding Queue Length Threshold

Consider, Geo/Geo/1, a single server queueing system with arrival parameter a and service parameter b. We want to ensure that the mean waiting time in the system for the customers is no more than w^*, or the probability of the number in the system exceeding L is less than some parameter σ. Suppose we want to achieve this by controlling the percentage of customers allowed into the system after the number in the system is at least $K < \infty$. The question is to determine what the threshold level K should be at which only a portion γ, $0 < \gamma < 1$ of customers are allowed into the system. This is a very simple control problem. The problem can be stated as follows. In general, the arrival probability when there are $i < K$ in the system is a, whereas when the number in the system is at least K the arrival probability is $a^* = \gamma a$, with $\bar{a}^* = 1 - a^*$. If we let X_n be the number of customers in the system at time n we know this is a DTMC with the transition matrix

$$P = \begin{bmatrix} \bar{a} & a & & & & & \\ \bar{a}b & \bar{a}\bar{b}+ab & a\bar{b} & & & & \\ & \ddots & \ddots & \ddots & & & \\ & & \bar{a}b & \bar{a}\bar{b}+ab & a\bar{b} & & \\ & & & \bar{a}^*b & \bar{a}^*\bar{b}+a^*b & a^*\bar{b} & \\ & & & & \bar{a}^*b & \bar{a}^*\bar{b}+a^*b & a^*\bar{b} \\ & & & & & \ddots & \ddots & \ddots \\ & & & & & & \ddots & \ddots & \ddots \end{bmatrix}.$$

Here the matrix is repeating from level 1 until level $K-1$ and then changes from level K when the threshold is arrived at and the arrival process changes. Essentially what we have is that the elements $p_{i,j}$ of matrix P

$$p_{i,i+1} = \begin{cases} a\bar{b}, & i \leq K-1, \\ a^*\bar{b}, & i \geq K, \end{cases}$$

$$p_{i,i-1} = \begin{cases} \bar{a}b, & i \leq K-1, \\ \bar{a}^*b, & i \geq K, \end{cases}$$

and

$$p_{i,i} = \begin{cases} \bar{a}\bar{b}+ab, & i \leq K-1, \\ \bar{a}^*\bar{b}+a^*b, & i \geq K. \end{cases}$$

This DTMC is about the same as the standard Geo/Geo/1 discussed in Chapter 5, and it is analyzed in the same manner. The only difference is that this DTMC has larger boundary. Letting $x_k(K) = Pr\{X_n = k\}|_{n\to\infty}$ we define

$$y_L(K) = Pr\{X_n \geq L\}|_{n\to\infty} = \sum_{k=L}^{\infty} x_k(K).$$

Our goal here is to select K, for given γ and σ, such that

$$y_L(K) \leq \sigma.$$

We know that, for a given set of parameters, $y_L(k+1) \geq y_L(k)$, hence this problem can be solved in a naive greedy approach. We are seeking a $K = k^*$ such that

$$y_L(k^*) \leq \sigma, \text{ and } y_L(k^*+1) > \sigma.$$

13.2.1.2 Finding Percentage Reduction in Arrival Rate

In this case K and σ are fixed and we wish to find γ that ensures that $y_L(k,\gamma) \leq \sigma$. So we vary γ from just above 0 to 1, i.e. $0 < \gamma \leq 1$, and select σ^* such that

$$y_L(K,\gamma^*) \leq \sigma \text{ and } y_L(K,\gamma^*+\delta) > \sigma, \ \forall (0 < \delta < 1-\gamma^*).$$

13.2.2 Two Queue Length Thresholds

13.2.2.1 Finding the Two Thresholds

An extension of the last example is where we have two thresholds, K_1 and K_2 such that $1 \leq K_1 \leq K_2 < \infty$. Consider a single server queueing system with arrival parameter a and service parameter b. Again we want to ensure that the mean waiting time in the system for the customers is no more than w^*, or the probability of the number in the system exceeding L is less than some parameter σ. Suppose we want to achieve this by not allowing a percentage or ratio of customers into the system after the number in the system increases past the level K_2. Let that ratio be $1-\gamma$, $0 \leq \gamma < 1$. However, as the number in the system drops we still do not allow more than γ ratio of them into the system until the number in the system drops below K_1. The question is to determine what the threshold levels K_1 and K_2 should be. This is an advance control problem because we now have two thresholds. The problem can be stated as follows. In general, arrival probability when there are $i < K_1$ in the system is a, and it is still a before the number in the system reaches K_2 as long as the growth was on its way up from K_1, until it reaches K_2. The arrival rate is then a^* as soon as the number in the system reaches and exceeds K_2. As the the number in the system decreases from K_2 on the way down, the arrival rate remains at a^* until the number in the system drops below K_1, when it goes up again to a. In this system the state space is as follows $\{0,1,\cdots,K_1,K_1^*,K_1+1,(K_1+1)^*,K_1+2,(K_1+2)^*,\cdots,K_2-1,(K_2-1)^*,K_2,K_2+1,\cdots\}$. Here we let $(K_1+j)^*$, $1 \leq j \leq K_2-K_1-1$ be the states corresponding to having K_1+j in the system while the queue length is on its way down from level K_2 without crossing level K_1 yet, whereas K_1+j, $1 \leq j \leq K_2-K_1-1$ states corresponding to the state of having K_1+j in the system when the system has not crossed level K_2 yet and is on its way up. Letting $p_{i,j}$ represent the probability of finding j in the system at time $n+1$, given that there were i in the system at time n, we have

$$p_{i,i+1} = \begin{cases} a\bar{b}, & i \le K_2 - 1, \\ a^*\bar{b}, & i \ge K_2, \end{cases}$$

$$p_{i,i-1} = \begin{cases} \bar{a}b, & i \le K_2 - 1, \\ \bar{a}^*b, & i \ge K_2 + 1, \end{cases}$$

$$p_{i^*,(i+1)^*} = a^*\bar{b},$$

$$p_{i^*,(i-1)^*} = \bar{a}^*b,$$

$$p_{(K_1)^*,K_1-1} = \bar{a}^*b,$$

$$p_{K_2,(K_2-1)^*} = \bar{a}^*b,$$

Every element $p_{i,i} = 1 - [\sum_j p_{i,j}]$ and also $p_{i^*,^*i} = 1 - [\sum_j p_{i^*,j}]$. This situation is represented in Fig. 13.1, with the bottom half being the K^* states.

If we let X_n be the number of customers in the system at time n we know this is a DTMC with a transition matrix P which can be constructed, however not easy to display here. The matrix P will be of the QBD type after a very large boundary set of states.

Letting $x_k(K_1,K_2) = Pr\{X_n = k\}|_{n\to\infty}$ we define

$$y_L(K_1,K_2) = Pr\{X_n \ge L\}|_{n\to\infty} = \sum_{k=L}^{\infty} x_k(K_1,K_2).$$

Our goal here is to select K_1 and K_2 such that

$$y_L(K_1,K_2) \le \sigma.$$

We know that $y_L(k_1,k_2) \ge y_L(k_1,k_2+1)$, hence this problem can be solved in a naive greedy approach by searching in two directions; first we fix K_1 and then search

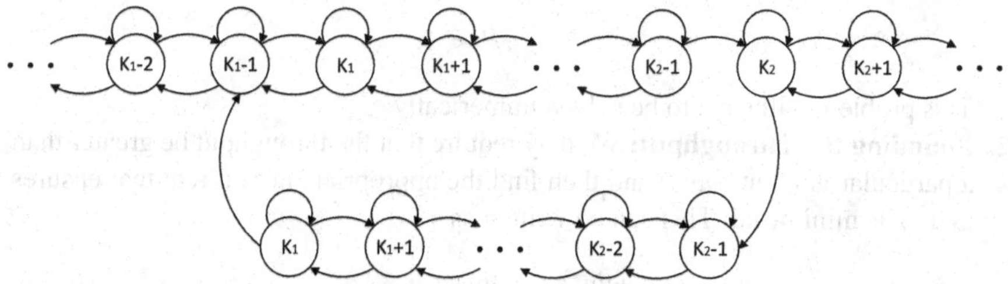

Fig. 13.1 Two Threshold Control

for K_2. Then we change K_1 and search again. We are seeking a the pair $(K_1, K_2) = (k_1, k_2)$ such that

$$y_L(k_1,k_2) < \sigma, \;\; y_L(k_1,k_2+1) \geq \sigma, \;\; y_L(k_1+1,k_2) \geq \sigma, \text{and} \;\; y_L(k_1+1,k_2+1) \geq \sigma.$$

13.2.2.2 Finding the Percentage Reduction in Arrival Rate

In this case K_1 and K_2 are fixed and we try to find γ such that $y_L(K_1, K_2, \gamma) \leq \sigma$. We apply the same idea as in the case of one threshold.

13.2.3 Maximizing Throughput, While Minimizing Loss Probability in Finite Buffer Systems

Consider a finite buffer queueing system with buffer size $K < \infty$. Suppose for a given service rate, we want to select an arrival rate that optimizes an objective function, where our function is related to the loss probability and throughput. We focus on the Geo/Geo/1/K system, with arrival and service parameters a and b, respectively. For this type of system it can be shown that the loss probability $p_L = ax_{K+1}$ and throughput is $\mu_T = a(1-p_L)$, where x_{K+1} is the probability that there are $K+1$ in the system.

There are three possible ways to select the appropriate arrival rate and they are:

1. **Bounding the Loss rate:** We can set the loss rate $p_L \leq \sigma$ and then find the appropriate arrival rate that ensures that μ_T is maximized. The problem can be written as

$$\begin{aligned} max \;\; z_a &= \mu_T(a,b,K), \\ s.t. \;\; ap_L(a,b,K) &\leq \sigma, \\ 0 < a &< 1, \\ a/b &< 1. \end{aligned}$$

   This problem will have to be solved numerically.

2. **Bounding the Throughput:** We may require that the throughput be greater than a particular amount, say γ and then find the appropriate arrival rate that ensures that p_L is minimized. This can be written as

$$\begin{aligned} min \;\; z_a &= ap_L(a,b,K), \\ s.t. \;\; \mu_T(a,b,K) &\geq \gamma, \\ 0 < a &< 1, \\ a/b &< 1. \end{aligned}$$

This problem will also have to be solved numerically.

3. **Combination Approach Using Cost Parameters:** This is usually a compromise type of approach where the interest is weighted by the system manager. They may place a cost of c_L to loss rate and b_T as the benefit (revenue) associated with each throughput unit. The problem can then be set up as

$$min\ z_a = c_L a p_L - b_T \mu_T = a[c_L p_L(a,b,K) - b_T(1 - p_L(a,b,K))],$$

$$s.t.\ \ 0 < a < 1,$$

$$a/b < 1.$$

This same idea can be used to find the optimal service rate given a fixed arrival rate.

13.2.4 *Finding the Maximum Profit, When Costs and Revenues are Associated*

For this class of problem we consider a modified special case of the Example 2.1.1 in Sennott (1999). Suppose there is a holding cost associated with each customer who waits in the queue every time epoch. Let $H(k)$ be the cost associated with k customers in the system at any time epoch. We also associate with the system a reward of B for accepting a customer and a cost C for rejecting a customer, i.e. not giving it service. The question then is that of making a decision at any time epoch whether to accept an arriving customer or reject them. Sennott (1999) discussed the case of batch arrivals, however in this chapter we focus on the single arrivals case. We show that this problem can be set up as a Markov decision process (MDP). We refer readers to classical MDP books where solution methods are discussed in details. Here we only show the set up and mention a technique that can be used to analyze it.

The decision making process here is whether to accept an arriving packet or not. If we accept it we say our decision is *YES* and represent it with $\mathscr{A}$. On the other hand if we do not accept it we say our decision is *NO* and represent it with $\mathscr{R}$. So our action space is

$$\mathscr{D} = \{\mathscr{A}, \mathscr{R}\}.$$

Let X_t be the number of items in the system at time t. So our state space is

$$\mathscr{S} = \{0, 1, 2, 3, \cdots\}.$$

The resulting random variable equation can thus be written as

$$X_{t+1} = (X_t - D_t)^+ + \begin{cases} A_t, & \mathscr{D} = \mathscr{A} \\ 0, & \mathscr{D} = \mathscr{R}, \end{cases}$$

where D_t is number of service completion at time t and in this example we let this number be either 1 or 0, and A_t is the number of arrivals at time t which we let equal 0 or 1. We do not discuss the case of non-batch arrivals here. Sennott (1999) in her book did present an example of batch arrivals.

When the system state is s and the action taken is $\mathscr{D}$, let the reward be $r(s,\mathscr{D})$, then we can write

$$r(s,\mathscr{D}) = \begin{cases} B - H(s), & \mathscr{D} = \mathscr{A}, \\ -C - H(s), & \mathscr{D} = \mathscr{R}, \end{cases}$$

Note that if an arrival is not accepted then $A_t = 0$ with probability 1. So if we let $p_{i,j}(\mathscr{D})$ be the transition probability from i in the system to j in the system with a particular decision $\mathscr{D}$, then we can write the transition probabilities of the resulting DTMC we have that

$$p_{0,0}(\mathscr{R}) = 1,$$

$$p_{0,0}(\mathscr{A}) = \bar{a},$$

$$p_{0,1}(\mathscr{R}) = a,$$

$$p_{i,i-1}(\mathscr{R}) = b,$$

$$p_{i,i-1}(\mathscr{A}) = b\bar{a},$$

$$p_{i,i}(\mathscr{R}) = \bar{b},\ i \geq 1,$$

$$p_{i,i}(\mathscr{A}) = ba + \bar{b}\bar{a},\ i \geq 1,$$

$$p_{i,i+1}(\mathscr{A}) = \bar{b}a,\ i \geq 1.$$

We see right away that we practically have two transition matrices $P(\mathscr{R})$ and $P(\mathscr{A})$, which refer to transitions associated to Decisions n or y, respectively, where

$$P(\mathscr{R}) = \begin{bmatrix} 1 & & & & \\ b & \bar{b} & & & \\ & b & \bar{b} & & \\ & & b & \bar{b} & \\ & & & \ddots & \ddots \end{bmatrix},\quad P(\mathscr{A}) = \begin{bmatrix} \bar{a} & a & & & \\ \bar{a}b & \bar{a}\bar{b} + ab & a\bar{b} & & \\ & \bar{a}b & \bar{a}\bar{b} + ab & a\bar{b} & \\ & & \bar{a}b & \bar{a}\bar{b} + ab & a\bar{b} \\ & & & \ddots & \ddots & \ddots \end{bmatrix}.$$

Consider a stationary policy π, i.e. a decision or action that depends on the states only. With this type of policy there is an associated value function $V^{\pi}(s)$ giving us the expected cumulative reward achieved by π, when starting from state s. We use the non-discounted average reward which is given as

$$V_N^{\pi} = E^{\pi}[\sum_{t=1}^{N} r(s_t, \mathscr{D}_t)],$$

where s_t and $\mathscr{D}_t$ are the state and decision, respectively at time t. Note that we could have used the discounted reward as an alternative. Let $v(s) = V_N^{\pi}|_{N\to\infty}$. It has been shown in Puterman (2005) that for each state s and stationary policy we have $v(s)$ being independent of t, hence we have the recursive equation

$$v(s) = max_{\mathscr{D}\in\{\mathscr{A},\mathscr{R}\}}\left\{ r(s,\mathscr{D}) + \sum_{j\in\mathscr{S}} p(s,j)(\mathscr{D})v(j) \right\}. \tag{13.1}$$

This we can solve iteratively as

$$_{k+1}v(s) = max_{\mathscr{D}\in\{\mathscr{A},\mathscr{R}\}}\left\{ r(s,\mathscr{D}) + \sum_{j\in\mathscr{S}} p(s,j)(\mathscr{D})(_kv(j)) \right\}. \tag{13.2}$$

We have the choice of using a finite or infinite planning horizon. We elect to go with infinite. Given infinite horizon we have a choice of discounted benefits or average reward and here we opt for the average reward. With all these assumptions now we can apply the value iteration approach.

Given all this information this problem can now be set up as stochastic dynamic programming problem as shown by Sennott (1999). However, because the state space is infinite one needs the Approximation Sequence Method to make it finite in the process of the computations. This is beyond the scope of this book and the reader is referred to Sennott (1999). After that the problem can be solved using linear programming approaches.

13.3 Controlling the Service Process

All the techniques shown in the previous sections of this chapter for controlling arrivals can equally be used to control service rates. The main difference is that instead of controlling a in order to achieve our goal we control b for the same goal. However, in these cases we are limited to how b can increase or decrease, keeping in mind that $0 < b < 1$ and also that $a/b < 1$. We however have more flexibility when dealing with batch service, as the mean batch size served can be greater than 1. Instead we focus on the case where we can select the number of servers in order to achieve some goals. So we consider multi server case.

13.3.1 Selection of Appropriate Number of Servers

Here we consider a Geo/Geo/k system. Our goal is to select the k such that we meet some goals. We know that a manager incurs costs that depend on the number k; the higher k is the higher the cost. Equally, we know that the queue length gets longer if the number of servers is decreased and that would affect the revenue for the operator because of reduction in QoS for users. So for such a system let $\mu_L(k)$ be the mean number in the system when k servers are used and let $c_o(k)$ be the cost to the system operator when there are k servers. Also we let $c_L(k)$ be the cost associated with the number in the system when there are k servers where $c_L(k)$ is a function of $\mu_L(k)$. Our interest is to find the optimal number k at which the system should operate. This can be formulated as an optimization problem as follows:

$$min \;\; z(k) = c_o(k) + c_L(k),$$

$$s.t. \;\; a < bk,$$

$$k \geq 1, \text{ and integer.}$$

This optimization problem can be carried out numerically, by first selecting the smallest integer k that meets the first constraint. Let that be k_0. Unless this selected k_0 value is the optimal one, we expect that if we initially increase k to $k_0 + 1$ we will improve z. We keep increasing k until $z(k_{n+1}) \geq z(k_n)$ for the first time, at which point we know that k_n is the optimal number of servers required. The algorithm is as follows

The Algorithm:

- **Step 1**: Select k_0 such that $a < bk_0, \;\; n = 0$
- **Step 2**: Compute $z(k_0)$,
- **Step 3**: For $n \leftarrow n+1$, compute $z(k_{n+1}) := z(k_n + 1)$
- **Step 4**: If $z(k_{n+1}) > z(k_n)$, Stop, and k_n is the optimal k GOTO Step 5, Otherwise GOTO Step 3.
- **Step 5**: STOP.

13.4 Problems

1. Question 1: Consider a single server queue with geometric arrivals and service with parameters a and b, respectively. This system has unlimited buffer space. Suppose when the system becomes empty the server goes on a vacation. The vacation is only interrupted if there are N customers waiting in the system. Suppose there is a reward of $R(n)$ for a vacation of n units of duration, and a cost

$C(n)$ for each unit n a customer waits. Find a threshold N for which the server should return from a vacation in order to minimize its total cost. For specific values of $a, b,\ C(.)$ and $R(.)$ find the value of N.

2. Question 2: Consider Question 1. Suppose when the system becomes empty the server goes on a vacation of duration T. There is a reward of $R(T)$ for the vacation of T units of duration. However if there are customers waiting during the vacation period there is a cost $C(n)$ for each unit n a customer waits. Find the optimal value of T for which the server should be away on a vacation in order to minimize its total cost. For specific values of $a, b,\ C(.)$ and $R(.)$ find the value of T.

References

Bertsekas D (1995) Dynamic Programming and Optimal Control, vol 2. Athena Scientific, MA

Kitaev MY, Rykov VV (1995) Controlled Queueing Systems. CRC Press, Boca Raton

Puterman ML (2005) Markov Decision Processes: Discrete Stochastic, Dynamic Programming. John Wiley & Sons

Sennott LI (1999) Stochastic dynamic programming and the control of queueing systems. In: Wiley Series in Probability and Statistics, John Wiley, New York

Appendix A
Appendix

A.1 The z−Transform

The $z-$transform also known as the probability generating function (pgf) is intermittently used in this book. In this section we give a very brief introduction to it for clarity and consistency.

Consider a discrete random variable $\mathscr{A}$ which can assume values in the set $\{0,1,2,3,\cdots\}$. Let

$$a_i = Pr\{\mathscr{A} = i\}.$$

We define its $z-$transform or pgf as $A(z)$, $|z| \leq 1$ as

$$A(z) = \sum_{i=0}^{\infty} z^i a_i.$$

Some of the well known properties of this are:

1. $A(z)|_{z\to 1} = 1$
2. The coefficient of z^k is the same as $Pr\{\mathscr{A} = k\}$
3. Let $E[\mathscr{A}]$ be the mean of $\mathscr{A}$, then $E[\mathscr{A}] = \frac{dA(z)}{dz}|_{z\to 1}$
4. Generally, the rth factorial moment of $\mathscr{A}$ is given as $E^r[\mathscr{A}] = \frac{drA(z)}{dz^r}|_{z\to 1}$
5. Consider two discrete random variables $\mathscr{A}$ and $\mathscr{B}$ both assuming values in the set $\{0,1,2,3\cdots\}$, if they are independent then the $z-$transform their convolution sum is the product of their transforms, i.e. $A(z)B(z)$.

A.S. Alfa, *Applied Discrete-Time Queues*, DOI 10.1007/978-1-4939-3420-1

A.2 Kronecker Product

In this section we give a brief summary of the operator *Kronecker Product* and few of its very useful properties that we use in this book.

The symbol $\otimes$ will be referred to as the Kronecker product. Consider two matrices A and B of orders $n \times m$ and $p \times q$, respectively. Let the matrix C be defined as

$$C = A \otimes B,$$

then C is of order $np \times mq$ and it is given as

$$C = \begin{bmatrix} a_{11}B & a_{12}B & \cdots & a_{1m}B \\ a_{21}B & a_{22}B & \cdots & a_{2m}B \\ \vdots & \vdots & \cdots & \vdots \\ a_{n1}B & a_{n2}B & \cdots & a_{nm}B \end{bmatrix},$$

where a_{ij} is the $(ij)^{th}$ element of the matrix A.

Some very well known properties of this operator include:

1. $A \otimes (B + C) = A \otimes B + A \otimes C$
2. $(A + B) \otimes C = A \otimes C + B \otimes C$
3. Note that $A \otimes B \neq B \otimes A$ unless $A = B$
4. $(A \otimes B) \otimes C = A \otimes (B \otimes C)$
5. $C^T = (A \otimes B)^T = A^T \otimes B^T$
6. If $E = (A \otimes B)(C \otimes D)$ exists then we also have that $E = (AC) \otimes (BD)$ and vice versa
7. If $C^{-1} = (A \otimes B)^{-1}$ exists then we also have that $(A \otimes B)^{-1} = A^{-1} \otimes B^{-1}$.
8. For a matrix $D = [D_1\ D_2\ \cdots\ D_m]$, where D_k is its kth column let $vecD = [D_1^T,\ D_2^T,\ \cdots,\ D_m^T]^T$ we can write

$$vec(AXB) = (B^T \otimes A)vecX.$$

Index

A.S. Alfa, *Applied Discrete-Time Queues*, DOI 10.1007/978-1-4939-3420-1

Printed in the United States
By Bookmasters

Printed in the United States
By Bookmasters